INVESTIGATIONS GUIDE

Environments

Full Option Science System
Developed at the Lawrence Hall of Science, University of California, Berkeley
Published and Distributed by Delta Education

FOSS Lawrence Hall of Science Team
Larry Malone and Linda De Lucchi, FOSS Project Codirectors and Lead Developers
Kathy Long, FOSS Assessment Director; David Lippman, Program Manager; Carol Sevilla, Publications Design Coordinator; Susan Stanley, Illustrator; John Quick, Photographer
FOSS Curriculum Developers: Brian Campbell, Teri Lawson, Alan Gould, Susan Kaschner Jagoda, Ann Moriarty, Jessica Penchos, Kimi Hosoume, Virginia Reid, Joanna Snyder, Erica Beck Spencer, Joanna Totino, Diana Velez, Natalie Yakushiji
Susan Ketchner, Technology Project Manager
FOSS Technology Team: Dan Bluestein Christopher Cianciarulo, Matthew Jacoby, Kate Jordan, Frank Kusiak, Nicole Medina, Jonathan Segal, Dave Stapley, Shan Tsai

Delta Education Team
Bonnie A. Piotrowski, Editorial Director, Elementary Science
Project Team: Jennifer Apt, Sandra Burke, Joann Hoy, Kristen Mahoney, Jennifer McKenna, Angela Miccinello

Thank you to all FOSS Grades K-5 Trial Teachers
Heather Ballard, Wilson Elementary, Coppell, TX; Mirith Ballestas De Barroso, Treasure Forest Elementary, Houston, TX; Terra L. Barton, Harry McKillop Elementary, Melissa, TX; Rhonda Bernard, Frances E. Norton Elementary, Allen, TX; Theresa Bissonnette, East Millbrook Magnet Middle School, Raleigh, NC; Peter Blackstone, Hall Elementary School, Portland, ME; Tiffani Brisco, Seven Hills Elementary, Newark, TX; Darrow Brown, Lake Myra Elementary School, Wendell, NC; Heather Callaghan, Olive Chapel Elementary, Apex, NC; Katie Cannon, Las Colinas Elementary, Irving, TX; Elaine M. Cansler, Brassfield Road Elementary School, Raleigh, NC; Kristy Cash, Wilson Elementary, Coppell, TX; Monica Coles, Swift Creek Elementary School, Raleigh, NC; Shirley Conner, Ocean Avenue Elementary School, Portland, ME; Sally Connolly, Cape Elizabeth Middle School, Cape Elizabeth, ME; Melissa Cook-Airhart, Harry McKillop Elementary, Melissa, TX; Melissa Costa, Olive Chapel Elementary, Apex, NC; Hillary P. Croissant, Harry McKillop Elementary, Melissa, TX; Rene Custeau, Hall Elementary School, Portland, ME; Nancy Davis, Martha and Josh Morriss Mathematics and Engineering Elementary School, Texarkana, TX; Nancy Deveneau, Wilson Elementary, Coppell, TX; Karen Diaz, Las Colinas Elementary, Irving, TX; Marlana Dumas, Las Colinas Elementary, Irving, TX; Mary Evans, R.E. Good Elementary School, Carrollton, TX; Jacquelyn Farley, Moss Haven Elementary, Dallas, TX; Corinna Ferrier, Oak Forest Elementary, Humble, TX; Allison Fike, Wilson Elementary, Coppell, TX; Barbara Fugitt, Martha and Josh Morriss Mathematics and Engineering Elementary School, Texarkana, TX; Colleen Garvey, Farmington Woods Elementary, Cary, NC; Judy Geller, Bentley Elementary School, Oakland, CA; Erin Gibson, Las Colinas Elementary, Irving, TX; Kelli Gobel, Melissa Ridge Intermediate School, Melissa, TX; Dollie Green, Melissa Ridge Intermediate School, Melissa, TX; Brenda Lee Harrigan, Bentley Elementary School, Oakland, CA; Cori Harris, Samuel Beck Elementary, Trophy Club, TX; Kim Hayes, Martha and Josh Morriss Mathematics and Engineering Elementary School, Texarkana, TX; Staci Lynn Hester, Lacy Elementary School, Raleigh, NC; Amanda Hill, Las Colinas Elementary, Irving, TX; Margaret Hillman, Ocean Avenue Elementary School, Portland, ME; Cindy Holder, Oak Forest Elementary, Humble, TX; Sarah Huber, Hodge Road Elementary, Knightdale, NC; Susan Jacobs, Granger Elementary, Keller, TX; Carol Kellum, Wallace Elementary, Dallas, TX; Jennifer A. Kelly, Hall Elementary School, Portland, ME; Brittani Kern, Fox Road Elementary, Raleigh, NC; Jodi Lay, Lufkin Road Middle School, Apex, NC; Melissa Lourenco, Lake Myra Elementary School, Wendell, NC; Ana Martinez, RISD Academy, Dallas, TX; Shaheen Mavani, Las Colinas Elementary, Irving, TX; Mary Linley McClendon, Math Science Technology Magnet School, Richardson, TX; Adam McKay, Davis Drive Elementary, Cary, NC; Leslie Meadows, Lake Myra Elementary School, Wendell, NC; Anne Mechler, J. Erik Jonsson Community School, Dallas, TX; Anne Miller, J. Erik Jonsson Community School, Dallas, TX; Shirley Diann Miller, The Rice School, Houston, TX; Keri Minier, Las Colinas Elementary, Irving, TX; Stephanie Renee Nance, T.H. Rogers Elementary, Houston, TX; Cynthia Nilsen, Peaks Island School, Peaks Island, ME; Elizabeth Noble, Las Colinas Elementary, Irving, TX; Courtney Noonan, Shadow Oaks Elementary School, Houston, TX; Sarah Peden, Aversboro Elementary School, Garner, NC; Carrie Prince, School at St. George Place, Houston, TX; Marlaina Pritchard, Melissa Ridge Intermediate School, Melissa, TX; Alice Pujol, J. Erik Jonsson Community School, Dallas, TX; Claire Ramsbotham, Cape Elizabeth Middle School, Cape Elizabeth, ME; Paul Rendon, Bentley Elementary, Oakland, CA; Janette Ridley, W.H. Wilson Elementary School, Coppell, TX; Kristina (Crickett) Roberts, W.H. Wilson Elementary School, Coppell, TX; Heather Rogers, Wendell Creative Arts & Science Magnet Elementary School, Wendell, NC; Alissa Royal, Melissa Ridge Intermediate School, Melissa, TX; Megan Runion, Olive Chapel Elementary, Apex, NC; Christy Scheef, J. Erik Jonsson Community School, Dallas, TX; Samrawit Shawl, T.H. Rogers School, Houston, TX; Nicole Spivey, Lake Myra Elementary School, Wendell, NC; Ashley Stephenson, J. Erik Jonsson Community School, Dallas, TX; Jolanta Stern, Browning Elementary School, Houston, TX; Gale Stimson, Bentley Elementary, Oakland, CA; Ted Stoeckley, Hall Middle School, Larkspur, CA; Cathryn Sutton, Wilson Elementary, Coppell, TX; Camille Swander, Ocean Avenue Elementary School, Portland, ME; Brandi Swann, Westlawn Elementary School, Texarkana, TX; Robin Taylor, Arapaho Classical Magnet, Richardson, TX; Michael C. Thomas, Forest Lane Academy, Dallas, TX; Jomarga Thompkins, Lockhart Elementary, Houston, TX; Mary Timar, Madera Elementary, Lake Forest, CA; Helena Tongkeamha, White Rock Elementary, Dallas, TX; Linda Trampe, J. Erik Jonsson Community School, Dallas, TX; Charity VanHorn, Fred A. Olds Elementary, Raleigh, NC; Kathleen VanKeuren, Lufkin Road Middle School, Apex, NC; Valerie Vassar, Hall Elementary School, Portland, ME; Megan Veron, Westwood Elementary School, Houston, TX; Mary Margaret Waters, Frances E. Norton Elementary, Allen, TX; Stephanie Robledo Watson, Ridgecrest Elementary School, Houston, TX; Lisa Webb, Madisonville Intermediate, Madisonville, TX; Matt Whaley, Cape Elizabeth Middle School, Cape Elizabeth, ME; Nancy White, Canyon Creek Elementary, Austin, TX; Barbara Yurick, Oak Forest Elementary, Humble, TX; Linda Zittel, Mira Vista Elementary, Richmond, CA

Photo Credits: © iStockphoto/Ken Canning (cover); © Anton Foltin/Shutterstock; © Laurie Meyer; © John Quick

Published and Distributed by Delta Education, a member of the School Specialty Family
The FOSS program was developed in part with the support of the National Science Foundation grant nos. MDR-8751727 and MDR-9150097. However, any opinions, findings, conclusions, statements, and recommendations expressed herein are those of the authors and do not necessarily reflect the views of NSF. FOSSmap was developed in collaboration between the BEAR Center at UC Berkeley and FOSS at the Lawrence Hall of Science.

Copyright © 2019 by The Regents of the University of California

Standards cited herein from NGSS Lead States. 2013. *Next Generation Science Standards: For States, By States.* Washington, DC: The National Academies Press. Next Generation Science Standards is a registered trademark of Achieve. Neither Achieve nor the lead states and partners that developed the Next Generation Science Standards was involved in the production of, and does not endorse, this product.

All rights reserved. Any part of this work may not be reproduced or transmitted in any form or by any means, electronic or mechanical, including photocopying and recording, or by an information storage or retrieval system without prior written permission. For permission please write to: FOSS Project, Lawrence Hall of Science, University of California, Berkeley, CA 94720 or foss@berkeley.edu.

Environments
Investigations Guide, 1487573
978-1-62571-333-9
Printing 6 – 12/2017
Webcrafters, Madison, WI

Welcome to FOSS® Next Generation™

Getting Started with FOSS Next Generation for Grades 3–5

Whether you're new to hands-on science or a FOSS veteran, you'll be up and running in no time and ready to lead your students on a fantastic voyage through the wonders of the natural and designed world.

Watch our short video series or browse the next few pages to get started!

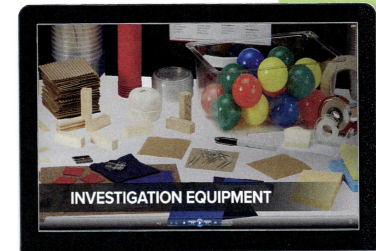

Getting Started with FOSS: Meet Your Module video

Scan here or visit deltaeducation.com/goFOSS

Three-Dimensional Active Science

It's time to experience the three dimensions of the NGSS— **disciplinary core ideas**, **crosscutting concepts**, and **science and engineering practices**. Engage in rich investigations that immerse your students in real-world applications of important scientific phenomena, supported by just-in-time teaching tips and strategies.

Getting Started with Your Equipment Kit

Meet Your FOSS Module!

Your FOSS module includes one or more large boxes, called drawers, and two smaller boxes for the Teacher Toolkit, student books, and other equipment. Each drawer has a label on the front listing its contents. Your packing list is always in Drawer 1.

Permanent Equipment

Your equipment kit includes enough permanent equipment for up to 8 groups (32 students). This equipment is classroom-tested and expected to last 7–10 years.

Consumable Equipment

Your kit also includes consumable materials for three class uses. Convenient refill kits provide materials for three additional uses and are available through Delta Education.

Easy Set-up and Clean-up!

FOSS Next Generation equipment drawers are packed by investigation to facilitate prep and to make packing up for the next use a snap!

Drawer sections include:

- Unique materials needed for one investigation
- Common equipment used in multiple investigations
- Consumable materials—when it's empty you know it's time to refill!

Order Refills Online

deltaeducation.com/refillcenter

Live Organisms

Some investigations require live organisms. Schools are encouraged to purchase these organisms from a local biological supply company to minimize both transit time and the impact of adverse weather on the health of the organisms.

If living material cards are purchased from Delta Education, they will be shipped separately in a green and white envelope. Keep these cards in a safe place until it's time to redeem them for the investigation.

Call Delta Education at 800-258-1302 at least three weeks before you need your organisms.

Premium Student eBook Access

If your school purchased a premium class license for the *FOSS Science Resources* student eBook, your access codes will be shipped separately in a blue and white striped envelope. Use this access code on FOSSweb to unlock student eBook access.

Getting Started with Your Teacher Toolkit

The Teacher Toolkit is the most important part of the FOSS program. There are three parts of the Teacher Toolkit—the **Investigations Guide**, **Teacher Resources**, and the student **Science Resources** book. It's here that all the wisdom and experience from years of research and classroom development comes together to support teachers with lesson facilitation and in-depth strategies for taking investigations to the next level.

1. Investigations Guide

The **Investigations Guide** is your roadmap to prepare for and lead the FOSS investigations. Chapters are tabbed for easy access to important module information.

The module **Overview** gives you a high-level look at the 10–12 weeks of instruction in each module including a summary matrix, schedule for the module, and product support contacts.

Framework and the NGSS provides a complete overview of NGSS connections, learning progressions, and background to support the conceptual framework for the module.

The **Materials** chapter is a must-read resource that helps you get your student equipment ready for first-time use and shares helpful tips for getting your classroom ready for FOSS.

The **Technology** chapter provides an overview for each digital resource in the module and gets you up and running on FOSSweb.com, complete with technical support.

Each **Investigation** includes an At-a-Glance overview, science background content with NGSS connections, and in-depth guidance for preparing and facilitating instruction.

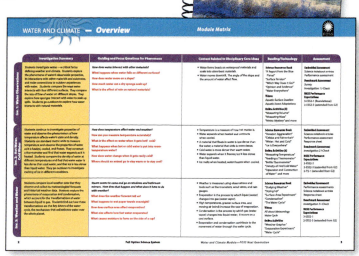

Module matrix

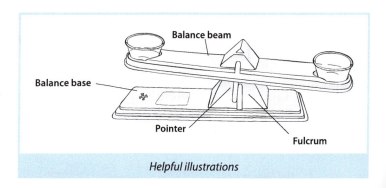

Helpful illustrations

The At-a-Glance chart includes:
- Summaries and pacing for investigation scheduling
- Focus questions for investigative phenomena
- Connections to disciplinary core ideas
- Reading, writing, and technology integration opportunities
- Embedded and benchmark assessments

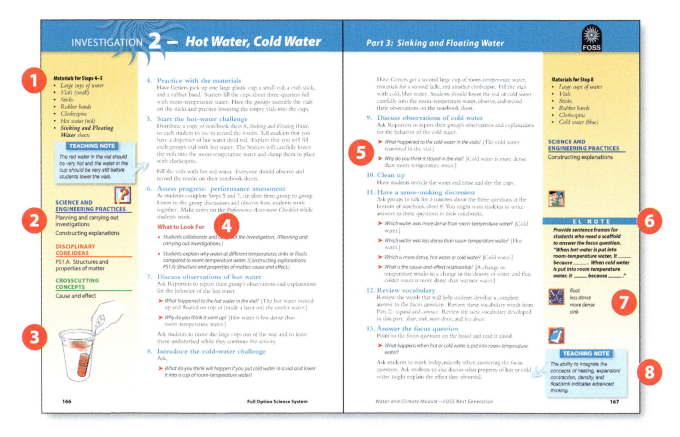

FOSS investigations provide the right support, when you need it with point-of-use guidance.

1. Materials used in the current steps
2. Key three-dimensional highlights
3. Helpful drawings and diagrams
4. Embedded assessment "What to Look For"
5. Guiding questions and expected responses
6. Strategies to support English learners
7. Vocabulary review
8. Teaching notes from real classrooms

The **Assessment** chapter gives you an in-depth look at the research-based components of the FOSS Assessment System, guidance on assessing for the NGSS, and generalized next-step strategies to use in your classroom. Find duplication masters, assessment charts, coding guides, and specific next-step strategies on FOSSweb.com.

Getting Started with Your Teacher Toolkit

2. Teacher Resources

Your *Investigations Guide* tells you how to facilitate each investigation of a module. The ***Teacher Resources*** provides guidance on how to do it at your grade level across three modules throughout the year with effective practices and strategies derived from extensive field-testing.

A grade-level **Planning Guide** provides an overview to your three modules and an introduction to three-dimensional teaching and learning.

The **Science Notebooks** chapter provides age-appropriate methods to support students in developing productive science notebooks. Access powerful research-based next-step strategies to maximize the effectiveness of the notebook as a formative assessment tool.

Science-Centered Language Development is a collection of standards-aligned strategies to support and enhance literacy development in the context of science—reading, writing, speaking, listening, and vocabulary development.

In **Taking FOSS Outdoors**, find guidance for managing the space, time, and materials needed to provide authentic, real-world learning experiences in students' local communities.

Teacher Resources also includes:

- Grade-level connections to Common Core ELA and Math standards
- Module-specific notebook, teacher, and assessment blackline masters.

Check FOSSweb for the latest updates to chapters in *Teacher Resources*.

3. FOSS Science Resources Student Book

The Teacher Toolkit includes one copy of the student book. Reading is an integral part of science learning. Reading informational text critically and effectively is an important component of today's ELA standards. Once students have engaged with phenomena firsthand, they go more in-depth with articles in *FOSS Science Resources*.

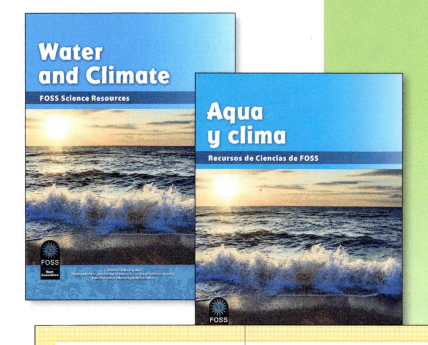

Articles from FOSS *Science Resources* complement and enhance the active investigations, giving students opportunities to:

- Ask and answer questions
- Use evidence to support their ideas
- Use text to acquire information
- Draw information from multiple sources
- Interpret illustrations to build understanding

Interactive eBooks

FOSS Science Resources is available as a convenient, platform-neutral interactive student eBook with integrated audio, highlighted text, and links to videos and online activities. Student access to eBooks is available as an additional purchase.

Getting Started with Technology

FOSSweb.com

Easy access to program support resources

FOSSweb.com is your home for accessing the complete portfolio of digital resources in the FOSS program. Easily manage each of your modules, create class pages, and keep helpful references at your fingertips.

eInvestigations Guide

This easy-to-use interactive version of the *Investigations Guide* is mobile-friendly and offers simplified navigation, collapsible sections, and the ability to add customized notes.

Resources by Investigation

Easily access the duplication masters, online activities, and streaming videos needed for the current investigation part.

Teacher Preparation Videos

Videos provide helpful equipment setup instructions, safety information, and a summary of what students will do and learn throughout a part.

Interactive Whiteboard Lessons

Developed for SMART™ or Promethean boards, these resources help you facilitate each part of every investigation and give the class a visual reference.

Online Activities for Differentiating Instruction

FOSSweb digital resources provide engaging, interactive virtual investigations and tutorials that offer additional content and skill support for students. These experiences also help students who were absent catch up with class.

Streaming Videos

Videos are available on FOSSweb to support many investigations and often take students "on location" around the world or showcase experiments that would be too messy, expensive, or dangerous for the classroom.

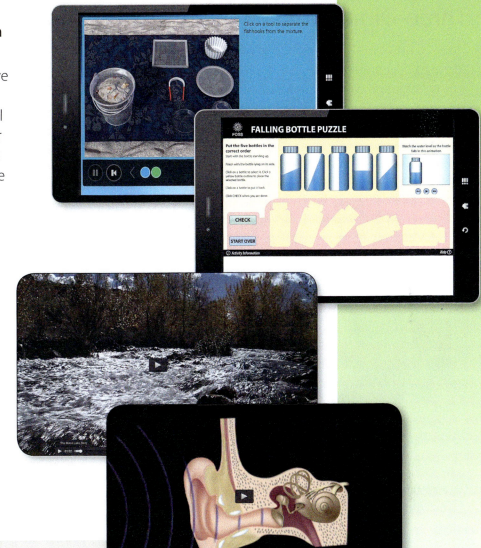

FOSSmap Online Assessment

Students in grades 3–5 can take benchmark assessments online with automatic coding of most responses to save you valuable time and provide instant feedback. A variety of student- and class-level reports help you identify where extra attention is needed and support communication with families and administration. FOSSmap is accessible through the FOSSweb portal.

Three-Dimensional Active Learning

The FOSS program has always placed student learning of science *practices* on equal footing with science *concepts and core ideas* and the NGSS and *Framework for K–12 Science Education* have provided a new language with which to articulate this. In each **FOSS Next Generation** investigation, students are engaged in the three dimensions of the NGSS to develop increasingly complex knowledge and understanding.

Science and engineering practices are the cognitive tools scientists and engineers use to answer questions and design solutions. FOSS students use these tools to gather evidence and to explain real-world phenomena.

Grade-level appropriate **disciplinary core ideas** are the concepts and established ideas of science. FOSS students develop these building blocks throughout investigations to make sense of phenomena.

Crosscutting concepts help students to connect the varied concepts and disciplines of science. FOSS students apply these concepts to different situations in order to make connections and develop comprehensive understanding.

FOSS Forward Thinking

The FOSS Vision

When the Full Option Science System (FOSS) began, the founders envisioned a science curriculum that was enjoyable, logical, and intuitive for teachers, and stimulating, provocative, and informative for students. Achieving this vision was informed by research in cognitive science, learning theory, and critical study of effective practice. The modular design of the FOSS product allowed users to select topics that aligned with district or state learning objectives, or simply resonated with their perception of comprehensive and reasonable science instruction. The original design of the FOSS Program was comprehensive in terms of coverage. FOSS was designed to provide real and meaningful student experience with important scientific ideas and to nurture developmentally appropriate knowledge of the objects, organisms, systems, and principles governing, the natural world.

The FOSS Next Generation Program

But the developers never envisioned FOSS to be a static curriculum, and now the Full Option Science System has evolved into a fully realized 21st century science program with authentic connection to the *Next Generation Science Standards (NGSS)*. The FOSS science curriculum is a comprehensive science program, featuring instructional guidance, student equipment, student reading materials, digital resources, and an embedded assessment system. The FOSS philosophy has always taken very seriously the teaching of good, comprehensive, accurate, science content using the methods of inquiry to advance that science knowledge. But the *Framework for K–12 Science Education*, on which the NGSS are based has allowed us to articulate our mission in a more coherent manner, using the vocabulary established by the authors of the *Framework*. The FOSS instructional design now strives to

a. communicate the disciplinary core ideas (content) of science, while

b. guiding and encouraging students to engage in or exercise the science and engineering practices (inquiry methods) to develop knowledge of the disciplinary core ideas, and

c. help students apprehend the crosscutting concepts (themes that unite core ideas, overarching concepts) that connect the learning experiences within a discipline and bridge meaningfully across disciplines as students gain more and more knowledge of the natural world.

> The Full Option Science System has evolved into a fully realized 21st century science program with authentic connection to the Next Generation Science Standards (NGSS).

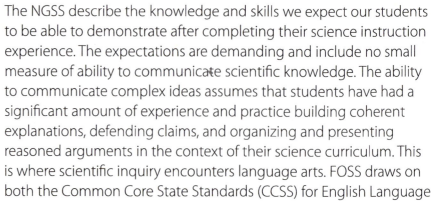

The NGSS describe the knowledge and skills we expect our students to be able to demonstrate after completing their science instruction experience. The expectations are demanding and include no small measure of ability to communicate scientific knowledge. The ability to communicate complex ideas assumes that students have had a significant amount of experience and practice building coherent explanations, defending claims, and organizing and presenting reasoned arguments in the context of their science curriculum. This is where scientific inquiry encounters language arts. FOSS draws on both the Common Core State Standards (CCSS) for English Language Arts and research data regarding the productive use of student science notebooks. FOSS developers realize that the most effective science program must seamlessly integrate science instruction goals and language arts skills. Science is one of the most engaging and productive arenas for introducing and exercising language arts skills: vocabulary, nonfiction (informational) reading, cause-and-effect relationships, on and on.

FOSS is strongly grounded in the realities of the classroom and the interests and experiences of the learners. The content in FOSS is teachable and learnable over multiple grade levels as students increase in their abilities to reason about and integrate complex ideas within and between disciplines.

FOSS is crafted with a structured, yet flexible, teaching philosophy that embraces the much-heralded 21st century skills; collaborative teamwork, critical thinking, and problem solving. The FOSS curriculum design promotes a classroom culture that allows both teachers and students to assume prominent roles in the management of the learning experience.

FOSS is built on the assumptions that understanding of core scientific knowledge and how science functions is essential for citizenship, that all teachers can teach science, and that all students can learn science. Formative assessment in FOSS creates a community of reflective practice. Teachers and students make up the community and establish norms of mutual support, trust, respect, and collaboration. The goal of the community is that everyone will demonstrate progress and will learn and grow.

WARNING — This set contains chemicals that may be harmful if misused. Read cautions on individual containers carefully. Not to be used by children except under adult supervision.

This warning label is required by the
U.S. Consumer Product Safety Commission.
The chemical in the FOSS Environments kit is
kosher salt.

INVESTIGATIONS GUIDE

Environments

TABLE OF CONTENTS

Overview 1
Framework and NGSS 31
Materials 53
Technology 67
Investigation 1: Environmental Factors 79
Part 1: Observing Mealworms 92
Part 2: Designing an Isopod Environment 117
Part 3: Leaf-Litter Critters 130
Investigation 2: Ecosystems 153
Part 1: Designing an Aquarium 164
Part 2: Food Chains and Food Webs 175
Part 3: Population Simulation 190
Part 4: Sound Off 202
Investigation 3: Brine Shrimp Hatching 223
Part 1: Setting Up the Experiment 234
Part 2: Determining Range of Tolerance 243
Part 3: Determining Viability 256
Part 4: Variation in a Population 264
Investigation 4: Range of Tolerance 281
Part 1: Water or Salt Tolerance and Plants 294
Part 2: Plant Patterns 314
Part 3: Plant Adaptations 325
Assessment 339

Overview

ENVIRONMENTS — *Overview*

INTRODUCTION

The study of the structures and behaviors of organisms and the relationships between one organism and its environment builds knowledge of all organisms. With this knowledge comes an awareness of limits. Such knowledge is important because humans can change environments.

The **Environments Module** has four investigations that focus on the anchor phenomenon that animals and plants interact with their environment and with each other. The driving question for the module deals with structure and function—How do the structures of an organism allow it to survive in its environment? Students design investigations to study preferred environments, range of tolerance, and optimum conditions for growth and survival of specific organisms, both terrestrial and aquatic. Students conduct controlled experiments by incrementally changing specific environmental conditions to determine the range of tolerance for early growth of seeds and hatching of brine shrimp, and use these data to develop and use models to understand the impact of changes to the environment. Students explore how animals use their sense of hearing and develop models for detecting and interpreting sound. They graph and interpret data from multiple trials of experiments and build explanations from evidence. Students gain experiences that will contribute to the understanding of crosscutting concepts of patterns; cause and effect; scale, proportion, and quantity; systems and system models; energy and matter; structure and function; and stability and change.

Contents

Introduction	1
Module Matrix	2
FOSS Components	6
FOSS Instructional Design	10
Differentiated Instruction for Access and Equity	18
FOSS Investigation Organization	21
Establishing a Classroom Culture	24
Safety in the Classroom and Outdoors	28
Scheduling the Module	29
FOSS Contacts	30

The NGSS Performance Expectations bundled in this module include:

Life Sciences
4-LS1-1
4-LS1-2
3-LS4-2 (extended)
3-LS4-4 (extended)

Earth and Space Sciences
5-ESS3-1 (foundational)

▶ **NOTE**
The three modules for grade 4 in FOSS Next Generation are

Energy

Soils, Rocks, and Landforms

Environments

Full Option Science System

ENVIRONMENTS — *Overview*

Module Summary	Guiding and Focus Questions for Phenomena
Inv. 1: Environmental Factors Students observe terrestrial organisms as a phenomenon—mealworms and isopods in the classroom and leaf-litter critters on the schoolyard. They set up a mealworm environment at two temperatures and observe the life cycle over time. Students investigate how isopods respond to environmental factors such as water and light, and design an isopod environment. Students investigate small animals that live in leaf litter and study their structures. Students observe and describe the living and nonliving components (biotic and abiotic factors) in terrestrial environments and are introduced to the diverse environments of deserts and rain forests. Students organize information they gather about organisms through first-hand investigations, readings, and videos to understand how structures function to meet the needs of organisms in terrestrial environments.	*How do the structures of terrestrial organisms function to support the survival of the organisms in that environment?* How do mealworm structures and behaviors help them grow and survive? What moisture conditions do isopods prefer? What light conditions do isopods prefer? What are the characteristics of animals living in the leaf-litter environment?
Inv. 2: Ecosystems Students investigate the phenomenon of life in water and how organisms' needs are the same and different from life on land. Students set up a freshwater aquarium with different kinds of fish, plants, and other organisms. They monitor the environmental factors in the system and look for feeding interactions among the populations. They learn about the role of producers, consumers, and decomposers in food chains and food webs in terrestrial and aquatic systems. Through an outdoor simulation, students learn about how food affects a population's home range. Students compare the structures of land and water organisms and the ways structures function to meet the organisms' needs. Students gather and compare information on how different animals obtain one basic need—oxygen. Students explore how animals receive information from their environment through their sensory system and use the information to guide their actions. Through simulations and research, students model how animals use their sense of hearing.	*How are the structures of aquatic organisms similar and different from land animals?* *How do organisms sense and interact with their environment?* What are the environmental factors in an aquatic system? What are the roles of organisms in a food chain? How does food affect a population in its home range? How do animals use their sense of hearing?

Full Option Science System

Module Matrix

Content Related to Disciplinary Core Ideas	Reading/Technology	Assessment
• An environment is everything living and nonliving that surrounds and influences an organism. • A relationship exists between environmental factors and how well organisms grow. • Animals have structures and behaviors that function to support survival, growth, and reproduction. • Every organism has a set of preferred environmental conditions.	**Science Resources Book** "Two Terrestrial Environments" "Darkling Beetles" "Setting Up a Terrarium" "Isopods" "Amazon Rain Forest Journal" **Videos** *Deserts* *Animals of the Rain Forest* *Animal Needs*	**Embedded Assessment** Science notebook entry Performance assessment Response sheet **Benchmark Assessment** *Survey* *Investigation 1 I-Check* **NGSS Performance Expectations** 4-LS1-1 4-LS1-2
• Aquatic environments include living and nonliving factors (water and temperature). • Organisms that live in water have structures that function to meet their needs. Terrestrial and aquatic organisms have similar needs; while their structures are different, the functions are similar. • An ecosystem is the interactions of organisms with one another and with the nonliving environment. • Organisms have structures that allow them to interact in feeding relationships in ecosystems (food chains and food webs). • Producers make their own food, which is also used by animals (consumers); decomposers eat dead plant and animal materials and recycle the nutrients in the system; organisms may compete for resources in an ecosystem. • Decomposers recycle the nutrients in the ecosystem. • Animals communicate to warn others of danger, scare predators away, or locate others of their kind. Animals detect and interpret sounds, and act on them.	**Science Resources Book** "Freshwater Environments" "What Is an Ecosystem?" "Food Chains and Food Webs" "Human Activities and Aquatic Ecosystems" "Comparing Aquatic and Terrestrial Ecosystems" "Animal Sensory Sytems" "Saving Murrelets through Mimicry" **Videos** *Animal Language and Communication* *All about the Senses* **Online Activities** "Virtual Aquarium" "Virtual Terrarium"	**Embedded Assessment** Science notebook entries Response sheet **Benchmark Assessment** *Investigation 2 I-Check* **NGSS Performance Expectations** 4-LS1-1 4-LS1-2 3-LS4-4

Environments Module—FOSS Next Generation 3

ENVIRONMENTS — *Overview*

	Module Summary	Guiding and Focus Questions for Phenomena
Inv. 3: Brine Shrimp Hatching	Students are presented with an ecological problem related to water level fluctuations in an important migratory bird environment—Mono Lake. Brine shrimp are a critical factor in the food web of the lake ecosystem and the salinity of the lake may change due to human activities. Students conduct a controlled experiment to determine which of four salt concentrations allow brine shrimp eggs to hatch. Students determine range of tolerance and optimum conditions for brine shrimp hatching. They use the data to make a recommendation about managing the environment. Students, through an outdoor simulation, look at variation in a population, and consider how variation among individuals contributes to survival of a population.	*How is optimum environment related to organism and population survival?* How can we find out if salinity affects brine shrimp hatching? How does salinity affect the hatching of brine shrimp eggs? Does changing the salt environment allow the brine shrimp eggs to hatch? What are some benefits of having variation within a population?
Inv. 4: Range of Tolerance	Students return to the desert and rain forest environments they studied in Investigation 1 and engage with the phenomenon that different plants survive in each environment. Students set up and monitor controlled experiments to determine the range of tolerance of water for germination of four kinds of seeds: corn, pea, barley, and radish. In a second experiment, students test the effect of salinity on these seeds. Students study local plants by mapping schoolyard plants and relate plant a distribution to environmental factors. Students look at plant adaptations that allow the organisms to thrive in dry desert environments and wet tropical environments.	*What environmental conditions result in the best growth and survival of different plants?* *How do the structures of plants function to support the survival of the organisms in a particular environment?* How much water is needed for early growth of different kinds of plants? What is the salt tolerance of several common farm crops? How does mapping the plants in the schoolyard help us to investigate environmental factors? What are some examples of plant adaptations?

Module Matrix

Content Related to Disciplinary Core Ideas	Reading/Technology	Assessment
• Organisms have ranges of tolerance for environmental factors; there are optimum conditions that produce maximum growth. • Brine shrimp eggs can hatch in a range of salt concentrations, but more hatch in environments with optimum salt concentration. • When environments change, some organisms survive and reproduce; others move; some die. • Individuals of the same kind differ in their characteristics; differences may give individuals an advantage in surviving and reproducing. • Human activities impact environments. Communities are using science to help protect Earth's resources and environments.	**Science Resources Book** "Brine Shrimp" "The Mono Lake Story" "What Happens When Ecosystems Change?" "The Shrimp Club" "Variation and Selection" **Online Activities** "Food Webs" "Virtual Investigation: Trout Range of Tolerance"	**Embedded Assessment** Performance assessment Science notebook entry Response sheet **Benchmark Assessment** Investigation 3 I-Check **NGSS Performance Expectations** 4-LS1-1 3-LS4-2 3-LS4-4 5-ESS3-1
• Organisms have ranges of tolerance for environmental factors; there are optimum conditions that produce maximum growth. • Adaptations are structures and behaviors of an organism that help it survive and reproduce. • A relationship exists between environmental factors and how well organisms grow.	**Science Resources Book** "Environmental Scientists" "Range of Tolerance" "How Organisms Depend on One Another" "Animals from the Past" (optional) **Video** *All about Plant Adaptations* **Online Activity** "Tutorial: Analyzing Environmental Experiments"	**Embedded Assessment** Performance assessment Response sheet **Benchmark Assessment** *Posttest* **NGSS Performance Expectations** 4-LS1-1 3-LS4-4

Environments Module—FOSS Next Generation

ENVIRONMENTS — *Overview*

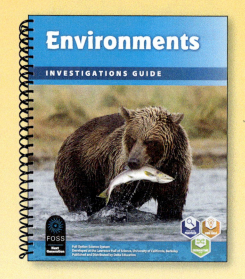

FOSS COMPONENTS

Teacher Toolkit for Each Module

The FOSS Next Generation Program has three modules for grade 4—Energy, Environments, and Soils, Rocks, and Landforms.

Each module comes with a *Teacher Toolkit* for that module. The *Teacher Toolkit* is the most important part of the FOSS Program. It is here that all the wisdom and experience contributed by hundreds of educators has been assembled. Everything we know about the content of the module, how to teach the subject, and the resources that will assist the effort are presented here. Each toolkit has three parts.

Investigations Guide. This spiral-bound document contains these chapters.

- Overview
- Framework and NGSS
- Materials
- Technology
- Investigations (four in this module)
- Assessment

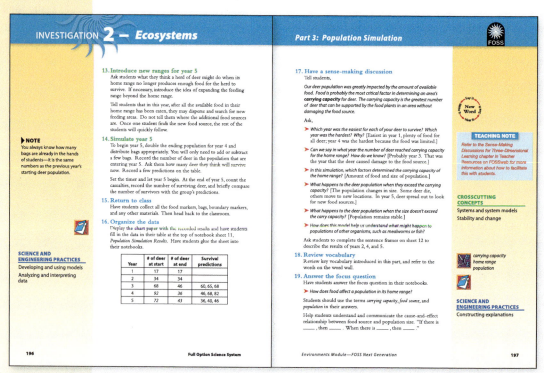

FOSS Components

FOSS Science Resources book. One copy of the student book of readings is included in the *Teacher Toolkit*.

Teacher Resources. These chapters can be downloaded from FOSSweb and are also in the bound *Teacher Resources* book.

- FOSS Program Goals
- Planning Guide—Grade 4
- Science and Engineering Practices—Grade 4
- Crosscutting Concepts—Grade 4
- Sense-Making Discussions for Three-Dimensional Learning—Grade 4
- Access and Equity
- Science Notebooks in Grades 3–5
- Science-Centered Language Development
- FOSS and Common Core ELA—Grade 4
- FOSS and Common Core Math—Grade 4
- Taking FOSS Outdoors
- Science Notebook Masters
- Teacher Masters
- Assessment Masters

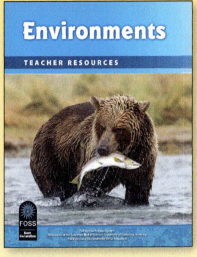

Equipment Kit for Each Module or Grade Level

The FOSS Program provides the materials needed for the investigations, in sturdy, front-opening drawer-and-sleeve cabinets. Inside, you will find high-quality materials packaged for a class of 32 students. Consumable materials are supplied for three uses before you need to resupply. Teachers may be asked to supply small quantities of common classroom materials.

Delta Education can assist you with materials management strategies for schools, districts, and regional consortia.

Environments Module—FOSS Next Generation

ENVIRONMENTS — Overview

FOSS Science Resources Books

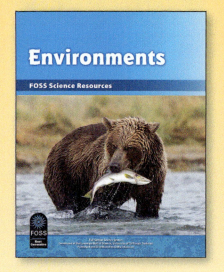

FOSS Science Resources: Environments is a book of original readings developed to accompany this module. The readings are referred to as articles in *Investigations Guide*. Students read the articles in the book as they progress through the module. The articles cover specific concepts, usually after the concepts have been introduced in the active investigation.

The articles in *FOSS Science Resources* and the discussion questions provided in *Investigations Guide* help students make connections to the science concepts introduced and explored during the active investigations. Concept development is most effective when students are allowed to experience organisms, objects, and phenomena firsthand before engaging the concepts in text. The text and illustrations help make connections between what students experience concretely and the ideas that explain their observations.

Animal Sensory Systems

How do you get the information you need to survive in your environment? You get information through your five **senses**. You use your sense of hearing, touch or feel, sight, smell, and taste.

Your senses make you aware of suitable food through smell and taste. They make you aware of things far away through sight and hearing. And they help you sense things that are close, through smell and touch or feel. The sense of touch or feel has many dimensions. Through touch, you can detect many kinds of input to your skin. Human touch can detect pressure, heat, cold, pain, tickle, itch, and textures such as smooth, rough, slippery, and sharp.

Sensory receptors on your body get information from the environment. Some of this information travels to your brain, which processes it. Then it sends information to your body to take action.

Animals use these same senses, or **variations** of these senses, to get the information they need to survive in their environments. Some animals use senses that are beyond the reach of humans.

Rattlesnakes are members of a family of snakes called pitvipers. Pitvipers have a sensory receptor on their face that detects heat. With this structure, rattlesnakes locate **prey** such as mice and rats. These small animals give off body heat. The snakes can sense the heat. Put a blindfold on a rattlesnake, and it can still sense and capture prey. A rattlesnake can find a meal even in complete darkness.

American pitvipers include 16 kinds of rattlesnake, the copperhead, and the cottonmouth. Other kinds of pitvipers live in Central and South America.

A western diamondback rattlesnake

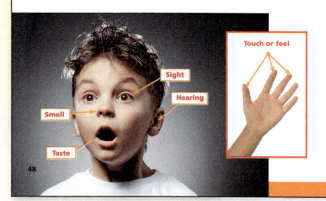

Close-up of a rattlesnake's head showing heat-sensing pit

FOSS Components

Technology

The FOSS website opens new horizons for educators, students, and families, in the classroom or at home. Each module has digital resources for students and families—interactive simulations, virtual investigations, and online activities. For teachers, FOSSweb provides online teacher *Investigations Guides*; grade-level planning guides (with connections to ELA and math); materials management strategies; science teaching and professional development tools; contact information for the FOSS Program developers; and technical support. In addition, FOSSweb provides digital access to PDF versions of the *Teacher Resources* component of the *Teacher Toolkit*, digital-only instructional resources that supplement the print and kit materials, and access to FOSSmap, the online assessment and reporting system for grades 3–8.

With an educator account, you can customize your homepage, set up easy access to the digital components of the modules you teach, and create class pages for your students with access to tutorials and online assessments.

▶ **NOTE**
To access all the teacher resources and to set up customized pages for using FOSS, log in to FOSSweb through an educator account. See the Technology chapter in this guide for more specifics.

Ongoing Professional Learning

The Lawrence Hall of Science and Delta Education strive to develop long-term partnerships with districts and teachers through thoughtful planning, effective implementation, and ongoing teacher support. FOSS has a strong network of consultants who have rich and experienced backgrounds in diverse educational settings using FOSS.

▶ **NOTE**
Look for professional development opportunities and online teaching resources on www.FOSSweb.com.

Environments Module—FOSS Next Generation

ENVIRONMENTS — *Overview*

FOSS INSTRUCTIONAL DESIGN

FOSS is designed around active investigation that provides engagement with science concepts and science and engineering practices. Surrounding and supporting those firsthand investigations are a wide range of experiences that help build student understanding of core science concepts and deepen scientific habits of mind.

The Elements of the FOSS Instructional Design

FOSS Instructional Design

Each FOSS investigation follows a similar design to provide multiple exposures to science concepts. The design includes these pedagogies.

- Active investigation in collaborative groups: firsthand experiences with phenomena in the natural and designed worlds
- Recording in science notebooks to answer a focus question dealing with the scientific phenomenon under investigation
- Reading informational text in *FOSS Science Resources* books
- Online activities to acquire data or information or to elaborate and extend the investigation
- Outdoor experiences to collect data from the local environment or to apply knowledge
- Assessment to monitor progress and inform student learning

In practice, these components are seamlessly integrated into a curriculum designed to maximize every student's opportunity to learn.

A **learning cycle** employs an instructional model based on a constructivist perspective that calls on students to be actively involved in their own learning. The model systematically describes both teacher and learner behaviors in a coherent approach to science instruction.

A popular model describes a sequence of five phases of intellectual involvement known as the 5Es: engage, explore, explain, elaborate, and evaluate. The body of foundational knowledge that informs contemporary learning-cycle thinking has been incorporated seamlessly and invisibly into the FOSS curriculum design.

Engagement with real-world **phenomena** is at the heart of FOSS. In every part of every investigation, the investigative phenomenon is referenced implicitly in the focus question that guides instruction and frames the intellectual work. The focus question is a prominent part of each lesson and is called out for the teacher and student. The investigation Background for the Teacher section is organized by focus question—the teacher has the opportunity to read and reflect on the phenomenon in each part in preparing for the lesson. Students record the focus question in their science notebooks, and after exploring the phenomenon thoroughly, explain their thinking in words and drawings.

In science, a phenomenon is a natural occurrence, circumstance, or structure that is perceptible by the senses—an observable reality. Scientific phenomena are not necessarily phenomenal (although they may be)—most of the time they are pretty mundane and well within the everyday experience. What FOSS does to enact an effective engagement with the NGSS is thoughtful selection of scientific phenomena for students to investigate.

▶ **NOTE**
The anchor phenomena establish the storyline for the module. The investigative phenomena guide each investigation part. Related examples of everyday phenomena are incorporated into the readings, videos, discussions, formative assessments, outdoor experiences, and extensions.

Environments Module—FOSS Next Generation

ENVIRONMENTS — Overview

Active Investigation

Active investigation is a master pedagogy. Embedded within active learning are a number of pedagogical elements and practices that keep active investigation vigorous and productive. The enterprise of active investigation includes

- context: sharing prior knowledge, questioning, and planning;
- activity: doing and observing;
- data management: recording, organizing, and processing;
- analysis: discussing and writing explanations.

Context: sharing, questioning, and planning. Active investigation requires focus. The context of an inquiry can be established with a focus question about a phenomenon or challenge from you or, in some cases, from students. (How can we find out if salinity affects the hatching of brine shrimp eggs?) At other times, students are asked to plan a method for investigation. This might start with a teacher demonstration or presentation. Then you challenge students to plan an investigation, such as to find out how much water is needed for early growth of plants. In either case, the field available for thought and interaction is limited. This clarification of context and purpose results in a more productive investigation.

Activity: doing and observing. In the practice of science, scientists put things together and take things apart, observe systems and interactions, and conduct experiments. This is the core of science—active, firsthand experience with objects, organisms, materials, and systems in the natural world. In FOSS, students engage in the same processes. Students often conduct investigations in collaborative groups of four, with each student taking a role to contribute to the effort.

The active investigations in FOSS are cohesive, and build on each other to lead students to a comprehensive understanding of concepts. Through investigations and readings, students gather meaningful data.

Data management: recording, organizing, and processing. Data accrue from observation, both direct (through the senses) and indirect (mediated by instrumentation). Data are the raw material from which scientific knowledge and meaning are synthesized. During and after work with materials, students record data in their science notebooks. Data recording is the first of several kinds of student writing.

Students then organize data so they will be easier to think about. Tables allow efficient comparison. Organizing data in a sequence (time) or series (size) can reveal patterns. Students process some data into graphs, providing visual display of numerical data. They also organize data and process them in the science notebook.

FOSS Instructional Design

Analysis: discussing and writing explanations. The most important part of an active investigation is extracting its meaning. This constructive process involves logic, discourse, and prior knowledge. Students share their explanations for phenomena, using evidence generated during the investigation to support their ideas. They conclude the active investigation by writing a summary of their learning as well as questions raised during the activity in their science notebooks.

Science Notebooks

Research and best practice have led FOSS to place more emphasis on the student science notebook. Keeping a notebook helps students organize their observations and data, process their data, and maintain a record of their learning for future reference. The process of writing about their science experiences and communicating their thinking is a powerful learning device for students. The science-notebook entries stand as credible and useful expressions of learning. The artifacts in the notebooks form one of the core exhibitions of the assessment system.

You will find the duplication masters for grades 1–5 presented in notebook format. They are reduced in size (two copies to a standard sheet) for placement (glue or tape) into a bound composition book. Full-sized masters for grades 3–5 that can be filled in electronically and are suitable for display are available on FOSSweb. Student work is entered partly in spaces provided on the notebook sheets and partly on adjacent blank sheets in the composition book. Look to the chapter in *Teacher Resources* called Science Notebooks in Grades 3–5 for more details on how to use notebooks with FOSS.

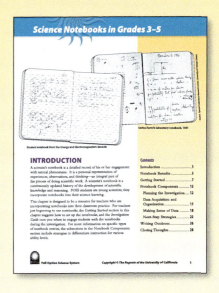

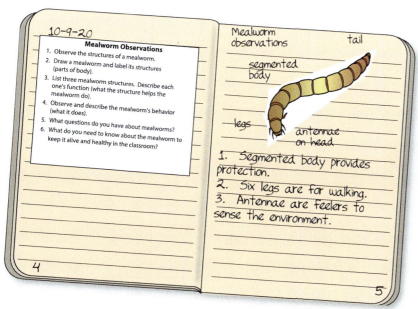

Environments Module—FOSS Next Generation

ENVIRONMENTS — Overview

Reading in FOSS Science Resources

The *FOSS Science Resources* books, available in print and interactive eBooks, are primarily devoted to expository articles and biographical sketches. FOSS suggests that the reading be completed during language-arts time to connect to the Common Core State Standards for ELA. When language-arts skills and methods are embedded in content material that relates to the authentic experience students have had during the FOSS active learning sessions, students are interested, and they get more meaning from the text material.

Recommended strategies to engage students in reading, writing, speaking, and listening around the articles in the *FOSS Science Resources* books are included in the flow of Guiding the Investigation. In addition, a library of resources is described in the Science-Centered Language Development chapter in *Teacher Resources*.

The FOSS and the Common Core ELA—Grade 4 chapter in *Teacher Resources* shows how FOSS provides opportunities to develop and exercise the Common Core ELA practices through science. A detailed table identifies these opportunities in the three FOSS modules for the fourth grade.

Engaging in Online Activities through FOSSweb

The simulations and online activities on FOSSweb are designed to support students' learning at specific times during instruction. Digital resources include streaming videos that can be viewed by the class or small groups. Resources may also include virtual investigations and tutorials that students can use to review the active investigations and to support students who need more time with the concepts or who have been absent and missed the active investigations.

The Technology chapter provides details about the online activities for students and the tools and resources for teachers to support and enrich instruction. There are many ways for students to engage with the digital resources—in class as individuals, in small groups, or as a whole class, and at home with family and friends.

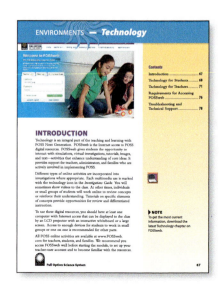

14

Full Option Science System

FOSS Instructional Design

Assessing Progress

The FOSS assessment system includes both formative and summative assessments. Formative assessment monitors learning during the process of instruction. It measures progress, provides information about learning, and is predominantly diagnostic. Summative assessment looks at the learning after instruction is completed, and it measures achievement.

Formative assessment in FOSS, called **embedded assessment**, is an integral part of instruction, and occurs on a daily basis. You observe action during class in a performance assessment or review notebooks after class. Performance assessments look at students' engagement in science and engineering practices or their recognition of crosscutting concepts, and are indicated with the second assessment icon. Embedded assessment provides continuous monitoring of students' learning and helps you make decisions about whether to review, extend, or move on to the next idea to be covered.

The embedded assessments are based on authentic work produced by students during the course of participating in the FOSS activities. Students do their science, and you observe the actions and look at their notebook entries. Bullet points in the Guiding the Investigation tell you specifically what students should know and be able to communicate.

Benchmark assessments include the *Survey*, I-Checks, *Posttest*, and interim assessments. The *Survey* is given before instruction begins. It provides information about students' prior knowledge. **I-Checks** are actually hybrid tools: they can provide summative information about students' achievement, but more importantly, they can be used formatively as well to provide diagnostic information. Reviewing specific items on an I-Check with the class provides additional opportunities for students to clarify their thinking. The *Posttest* is a summative assessment given after instruction is complete.

Interim assessments give students practice with items specifically designed to measure three-dimensional learning described by NGSS performance expectations. Interim assessment tasks generally begin with a scenario, and ask students to apply practices and crosscutting concepts as well as disciplinary core ideas to respond to the item. Interim assessment tasks can be administered during a module, at the end of the module, or as an end-of-year grade-level assessment.

All benchmark items are carefully designed to be valid, reliable, and accessible to all students. They focus on assessment *for* learning, and when accompanied by thoughtful self-assessment activities and feedback, contribute to the development of a growth mindset.

▶ **TECHNOLOGY COMPONENTS OF THE FOSS ASSESSMENT SYSTEM**

FOSSmap for teachers and online assessment for students are the technology components of the FOSS assessment system. Students in grades 3–5 can take assessments online. FOSSmap provides the tools for you to review those assessments online so you can determine next steps for the class or differentiated instruction for individual students based on assessment performance. See the Assessment chapter for more information on these technology components.

Environments Module—FOSS Next Generation

ENVIRONMENTS — Overview

Taking FOSS Outdoors

FOSS throws open the classroom door and proclaims the entire school campus to be the science classroom. The true value of science knowledge is its usefulness in the real world and not just in the classroom. Taking regular excursions into the immediate outdoor environment has many benefits. First of all, it provides opportunities for students to apply things they learned in the classroom to novel situations. When students are able to transfer knowledge of scientific principles to natural systems, they experience a sense of accomplishment.

In addition to transfer and application, students can learn things outdoors that they are not able to learn indoors. The most important object of inquiry outdoors is the outdoors itself. To today's youth, the outdoors is something to pass through as quickly as possible to get to the next human-managed place. For many, engagement with the outdoors and natural systems must be intentional, at least at first. With repeated visits to familiar outdoor learning environments, students may first develop comfort in the outdoors, and then a desire to embrace and understand natural systems.

The last part of most investigations is an outdoor experience. Venturing out will require courage the first time or two you mount an outdoor expedition. It will confuse students as they struggle to find the right behavior that is a compromise between classroom rigor and diligence and the freedom of recreation. With persistence, you will reap rewards. You will be pleased to see students' comportment develop into proper field-study habits, and you might be amazed by the transformation of students with behavior issues in the classroom who become your insightful observers and leaders in the schoolyard environment.

▶ **NOTE**
The kit includes a set of four *Conservation* posters so you can discuss the importance of natural resources with students.

Teaching outdoors is the same as teaching indoors—except for the space. You need to manage the same four core elements of classroom teaching: time, space, materials, and students. Because of the different space, new management procedures are required. Students can get farther away. Materials have to be transported. The space has to be defined and honored. Time has to be budgeted for getting to, moving around in, and returning from the outdoor study site. All these and more issues and solutions are discussed in the Taking FOSS Outdoors chapter in *Teacher Resources*.

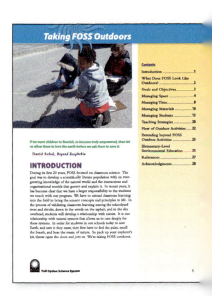

16 Full Option Science System

FOSS Instructional Design

Science-Centered Language Development and Common Core State Standards for ELA

The FOSS active investigations, science notebooks, *FOSS Science Resources* articles, and formative assessments provide rich contexts in which students develop and exercise thinking and communication. These elements are essential for effective instruction in both science and language arts—students experience the natural world in real and authentic ways and use language to inquire, process information, and communicate their thinking about scientific phenomena. FOSS refers to this development of language process and skills within the context of science as science-centered language development.

In the Science-Centered Language Development chapter in *Teacher Resources*, we explore the intersection of science and language and the implications for effective science teaching and language development. Language plays two crucial roles in science learning: (1) it facilitates the communication of conceptual and procedural knowledge, questions, and propositions, and (2) it mediates thinking—a process necessary for understanding. For students, language development is intimately involved in their learning about the natural world. Science provides a real and engaging context for developing literacy and language-arts skills identified in contemporary standards for English language arts.

The most effective integration depends on the type of investigation, the experience of students, the language skills and needs of students, and the language objectives that you deem important at the time. The Science-Centered Language Development chapter is a library of resources and strategies for you to use. The chapter describes how literacy strategies are integrated purposefully into the FOSS investigations, gives suggestions for additional literacy strategies that both enhance students' learning in science and develop or exercise English-language literacy skills, and develops science vocabulary with scaffolding strategies for supporting all learners. We identify effective practices in language-arts instruction that support science learning and examine how learning science content and engaging in science and engineering practices support language development.

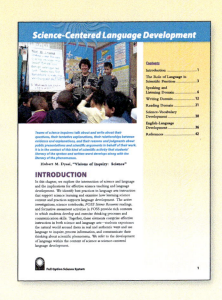

Specific methods to make connections to the Common Core State Standards for English Language Arts are included in the flow of Guiding the Investigation. These recommended methods are linked to the CCSS ELA through ELA Connection notes. In addition, the FOSS and the Common Core ELA chapter in *Teacher Resources* summarizes all of the connections to each standard at the given grade level.

Environments Module—FOSS Next Generation 17

ENVIRONMENTS — Overview

DIFFERENTIATED INSTRUCTION FOR ACCESS AND EQUITY

Learning from Experience

The roots of FOSS extend back to the mid-1970s and the Science Activities for the Visually Impaired and Science Enrichment for Learners with Physical Handicaps projects (SAVI/SELPH Program). As this special-education science program expanded into fully integrated (mainstreamed) settings in the 1980s, hands-on science proved to be a powerful medium for bringing all students together. The subject matter is universally interesting, and the joy and satisfaction of discovery are shared by everyone. Active science by itself provides part of the solution to full inclusion and provides many opportunities at the same time for differentiated instruction.

Many years later, FOSS began a collaboration with educators and researchers at the Center for Applied Special Technology (CAST), where principles of Universal Design for Learning (UDL) had been developed and applied. FOSS continues to learn from our colleagues about ways to use new media and technologies to improve instruction. Here are the UDL guiding principles.

Principle 1. Provide multiple means of representation. Give learners various ways to acquire information and demonstrate knowledge.

Principle 2. Provide multiple means of action and expression. Offer students alternatives for communicating what they know.

Principle 3. Provide multiple means of engagement. Help learners get interested, be challenged, and stay motivated.

> *"Active science by itself provides part of the solution to full inclusion and provides many opportunities at the same time for differentiated instruction."*

Differentiated Instruction for Access and Equity

FOSS for All Students

The FOSS Program has been designed to maximize the science learning opportunities for all students, including those who have traditionally not had access to or have not benefited from equitable science experiences—students with special needs, ethnically diverse learners, English learners, students living in poverty, girls, and advanced and gifted learners. FOSS is rooted in a 30-year tradition of multisensory science education and informed by recent research on UDL and culturally and linguistically responsive teaching and learning. Procedures found effective with students with special needs and students who are learning English are incorporated into the materials and strategies used with all students during the initial instruction phase. In addition, the **Access and Equity** chapter in *Teacher Resources* (or go to FOSSweb to download this chapter) provides strategies and suggestions for enhancing the science and engineering experiences for each of the specific groups noted above.

Throughout the FOSS investigations, students experience multiple ways of interacting with phenomena and expressing their understanding through a variety of modalities. Each student has multiple opportunities to demonstrate his or her strengths and needs, thoughts, and aspirations.

The challenge is then to provide appropriate follow-up experiences or enhancements appropriate for each student. For some students, this might mean more time with the active investigations or online activities. For other students, it might mean more experience and/or scaffolds for developing models, building explanations, or engaging in argument from evidence.

For some students, it might mean making vocabulary and language structures more explicit through new concrete experiences or through reading to students. It may help them identify and understand relationships and connections through graphic organizers.

For other students, it might be designing individual projects or small-group investigations. It might be more opportunities for experiencing science outside the classroom in more natural, outdoor environments or defining problems and designing solutions in their communities.

Environments Module—FOSS Next Generation

Assessment and Extensions

The next-step strategies suggested during the self-assessment sessions following I-Checks provide opportunities for differentiated instruction. These strategies can also be used to provide targeted and strategic instruction for students who need additional specific support. For more on next-step strategies, see the Assessment chapter.

There are additional approaches and strategies for ensuring access and equity by providing appropriate differentiated instruction. The FOSS Program provides tools and strategies so that you know what students are thinking throughout the module. Based on that knowledge and what you know about your students' cultural and linguistic background, as well as their individual strengths and needs, read through the extension activities for additional ways to enhance the learning experience for your students. Interdisciplinary extensions in the arts, social studies, math, and language arts, as well as more advanced projects are listed at the end of each investigation. Use these ideas to meet the individual needs and interests of your students. In addition, online activities including tutorials and virtual investigations are effective tools to provide differentiated levels of instruction.

English Learners

The FOSS Program provides a rich laboratory for language development for English learners. A variety of techniques are provided to make science concepts clear and concrete, including modeling, visuals, and active investigations in small groups. Instruction is guided and scaffolded through carefully designed lesson plans, and students are supported throughout.

Science vocabulary and language structures are introduced in authentic contexts while students engage in hands-on learning and collaborative discussion. Strategies for helping all students read, write, speak, and listen are described in the Science-Centered Language Development chapter. A specific section on English learners provides suggestions for both integrating English language development (ELD) approaches during the investigation and for developing designated (targeted and strategic) ELD-focused lessons that support science learning.

FOSS Investigation Organization

FOSS INVESTIGATION ORGANIZATION

Modules are subdivided into **investigations** (four in this module). Investigations are further subdivided into three to five **parts**. Each investigation has a general guiding question for the phenomenon students investigate, and each part of each investigation is driven by a **focus question**. The focus question, usually presented as the part begins, engages the student with the phenomenon and signals the challenge to be met, mystery to be solved, or principle to be uncovered. The focus question guides students' actions and thinking and makes the learning goal of each part explicit for teachers. Each part concludes with students recording an answer to the focus question in their notebooks.

The investigation is summarized for the teacher in the At-a-Glance chart at the beginning of each investigation.

Investigation-specific **scientific background** information for the teacher is presented in each investigation chapter organized by the focus questions.

The **Teaching Children about** section makes direct connections to the NGSS foundation boxes for the grade level—Disciplinary Core Ideas, Science and Engineering Practices, and Crosscutting Concepts. This information is later presented in color-coded sidebar notes to identify specific places in the flow of the investigation where connections to the three dimensions of science learning appear. The Teaching Children about section ends with information about teaching and learning and a conceptual-flow graphic of the content.

The **Materials** and **Getting Ready** sections provide scheduling information and detail exactly how to prepare the materials and resources for conducting the investigation.

Teaching notes and **ELA Connections** appear in blue boxes in the sidebars. These notes comprise a second voice in the curriculum—an educative element. The first (traditional) voice is the message you deliver to students. The second (educative) voice, shared as a teaching note, is designed to help you understand the science content and pedagogical rationale at work behind the instructional scene. ELA Connection boxes provide connections to the Common Core State Standards for English Language Arts.

FOCUS QUESTION
How do animals use their sense of hearing?

SCIENCE AND ENGINEERING PRACTICES
Developing and using models

DISCIPLINARY CORE IDEAS
LS1.A: Structure and function

CROSSCUTTING CONCEPTS
Systems and system models

TEACHING NOTE
This focus question can be answered with a simple yes or no, but the question has power when students support their answers with evidence. Their answers should take the form "Yes, because ___."

Environments Module—FOSS Next Generation

ENVIRONMENTS — Overview

The **Getting Ready** and **Guiding the Investigation** sections have several features that are flagged in the sidebars. These include several icons to remind you when a particular pedagogical method is suggested, as well as concise bits of information in several categories.

The **safety** icon alerts you to potential safety issues related to chemicals, allergic reactions, and the use of safety goggles.

The small-group **discussion** icon asks you to pause while students discuss data or construct explanations in their groups.

The **new-word** icon alerts you to a new vocabulary word or phrase that should be introduced thoughtfully.

The **vocabulary** icon indicates where students should review recently introduced vocabulary.

The **recording** icon points out where students should make a science-notebook entry.

The **reading** icon signals when the class should read a specific article in the *FOSS Science Resources* book.

The **technology** icon signals when the class should use a digital resource on FOSSweb.

Full Option Science System

FOSS Investigation Organization

The **assessment** icons appear when there is an opportunity to assess student progress by using embedded or benchmark assessments. Some are performance assessments—observations of science and engineering practices, indicated by a second icon which includes a beaker and ruler.

The **outdoor** icon signals when to move the science learning experience into the schoolyard.

The **engineering** icon indicates opportunities for an experience incorporating engineering practices.

The **math** icon indicates an opportunity to engage in numerical data analysis and mathematics practice.

The **crosscutting concepts** icon indicates an opportunity to expand on the concept by going to *Teacher Resources*, Crosscutting Concepts chapter. This chapter provides details on how to engage students with that concept in the context of the investigation.

The **EL note** provides a specific strategy to use to assist English learners in developing science concepts.

E L   N O T E

To help with pacing, you will see icons for **breakpoints**. Some breakpoints are essential, and others are optional.

POSSIBLE BREAKPOINT

Environments Module—FOSS Next Generation

ENVIRONMENTS — Overview

ESTABLISHING A CLASSROOM CULTURE

Working in Collaborative Groups

Collaboration is important in science. Scientists usually collaborate on research enterprises. Groups of researchers often contribute to the collection of data, the analysis of findings, and the preparation of the results for publication.

Collaboration is expected in the science classroom too. Some tasks call for everyone to have the same experience, either taking turns or doing the same things simultaneously. At other times, group members may have different experiences that they later bring together.

Research has shown that students learn better and are more successful when they collaborate. Working together promotes student interest, participation, learning, language development, and self-confidence. FOSS investigations use collaborative groups extensively. Here are a few general guidelines for collaborative learning that have proven successful over the years.

For most activities in upper-elementary grades, collaborative groups of four in which students take turns assuming specific responsibilities work best. Groups can be identified completely randomly (first four names drawn from a hat constitute group 1), or you can assemble groups to ensure diversity and inclusion. Thoughtfully constituted groups tend to work better.

Groups can be maintained for extended periods of time, or they can be reconfigured frequently. Six to nine weeks seems about optimum, so students might stay together throughout an entire module.

Functional roles within groups can be determined by the members themselves, or they can be assigned in one of several ways. Each member in a collaborative group can be assigned a number or a color. Then you need only announce which color or number will perform a certain task for the group at a certain time. Compass points can also be used: the person seated on the east side of the table will be the Reporter for this investigation.

The functional roles used in the investigations follow. If you already use other names for functional roles in your class, use them in place of those in the investigations.

The **Getter** is responsible for materials. One person from each group gets equipment from the materials station, and later returns the equipment.

Establishing A Classroom Culture

One person is the **Starter** for each task. This person supervises setting up the equipment and makes sure that everyone gets a turn and that everyone has an opportunity to contribute ideas to the investigation.

The **Recorder** makes sure that everyone has recorded information in his or her science notebook and, as appropriate, records the group data and ideas.

The **Reporter** reports group data to the class or transcribes it to the board or class chart, and shares the main points of the group discussion during class discussion.

Getting started with collaborative groups requires patience, but the rewards are great. Once collaborative groups are in place, you will be able to engage students in more meaningful conversations about science content. You are free to "cruise" the groups, to observe and listen to students as they work, and to interact with individuals and small groups as needed.

Norms for Sense-Making Discussions

Setting up norms for discussion and holding yourself and your students accountable is the first step toward creating a culture of productive talk in the classroom that supports engagement in the science and engineering practices. Students need to feel free to express their ideas, and to provide and receive criticism from others as they work toward understanding of the disciplinary core ideas of science and methods of engineering.

Establish norms at the beginning of the school year. It is recommended that this be done together as a class activity. However, presenting a poster of norms to students and asking them to discuss why each one is important can also be effective. Before each sense-making discussion, review the norms. Review what it will look like, sound like, and feel like when everyone is following the agreements. You might have students work on one or two at a time as they are developing their oral discourse skills. After discussion, save a few minutes for reflection on how well the group or the class adhered to the norms and what they can do better next time. More strategies for supporting academic discourse can be found in the Sense-Making Discussions for Three-Dimensional Learning and Science-Centered Language Development chapters in *Teacher Resources* (also available as downloadable PDFs on FOSSweb).

This poster is an example of student responsibilities that the class discussed and adopted as their norms.

Environments Module—FOSS Next Generation

ENVIRONMENTS — Overview

Managing Materials

The Materials section lists the items in the equipment kit and any teacher-supplied materials. It also describes things to do to prepare a new kit and how to check and prepare the kit for your classroom. Classroom volunteers and helpful students can also assist in setting up and preparing the materials.

Individual photos of each piece of FOSS equipment are available for printing from FOSSweb, and can help students and you identify each item. They can be used to support emerging readers and English language learners to acquire and use new vocabulary words necessary to engage in the investigations. (Photo equipment cards are available in English and Spanish formats.)

The FOSS Program designers suggest using a central materials distribution system. You organize all the materials for an investigation at a single location called the materials station. As the investigation progresses, one member of each group gets materials as they are needed, and another returns the materials when the investigation is complete. You place the equipment and resources at the station, and students do the rest. Students can also be involved in cleaning and organizing the materials at the end of a session. Be sure to work with students to Reduce, Reuse, and Recycle the materials used in science.

When Students Are Absent

When a student is absent for a session, give him or her a chance to spend some time with the materials at a center. Another student might act as a peer tutor. Allow the student to bring home a *FOSS Science Resources* book to read with a family member. There are suggested interactive reading strategies for each article as well as embedded review items that the student can respond to verbally or in writing.

There is a set of two or three virtual investigations for each FOSS module for grades 3–5. Students who have been absent from certain investigations can access these simulations online through FOSSweb. The virtual investigations require students to record data and answer concluding questions in their science notebooks. Sometimes the notebook sheet that was used in the classroom investigation is also used for the virtual investigation.

Establishing A Classroom Culture

Collaborative Teaching and Learning

Collaborative learning requires a collective as well as individual growth mindset. A growth mindset is when people believe that their most basic abilities can be developed through dedication and hard work (see the research of Carol Dweck and her book *Mindset: The Psychology of Success*). As students work together to make sense of phenomena and develop their inquiry and discourse skills, it's important to recognize and value their efforts to try new approaches and their willingness to make their thinking visible. Remind students that everyone in the classroom, including you the teacher, will be learning new ideas and ways to think about the world. Where there is productive struggle, there is learning. Here are a few ways to help students develop a growth mindset for science and engineering.

- **Praise effort, not right answers**. When students are successful at a task, provide positive feedback about their level of engagement and effort in the practices, e.g., the efforts they put into careful observations, how well they organized and interpreted their data, the relevancy of their questions, how well they connected or applied new concepts, and their use of precise vocabulary, etc. Also, try to provide feedback that encourages students to continue to improve their learning and exploring, e.g., is there another way to approach this question? Have you thought about _____? What evidence is there to support _____?

- **Foster and validate divergent thinking**. During sense-making discussions, continually emphasize how important it is to share emerging ideas and to be open to the ideas of others in order to build understanding. Model for students how you refine and revise your thinking based on new information. Make it clear to students that the point is not for them to show they have the right answer, but rather to help each other arrive at new understandings. Point out positive examples of students expressing and revising their ideas.

Establishing a classroom culture that supports three-dimensional teaching and learning centers on collaboration. Collaborative groupings, materials management, and norms are structures you can put into place to foster collaboration. These structures along with the expectations that students will be negotiating meaning together as a community of learners, creates a learning environment where students are compelled to work, think, and communicate like scientists and engineers to help one another learn.

Environments Module—FOSS Next Generation

ENVIRONMENTS — *Overview*

SAFETY IN THE CLASSROOM AND OUTDOORS

Following the procedures described in each investigation will make for a very safe experience in the classroom. You should also review your district safety guidelines and make sure that everything you do is consistent with those guidelines. Two posters are included in the kit: *FOSS Science Safety* for classroom use and *FOSS Outdoor Safety* for outdoor activities.

Look for the safety icon in the Getting Ready and Guiding the Investigation sections that will alert you to safety considerations throughout the module.

Safety Data Sheets (SDS) for materials used in the FOSS Program can be found on FOSSweb. If you have questions regarding any SDS, call Delta Education at 1-800-258-1302 (Monday–Friday, 8 a.m.–5 p.m. ET).

Science Safety in the Classroom

General classroom safety rules to share with students are listed here.

1. Listen carefully to your teacher's instructions. Follow all directions. Ask questions if you don't know what to do.
2. Tell your teacher if you have any allergies.
3. Never put any materials in your mouth. Do not taste anything unless your teacher tells you to do so.
4. Never smell any unknown material. If your teacher tells you to smell something, wave your hand over the material to bring the smell toward your nose.
5. Do not touch your face, mouth, ears, eyes, or nose while working with chemicals, plants, or animals.
6. Always protect your eyes. Wear safety goggles when necessary. Tell your teacher if you wear contact lenses.
7. Always wash your hands with soap and warm water after handling chemicals, plants, or animals.
8. Never mix any chemicals unless your teacher tells you to do so.
9. Report all spills, accidents, and injuries to your teacher.
10. Treat animals with respect, caution, and consideration.
11. Clean up your work space after each investigation.
12. Act responsibly during all science activities.

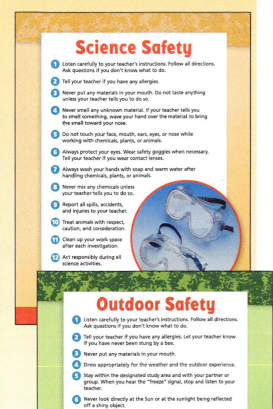

Scheduling the Module

SCHEDULING THE MODULE

Below is a suggested teaching schedule for the module. The investigations are numbered and should be taught in order, as the concepts build upon each other from investigation to investigation. We suggest that a minimum of 9 weeks be devoted to this module.

Active-investigation (A) sessions include hands-on work with materials and organims, active thinking about experiences, small-group discussion, writing in science notebooks, and learning new vocabulary in context.

Reading (R) sessions involve reading *FOSS Science Resources* articles. Reading can be completed during language-arts time to make connections to Common Core State Standards for ELA (CCSS ELA).

During **Wrap-Up/Warm-Up (W)** sessions, students share notebook entries and engage in connections to CCSS ELA. These sessions can also be completed during language-arts time.

I-Checks are short summative assessments at the end of each investigation. Students have a short notebook review session the day before and a self-assessment of selected items the following day. (See the Assessment chapter for the next-step strategies for self-assessment.)

Week	Day 1	Day 2	Day 3	Day 4	Day 5
1	Survey START Inv. 1 Part 1 A (mealworms)	(Dry soil) R/W	START Inv. 1 Part 2 A/R	A/R	A
2	A/W	START Inv. 1 Part 3 A/R	R	R/Review	(Start aging water) I-Check 1
3	(Observe mealworms) Self-assess	START Inv. 2 Part 1 A	A/R/W	START Inv. 2 Part 2 A	A/R
4	(Test shrimp eggs) R/W	START Inv. 2 Part 3 A/R	R/W	START Inv. 2 Part 4 A	A/R
5	R	Review	I-Check 2	Self-assess	(Observe mealworms)
6	START Inv. 3 Part 1 A/W	START Inv. 3 Part 2 A	A/R	A/R/W	START Inv. 3 Part 3 A
7	(Dry soil) A/R/W	START Inv. 3 Part 4 A/R	Review	START Inv. 4 Part 1 A (planting)	I-Check 3
8	Self-assess 3	(Day 5 observe) A/R	R/W	START Inv. 4 Part 2 A	(Day 8 observe) A/R/W
9	START Inv. 4 Part 3 A	("Darkling Beetles") R	(Day 13 observe) A	Review	Posttest

> **NOTE**
> Read through Investigation 1, Part 1, so you know how to plan the schedule for observing the life cycle stages of the darkling beetle (mealworm). It will take about 6 weeks for the mealworms to become adult beetles. Students are introduced to the reading, "Darkling Beetles", in week 9, day 2, after they have observed the entire life cycle.
>
> It is important to start Investigation 3, Part 1, with brine shrimp eggs on a Monday, have students make observations each day, and set up Part 3 on Friday.
>
> In Investigation 4, Part 1, students set up two plant experiments (half the class sets up the water tolerance experiment and the other half sets up the salt tolerance experiment). Students observe and record plant changes on days 5, 8, and 13. On day 13 or 14 the experiments come to an end; students make final observations and record. Plan to go on to Investigation 4, Part 2 and Part 3, before finishing Part 1 of that investigation.
>
> Be prepared—read the Getting Ready section thoroughly and review the teacher preparation video on FOSSweb.

Environments Module—FOSS Next Generation

ENVIRONMENTS — Overview

FOSS CONTACTS

General FOSS Program information

www.FOSSweb.com

www.DeltaEducation.com/FOSS

Developers at the Lawrence Hall of Science

foss@berkeley.edu

Customer Service at Delta Education

www.DeltaEducation.com/contact.aspx

Phone: 1-800-258-1302, 8:00 a.m.–5:00 p.m. ET

FOSSmap (online component of FOSS assessment system)

http://FOSSmap.com/

FOSSweb account questions/access codes/help logging in

techsupport.science@schoolspecialty.com

Phone: 1-800-258-1302, 8:00 a.m.–5:00 p.m. ET

School Specialty online support

loginhelp@schoolspecialty.com

Phone: 1-800-513-2465, 8:30 a.m.–6:00 p.m. ET

FOSSweb tech support

support@fossweb.com

Professional development

www.FOSSweb.com/Professional-Development

Safety issues

www.DeltaEducation.com/SDS

Phone: 1-800-258-1302, 8:00 a.m.–5:00 p.m. ET

For chemical emergencies, contact Chemtrec 24 hours per day.
Phone: 1-800-424-9300

Sales and replacement parts

www.DeltaEducation.com/FOSS/buy

Phone: 1-800-338-5270, 8:00 a.m.–5:00 p.m. ET

Framework and NGSS

ENVIRONMENTS — Framework and NGSS

INTRODUCTION TO PERFORMANCE EXPECTATIONS

"The NGSS are standards or goals, that reflect what a student should know and be able to do; they do not dictate the manner or methods by which the standards are taught. . . . Curriculum and assessment must be developed in a way that builds students' knowledge and ability toward the PEs [performance expectations]" (*Next Generation Science Standards*, 2013, page xiv).

This chapter shows how the NGSS Performance Expectations were bundled in the **Environments Module** to provide a coherent set of instructional materials for teaching and learning.

This chapter also provides details about how this FOSS module fits into the matrix of the FOSS Program (page 39). Each FOSS module K–5 and middle school course 6–8 has a functional role in the FOSS conceptual frameworks that were developed based on a decade of research on science education and the influence of *A Framework for K–12 Science Education* (2012) and *Next Generation Science Standards* (NGSS, 2013).

The FOSS curriculum provides a coherent vision of science teaching and learning in the three ways described by the NRC *Framework*. First, FOSS is designed around learning as a developmental progression, providing experiences that allow students to continually build on their initial notions and develop more complex science and engineering knowledge. Students develop functional understanding over time by building on foundational elements (intermediate knowledge). That progression is detailed in the conceptual frameworks.

Second, FOSS limits the number of core ideas, choosing depth of knowledge over broad shallow coverage. Those core ideas are addressed at multiple grade levels in ever greater complexity. FOSS investigations at each grade level focus on elements of core ideas that are teachable and learnable at that grade level.

Third, FOSS investigations integrate engagement with scientific ideas (content) and the practices of science and engineering by providing firsthand experiences.

Teach the module with the confidence that the developers have carefully considered the latest research and have integrated into each investigation the three dimensions of the *Framework* and NGSS, and have designed powerful connections to the Common Core State Standards for English Language Arts.

Contents

Introduction to Performance Expectations	31
FOSS Conceptual Framework	38
Background for the Conceptual Framework in Environments	40
Connections to NGSS by Investigation	48
Recommended FOSS Next Generation K–8 Scope and Sequence	52

The NGSS Performance Expectations bundled in this module include:

Life Sciences
4-LS1-1
4-LS1-2
3-LS4-2 extended from G3
3-LS4-4 extended from G3

Earth and Space Sciences
5-ESS3-1 foundational

Full Option Science System

ENVIRONMENTS — Framework and NGSS

DISCIPLINARY CORE IDEAS

A Framework for K–12 Science Education has four core ideas in life sciences.

- LS1: From molecules to organisms: Structures and processes
- LS2: Ecosystems: Interactions, energy, and dynamics
- LS3: Heredity: Inheritance and variation of traits
- LS4: Biological evolution: Unity and diversity

The questions and descriptions of the core ideas in the text on these pages are taken from the NRC *Framework* for the grades 3–5 grade band to keep the core ideas in a rich and useful context.

The performance expectations related to each core idea are taken from the NGSS for grades 3–4.

Disciplinary Core Ideas Addressed

The **Environments Module** connects with the NRC *Framework* for the grades 3–5 grade band and the NGSS performance expectations for grade 4. The module focuses on core ideas for life sciences.

Life Sciences

Framework core idea LS1: From molecules to organisms: structures and processes—How do organisms live, grow, respond to their environment, and reproduce?

- **LS1.A: Structure and function**
  How do the structures of organisms enable life's functions? [Plants and animals have both internal and external structures that serve various functions in growth, survival, behavior, and reproduction.]

- **LS1.D: Information processing**
  How do organisms detect, process, and use information about the environment? [Different sense receptors are specialized for particular kinds of information, which may then be processed and integrated by an animal's brain, with some information stored as memories. Animals are able to use their perceptions and memories to guide their actions. Some responses to information are instinctive—that is, animals' brains are organized so that they do not have to think about how to respond to certain stimuli.]

The following NGSS Performance Expectations for LS1 is derived from the Framework disciplinary core ideas above.

- **4-LS1-1.** Construct an argument that plants and animals have internal and external structures that function to support survival, growth, behavior, and reproduction. [Clarification Statement: Examples of structures could include thorns, stems, roots, colored petals, heart, stomach, lung, brain, and skin.] [Assessment Boundary: Assessment is limited to macroscopic structures within plant and animal systems.]

- **4-LS1-2.** Use a model to describe that animals receive different types of information through their senses, process the information in their brain, and respond to the information in different ways. [Clarification Statement: Emphasis is on systems of information transfer.] [Assessment Boundary: Assessment does not include the mechanisms by which the brain stores and recalls information or the mechanisms of how sensory receptors function.]

Introduction to Performance Expectations

Framework core idea LS2: Ecosystems: Interactions, energy, and dynamics—How and why do organisms interact with their environment and what are the effects of those interactions?

- **LS2.C: Ecosystem dynamics, functioning, and resilience**
  What happens to ecosystems when the environment changes? [When the environment changes in ways that affect a place's physical characteristics, temperature, or availability of resources, some organisms survive and reproduce, others move to new locations, yet others move into the transformed environment, and some die.]

No separate performance expectation primary for core idea LS2, only secondary to 3-LS4-4 on next page.

Framework core idea LS4: Biological evolution: Unity and diversity—How can there be so many similarities among organisms yet so many different kinds of plants, animals, and microorganisms? How does biodiversity affect humans?

- **LS4.A: Evidence of common ancestry and diversity**
  What evidence shows that different species are related? [Fossils provide evidence about the types of organisms (both visible and microscopic) that lived long ago and also about the nature of their environments. Fossils can be compared with one another and to living organisms according to their similarities and differences.]

- **LS4.B: Natural selection**
  How does genetic variation among organisms affect survival and reproduction? [Sometimes the differences in characteristics between individuals of the same species provide advantages in surviving, finding mates, and reproducing.]

- **LS4.D: Biodiversity and humans**
  What is biodiversity, how do humans affect it, and how does it affect humans? [Scientists have identified and classified many plants and animals. Populations of organisms live in a variety of habitats, and change in those habitats affects the organisms living there. Humans, like all other organisms, obtain living and nonliving resources from their environment.]

Environments Module—FOSS Next Generation

ENVIRONMENTS — Framework and NGSS

The following NGSS Performance Expectations for LS4 are derived from the Framework disciplinary core ideas above.

- **3-LS4-2.** Use evidence to construct an explanation for how the variations in characteristics among individuals of the same species may provide advantages in surviving, finding mates, and reproducing. [Clarification Statement: Examples of cause-and-effect relationships could be plants that have larger thorns than other plants may be less likely to be eaten by predators; and, animals that have better camouflage coloration than other animals may be more likely to survive and therefore more likely to leave offspring.]

- **3-LS4-4.** Make a claim about the merit of a solution to a problem caused when the environment changes and the types of plants and animals that live there may change. [Clarification Statement: Examples of environmental changes could include changes in land characteristics, water distribution, temperature, food, and other organisms.] [Assessment Boundary: Assessment is limited to a single environmental change. Assessment does not include the greenhouse effect or climate change.]

Earth and Space Sciences

Framework core idea ESS3: Earth and human activity—How do Earth's surface processes and human activities affect each other?

- **ESS3.C: Human impact on earth systems**
  How do humans change the planet? [Human activities in agriculture, industry and everyday life have had major effects on the land, vegetation, streams, ocean, air, and even outer space. But individuals and communities are doing things to protect Earth's resources and environments. For example, they are treating sewage, reducing the amounts of materials they use, and regulating sources of pollution such as emissions from factories and power plants or the runoff from agricultural activities.]

- **5-ESS3-1.** Obtain and combine information about ways individual communities use science ideas to protect the Earth's resources and environment.

DISCIPLINARY CORE IDEAS

A Framework for K–12 Science Education has three core ideas in Earth and space sciences.

ESS1: Earth's place in the universe

ESS2: Earth's systems

ESS3: Earth and human activity

The questions and descriptions of the core ideas in the text on this page is taken from the NRC *Framework* for the grades 3–5 grade band to keep the core ideas in a rich and useful context.

The performance expectations related to each core idea are taken from the NGSS for grade 4.

Introduction to Performance Expectations

Science and Engineering Practices Addressed

1. **Asking questions and defining problems**
   - Ask questions that can be investigated based on patterns such as cause-and-effect relationships.

2. **Developing and using models**
   - Use a model to test cause-and-effect relationships or interactions concerning the functioning of a natural system.

3. **Planning and carrying out investigations**
   - Plan and conduct an investigation collaboratively to produce data to serve as the basis for evidence, using fair tests in which variables are controlled and the number of trials considered.
   - Evaluate appropriate methods and/or tools for collecting data.
   - Make observations and/or measurements to produce data to serve as the basis for evidence for an explanation of a phenomenon or test a design solution.
   - Make predictions about what would happen if a variable changes.

4. **Analyzing and interpreting data**
   - Represent data in tables and various graphical displays to reveal patterns that indicate relationships.
   - Analyze and interpret data to make sense of phenomena using logical reasoning.
   - Compare and contrast data collected by different groups in order to discuss similarities and differences in their findings.

5. **Using mathematics and computational thinking**
   - Organize simple data sets to reveal patterns that suggest relationships.

6. **Constructing explanations and designing solutions**
   - Construct an explanation of observed relationships (e.g., the distribution of plants in the backyard).
   - Use evidence (e.g., measurements, observations, patterns) to construct or support an explanation or design a solution to a problem.
   - Identify the evidence that supports particular points in an explanation.

SCIENCE AND ENGINEERING PRACTICES

A Framework for K–12 Science Education (National Research Council, 2012) describes eight science and engineering practices as essential elements of a K–12 science and engineering curriculum. All eight practices are incorporated into the learning experiences in the **Environments Module**.

The learning progression for this dimension of the framework is addressed in *Next Generation Science Standards* (National Academies Press, 2013), volume 2, appendix F. Elements of the learning progression for practices recommended for grade 4 as described in the performance expectations appear in bullets below each practice.

ENVIRONMENTS — Framework and NGSS

7. Engaging in argument from evidence

- Respectfully provide and receive critiques from peers about a proposed procedure, explanation, or model by citing relevant evidence and posing specific questions.
- Construct an argument with evidence, data, and/or a model.
- Use data to evaluate claims about cause and effect.

8. Obtaining, evaluating, and communicating information

- Read and comprehend grade-appropriate complex texts and/or other reliable media to summarize and obtain scientific and technical ideas and describe how they are supported by evidence.
- Obtain and combine information from books and/or other reliable media to explain phenomena or solutions to a design problem.
- Communicate scientific and/or technical information orally and/or in written formats, including various forms of media, and may include tables, diagrams, and charts.

Crosscutting Concepts Addressed

Patterns
- Similarities and differences in patterns can be used to sort and classify natural phenomena. Patterns of change can be used to make predictions.

Cause and effect
- Cause-and-effect relationships are routinely identified and used to explain change.

Scale, proportion, and quantity
- Standard units are used to measure and describe quantities such as weight, time, temperature, and volume.

Systems and system models
- A system can be described in terms of its components and their interactions.

Energy and matter
- Energy can be transferred in various ways and between objects.

CROSSCUTTING CONCEPTS

A Framework for K–12 Science Education describes seven crosscutting concepts as essential elements of a K–12 science and engineering curriculum. The crosscutting concepts listed here are those recommended for grade 4 in the NGSS and are incorporated into the learning opportunities in the **Environments Module**.

The learning progression for this dimension of the framework is addressed in volume 2, appendix G, of the NGSS. Elements of the learning progression for crosscutting concepts recommended for grade 4, as described in the performance expectations, appear after bullets below each concept.

Introduction to Performance Expectations

Structure and function
- Substructures have shapes and parts that serve functions.

Stability and change
- Change is measured in terms of differences over time and may occur at different rates.

Connections: Understandings about the Nature of Science

Scientific investigations use a variety of methods.
- Scientific methods are determined by questions.
- Scientific investigations use a variety of methods, tools, and techniques.

Scientific knowledge is based on empirical evidence.
- Science findings are based on recognizing patterns. Scientists use tools and technologies to make accurate measurements and observations.

Science is a way of knowing.
- Science is both a body of knowledge and processes that add new knowledge. Science is a way of knowing that is used by many people.

Science is a human endeavor.
- Men and women from all cultures and backgrounds choose careers as scientists and engineers. Most scientists and engineers work in teams. Science affects everyday life. Creativity and imagination are important to science.

CONNECTIONS

See volume 2, appendix H and appendix J, in the NGSS for more on these connections.

For details on learning connections to Common Core State Standards English Language Arts and Math, see the chapters FOSS and Common Core ELA—Grade 4 and FOSS and Common Core Math—Grade 4 in *Teacher Resources*.

ENVIRONMENTS – Framework and NGSS

FOSS CONCEPTUAL FRAMEWORK

In the last half decade, teaching and learning research has focused on learning progressions. The idea behind a learning progression is that **core ideas** in science are complex and wide-reaching, requiring years to develop fully—ideas such as the structure of matter or the relationship between the structure and function of organisms. From the age of awareness throughout life, matter and organisms are important to us. There are things students can and should understand about these core ideas in primary school years, and progressively more complex and sophisticated things they should know as they gain experience and develop cognitive abilities. When we as educators can determine those logical progressions, we can develop meaningful and effective curriculum for students.

FOSS has elaborated learning progressions for core ideas in science for kindergarten through grade 8. Developing a learning progression involves identifying successively more sophisticated ways of thinking about a core idea over multiple years.

If mastery of a core idea in a science discipline is the ultimate educational destination, then well-designed learning progressions provide a map of the routes that can be taken to reach that destination. . . . Because learning progressions extend over multiple years, they can prompt educators to consider how topics are presented at each grade level so that they build on prior understanding and can support increasingly sophisticated learning. (National Research Council, *A Framework for K–12 Science Education*, 2012, p. 26)

> **TEACHING NOTE**
>
> FOSS has conceptual structure at the module and strand levels. The concepts are carefully selected and organized in a sequence that makes sense to students when presented as intended.

The FOSS modules are organized into three domains: physical science, earth science, and life science. Each domain is divided into two strands, as shown in the table "FOSS Next Generation—K–8 Sequence." Each strand represents a core idea in science and has a conceptual framework.

- Physical Science: matter; energy and change
- Earth and Space Science: dynamic atmosphere; rocks and landforms
- Life Science: structure and function; complex systems

The sequence in each strand relates to the core ideas described in the NRC *Framework*. Modules at the bottom of the table form the foundation in the primary grades. The core ideas develop in complexity as you proceed up the columns.

FOSS Conceptual Framework

Information about the FOSS learning progression appears in the **conceptual framework** (page 41), which shows the structure of scientific knowledge taught and assessed in this module, and the **content sequence** (pages 46–47), a graphic and narrative description that puts this single module into a K–8 strand progression.

FOSS is a research-based curriculum designed around the core ideas described in the NRC *Framework*. The FOSS module sequence provides opportunities for students to develop understanding over time by building on foundational elements or intermediate knowledge leading to the understanding of core ideas. Students develop this understanding by engaging in appropriate science and engineering practices and exposure to crosscutting concepts. The FOSS conceptual frameworks therefore are *more detailed* and *finer-grained* than the set of goals described by the NGSS performance expectations (PEs). The following statement reinforces the difference between the standards as a blueprint for assessment and a curriculum, such as FOSS.

Some reviewers of both public drafts [of NGSS] requested that the standards specify the intermediate knowledge necessary for scaffolding toward eventual student outcomes. However, the NGSS are a set of goals. They are PEs for the end of instruction—not a curriculum. Many different methods and examples could be used to help support student understanding of the DCIs and science and engineering practices, and the writers did not want to prescribe any curriculum or constrain any instruction. It is therefore outside the scope of the standards to specify intermediate knowledge and instructional steps. (Next Generation Science Standards, 2013, volume 2, p. 342)

FOSS Next Generation—K–8 Sequence

	PHYSICAL SCIENCE		EARTH SCIENCE		LIFE SCIENCE	
	MATTER	ENERGY AND CHANGE	ATMOSPHERE AND EARTH	ROCKS AND LANDFORMS	STRUCTURE/ FUNCTION	COMPLEX SYSTEMS
6–8	Waves; Gravity and Kinetic Energy; Chemical Interactions; Electromagnetic Force		Planetary Science; Earth History; Weather and Water		Heredity and Adaptation; Populations and Ecosystems; Diversity of Life; Human Systems Interactions	
5	Mixtures and Solutions		Earth and Sun		Living Systems	
4		Energy		Soils, Rocks, and Landforms	Environments	
3	Motion and Matter		Water and Climate		Structures of Life	
2	Solids and Liquids			Pebbles, Sand, and Silt	Insects and Plants	
1		Sound and Light	Air and Weather		Plants and Animals	
K	Materials and Motion		Trees and Weather		Animals Two by Two	

ENVIRONMENTS — Framework and NGSS

BACKGROUND FOR THE CONCEPTUAL FRAMEWORK
in Environments

The Ancient Environment

So far as we know, the environment we enjoy here on Earth is unique in the universe. There may be no other planet like it. Earth is a water planet with a thin shell of gases and vapors surrounding it. Islands of rock (some of them quite large!) rise above the surface of the water, providing a place to dry out. The overall Earth environment is rather temperate; it rarely gets much below −75°C, and hardly ever gets above 50°C at the hottest. The atmosphere is surprisingly homogeneous in composition, and the seas are even more uniform. Even so, you can see multitudes of localized environments, each with its own association of organisms. Conditions and life on Earth are marvelously diverse.

But it wasn't always so. Paleobiologists, scientists who study the biology of ancient Earth from the fossil record, tell us that when life emerged on Earth, the atmosphere was lethal. There was no oxygen in the atmosphere, and gases such as methane, ammonia, hydrogen sulfide, chlorine, and carbon monoxide predominated—all poisons to life as we know it today. Furthermore, there was no ozone layer high in the atmosphere to filter out the killing ultraviolet radiation of the Sun.

The first life-forms were probably single-celled organisms similar to present-day yeasts, living in the seas where water provided a shield against the Sun and isolation from the toxic atmosphere. Even so, life must have been a precarious affair for several million years. But life prevailed, reproduced, and changed over time. By 3 billion years ago an important advance had developed: chlorophyll. Chlorophyll provided the primitive green plants with a means for synthesizing food from carbon dioxide and sunlight. And, much to our benefit today, one of the by-products of this process is oxygen.

As oxygen increasingly invaded the atmosphere, the environment on Earth changed radically. Oxygen formed ozone, and a protective screen developed, perhaps 600 million years ago, allowing life to survive at the surface and edges of the seas. During the next several tens of millions of years, plants proliferated, producing oxygen much faster than it was consumed. As oxygen increased, so did animal life. Plants invaded the land 500 million years ago, and soon after (maybe 100 million years later) animals followed. The atmospheric oxygen reached its present-day level of 20 percent about 350 million years ago. Since that time it has fluctuated extensively, creating environments that favored first one kind of organism and then another. And so it goes still.

FOSS Conceptual Framework

CONCEPTUAL FRAMEWORK
Life Science, Focus on Complex Systems:
Environments

Structures and Function

Concept A All living things need food, water, a way to dispose of waste, and an environment in which they can live.

- Animals and plants have structures and behaviors that serve various functions in growth, survival, and reproduction.

Concept C Animals detect, process, and use information about their environment to survive.

- Different sense receptors are specialized for particular kinds of information gathering.

Complex Systems

Concept A Organisms and populations of organisms are dependent on their environmental interactions both with other living things and with nonliving factors.

- An ecosystem is the interactions of organisms with one another and the abiotic environments.
- Organisms have ranges of tolerance for environmental factors.
- Organisms interact in feeding relationships (food chains and food webs).

Concept B Ecosystems are dynamic and change over time.

- When the environment changes, some organisms and populations of organisms survive, thrive, and reproduce; others move, decline, or die.

Concept D Biological evolution, the process by which all living things have evolved over many generations from common ancestors, explains both the unity and diversity of species.

- Different organisms can live in different environments; organisms have adaptations that allow them to survive in that environment.
- Individuals of the same kind differ in their characteristics and sometimes the differences give individuals an advantage in surviving and reproducing in changing environmental conditions.

ENVIRONMENTS — Framework and NGSS

This is an extremely abbreviated history of Earth's atmosphere. The unabridged version is fascinating. It gives tremendous dimension to the story of life on Earth and gives one respect for the processes that weave the biological tapestry of the planet. It is, however, a story that is larger than most upper-elementary students can appreciate, so we recommend that they start by developing an understanding of the fundamental interactions between organisms and the physical environment and then between organisms.

An ecosystem, the basic functional unit of nature, has four structural levels or components. One component is nonliving (abiotic), the physical and chemical conditions of the environment and the energy sources that drive the system. The other three components are living (biotic) components: the producers, chiefly the green plants, which fix the energy of the Sun and manufacture food from simple substances; the consumers, which use the food stored by the green plants; and the decomposers, chiefly bacteria and fungi, which break down complex compounds into raw materials to be recycled and used again by the plants. The biotic community not only depends on the environment for its well-being; it also influences the conditions that surround the organisms. The living and nonliving components interact constantly, creating an ecological system, or ecosystem.

While a study of the ecosystem focuses on the dynamic interrelationships between the biotic community and the abiotic conditions, the study of the environment focuses on the specific factors that exert an influence on individuals, populations, or communities of organisms. Environment is defined as the objects, organisms, and conditions that exert an influence on an organism. Any one part of the environment is an environmental factor or component and can be biotic or abiotic.

Range of Tolerance

In 1840 a chemist, Justus von Liebig (1803–1873), contributed to the understanding of environmental factors in his book called *Organic Chemistry and Its Application to Agriculture and Physiology*. In it he said, "The crops on a field diminish or increase in exact proportion to the diminution or increase of the mineral substances conveyed to it in manure." This seems like common knowledge to us today, but it was rather revolutionary for the time. He had found that each plant requires certain kinds and quantities of nutrients to survive. If one of these nutrients is absent, the plant will not survive. Further, if the nutrient is present in minimal amounts, the growth of the plant will be minimal. This came to be known as the law of the minimum.

FOSS Conceptual Framework

Later studies of plants and animals demonstrated that not only does too little of a substance or condition affect the survival of an organism, but an excess can also be harmful. Each organism can survive within a certain range of conditions for any one factor, and the range is bounded by minimum and maximum limits. In 1913, V. E. Shelford (1877–1968) expanded on the law of the minimum to describe the law of tolerance. Organisms can survive only within their range of tolerance for a particular environmental factor. Outside their range of tolerance they fail to reproduce or grow.

An organism can survive when it is within its range of tolerance, but at an optimum place in that range the organism thrives. Animals that are mobile can move to favorable conditions and can demonstrate environmental preferences through their behavior.

Environmental Factors

The three most fundamental factors that determine the well-being of an organism are water, food, and oxygen. The most important is water. Without water there is no life. In fact, the cells of all living organisms are bathed in an internal solution that is remarkably like seawater, where life is thought to have first evolved.

For aquatic organisms water is no problem unless the well runs dry. There is essentially one acceptable condition of water: enough to submerge in. Other organisms must have strategies for obtaining water. Animals and plants have countless mechanisms for obtaining and conserving water within their range of tolerance in just about every environment on Earth. Some plants and animals in dry regions complete their life cycle in a short period immediately after a rain. They survive the dry periods as seeds or drought-resistant eggs. Some plants, such as succulents, store water in their tissues and use it during dry periods. There are structural plant adaptations to withstand desiccation: thick waxy leaves, small leaves, and long, complex root systems.

Food is important for two reasons. It provides building blocks for growth, development and system repair in organisms. And food is the source of energy that organisms need to live. All animals, fungi, and many bacteria consume other organisms to get the nutrtion they need to survive. Plants, algae, and some bacteris produce their own food.

The basis unit of life is the cell and all organisms are made of living cells. The simplest organisms are made of one cell and they need water, food, and gases to survive. Many organisms are multicelluar, made of many cells that form specialized structures to transport resources to each cell. It is in the cell where food (sugar) is broken down to get energy.

Environments Module—FOSS Next Generation 43

ENVIRONMENTS — Framework and NGSS

Cells need oxygen to do the job. One of the by-products of the sugar breakdown is the waste gas, carbon dioxide. If cells don't get oxygen they will die. If cells don't get rid of the carbon dioxide, they will die. In humans and many other animals, blood, which is mostly water, is pumped through blood vessels to all the cells. The blood carries food and oxygen to the cells, and carries away wastes, including carbon dioxide. Terrestial organisms get their oxygen from the air while aquatic organisms get oxygen from the water. Different structures serve similar functions.

Organisms' tolerance to extremes in temperature varies widely, but there are limits above and below which life cannot exist. The upper limit for most sustainable animal life is 52°C. Some organisms, such as hot-water algae, can live in water up to 77°C, while others, such as arctic algae, can complete their life cycle in temperatures close to 0°C. But all organisms seem to have a temperature range outside of which they fail to grow or reproduce. Within the range of tolerance, organisms have an optimum or preferred temperature at which they can best maintain their life processes. This temperature might vary throughout the life cycle of the organism. For instance, the seeds of some plants will not germinate nor will the eggs and pupae of some insects hatch or develop until chilled. Brook trout grow best at 13°C to 16°C, but the eggs develop best at 8°C.

Light is absolutely essential to green plants. Each plant has a minimum intensity level below which sufficient photosynthesis to maintain the plant is not possible. At this point photosynthesis balances respiration; plants are just able to replace energy used in respiration. Some plants are more shade tolerant than others. For example, sugar maple, white cedar, and hemlock can grow under a dense forest canopy at low light intensities, but they are not as vigorous as similar plants in brighter sunlight. In aquatic systems, light penetration is often a limiting factor for rooted plants. When nutrients are plentiful, light can stimulate the growth of algae at the surface of a pond to the point that the plants below are completely shaded out.

FOSS Conceptual Framework

Living Organisms Need Energy and Matter

An ecosystem is a community of living things, all the nonliving things that surround it, and the relationships among them. Ecology is the study of the interactions between a community of organisms and the environment in which the community lives.

One of the major ideas presented in this context is the mechanisms by which organisms acquire the material resources (chemicals in the form of minerals, gases, and carbohydrates) and energy to conduct their lives. In most ecosystems on Earth, energy enters the ecosystem as light from the Sun and is made available for life through the complex process of photosynthesis performed by plants, algae, some bacteria, and some protists. Once plants (producers) have synthesized carbon dioxide and water into energy-rich carbohydrates, the energy and material in those carbohydrates is available to the consumers (the animals, fungi, bacteria, and protists) in the ecosystem, which cannot make their own food through photosynthesis. Energy captured by plants is the energy available to run the whole ecosystem. The consumers can acquire energy they need only by consuming other organisms.

ENVIRONMENTS — Framework and NGSS

Life Science Content Sequence

This table shows most of the modules in the FOSS content sequence for Life Science. The supporting elements in these modules (somewhat abbreviated) are listed. The elements for the **Environments Module** are expanded to show how they fit into the sequence.

Module or course	LIFE SCIENCE — Structure and Function	Complex Systems
Diversity of Life (middle school)	• All living things are made of cells (unicellular or multicellular). Special structures within cells are responsible for various functions. • Cells have the same needs and perform the same functions as more complex organisms. • All living things need food, water, a way to dispose of waste, and an environment in which they can live (macro and micro level). • Plants reproduce in a variety of ways, sometimes depending on animal behaviors and specialized features for reproduction.	• Adaptations are structures or behaviors of organisms that enhance their chances to survive and reproduce in their environment. • Sexual reproduction results in offspring with genetic variation, similar to parents but not identical. Asexual reproduction results in offspring with identical genetic information. • Biodiversity is the wide range of existing life-forms that have adapted to the variety of conditions on Earth, from terrestrial to marine ecosystems.
Living Systems (grade 5)	• Food provides animals with the materials they need for body repair and growth and is digested to release the energy they need to maintain body warmth and for motion. • Reproduction is essential to the continued existence of every kind of organism. • Humans and other animals have systems specialized for particular body functions. • Animals detect, process, and use information about their environment to survive.	• Organisms obtain gases, water, and minerals from the environment and release waste matter back into the environment. • Matter cycles between air and soil, and among plants, animals, and microbes as these organisms live and die. • Organisms are related in food webs. • Some organisms, such as fungi and bacteria, break down dead organisms, operating as decomposers.
Environments (grade 4)		
Structures of Life (grade 3)	• Plants and animals have structures that function in growth, survival, and reproduction. • Reproduction is essential to the continued existence of every kind of organism. Organisms have diverse life cycles. • Plants and animals grow and change and have predictable characteristics at different stages of development. • Bones have several functions: support, protection, and movement.	• Different organisms can live in different environments; organisms have adaptations that allow them to survive in that environment. • Changes in an organism's habitat are sometimes beneficial to it and sometimes harmful. • Many characteristics of organisms are inherited from parents; other characteristics result from interaction with the environment.
Insects and Plants (grade 2)	• Insects need air, food, water, and space, including shelter, and different insects meet these needs in different ways. • Plants depend on the environment for water and light to grow. • Reproduction is essential to the continued existence of every kind of organism.	• Bees and other insects help some plants by moving pollen from flower to flower. • Animals interact with plants, using them as food. They also assist in plant reproduction. • There are many different kinds of living things and they exist in different habitats on land and in water.
Plants and Animals (grade 1)	• Animals and plants have structures that serve functions in growth and survival. • Plants can produce new plants in many ways. • Adult animals can have offspring; they have behaviors to help the offspring survive.	• Plants and animals are very much, but not exactly, like their parents and resemble other organisms of the same kind.
Animals Two by Two **Trees and Weather** (grade K)	• Animals have identifiable structures and behaviors. • Animals and plants have basic needs. • Trees are living plants and have structures. • Trees go through predictable stages.	• Living things can survive only when their needs are met. • Individuals of the same kind (plants or animals) are recognizable as similar but can also vary in many ways.

FOSS Conceptual Framework

	Structure and Function	Complex Systems
Environments	• Plants and animals have structures and behaviors that function in growth, survival, and reproduction. Terrestrial and aquatic organisms have similar needs but different structures. • Organisms have sensory systems to detect information about their environment and act on it. • Animals communicate to warn others of danger, scare predators away, or locate others of their kinds, including family members. • Producers make their own food. • Animals obtain food from eating plants or eating other animals.	• An ecosystem is the interactions of organisms with one another and the abiotic environment. • Organisms have ranges of tolerance for environmental factors. • A relationship exists between environmental factors and how well organisms grow. • Organisms interact in feeding relationships in ecosystems (food chains and food webs). • Individuals of the same kind differ in their characteristics; differences may give individuals an advantage in surviving and reproducing in changing environmental conditions.

> **NOTE**
> See the Assessment chapter at the end of this *Investigations Guide* for more details on how the FOSS embedded and benchmark assessment opportunities align to the conceptual frameworks and the learning progressions. In addition, the Assessment chapter describes specific connections between the FOSS assessments and the NGSS performance expectations.

The NGSS Performance Expectations addressed in this module include:

Life Sciences
4-LS1-1
4-LS1-2
3-LS4-2 extended from G3
3-LS4-4 extended from G3

Earth and Space Sciences
5-ESS3-1 foundational

See pages 32–34 in this chapter for more details on the Grade 4 NGSS Performance Expectations.

Environments Module—FOSS Next Generation 47

ENVIRONMENTS — Framework and NGSS

CONNECTIONS TO NGSS BY INVESTIGATION

	Science and Engineering Practices	Connections to Common Core State Standards—ELA
Inv. 1: Environmental Factors	Asking questions Developing and using models Planning and carrying out investigations Analyzing and interpreting data Constructing explanations Engaging in argument from evidence Obtaining, evaluating, and communicating information	RI 1: Refer to details/examples when explaining what the text says and when drawing inferences from text. RI 2: Determine the main idea of a text and explain how it is supported by key details; summarize the text. RI 3: Explain procedures or concepts in a scientific text. RI 4: Determine the meaning of general academic domain-specific words or phrases. RI 5: Describe the overall structure of information in a text. RI 6: Compare and contrast a firsthand and secondhand account of the same topic. RI 7: Interpret information presented visually, and explain how the information contributes to an understanding of the text. RI 9: Integrate information from two texts on the same topic. W 2: Write informative/explanatory text. W 5: Strengthen writing by revising. W 8: Gather relevant information from experiences and print, and categorize the information. SL 1: Engage in collaborative discussions. L 4: Determine or clarify the meaning of unknown and multiple-meaning words and phrases. L 5: Demonstrate understanding of word relationships.
Inv. 2: Ecosystems	Developing and using models Planning and carrying out investigations Analyzing and interpreting data Constructing explanations Engaging in argument from evidence Obtaining, evaluating, and communicating information	RI 1: Refer to details/examples when explaining what the text says and when drawing inferences from text. RI 3: Explain procedures or concepts in a scientific text. RI 4: Determine the meaning of general academic domain-specific words or phrases. RI 5: Describe the overall structure of information in a text. RI 7: Interpret information presented visually, and explain how the information contributes to an understanding of the text. RI 8: Explain how an author uses reasons and evidence to support particular points in a text. RI 9: Integrate information from two texts on the same topic. SL 1: Engage in collaborative discussions. SL 5: Add visual displays to presentations.

Full Option Science System

Connections to NGSS by Investigation

Disciplinary Core Ideas		Crosscutting Concepts
LS1.A: Structure and function • Plants and animals have both internal and external structures that serve various functions in growth, survival, behavior, and reproduction. (4-LS1-1) **LS1.D: Information processing** • Different sense receptors are specialized for particular kinds of information, which may then be processed by an animal's brain. Animals are able to use their perceptions and memories to guide their actions. (4-LS1-2)	**LS2.C: Ecosystem dynamics, functioning, and resilience** • When the environment changes in ways that affect a place's physical characteristics, temperature, or availability of resources, some organisms survive and reproduce, others move to new locations, yet others move into the transformed environment, and some die. (3-LS4-4, extended from grade 3) **LS4.D: Biodiversity and humans** • Populations live in a variety of habitats, and change in those habitats affects the organisms living there. (3-LS4-4, extended from grade 3)	Cause and effect Systems and system models Structure and function
LS1.A: Structure and function • Plants and animals have both internal and external structures that serve various functions in growth, survival, behavior, and reproduction. (4-LS1-1) **LS1.D: Information processing** • Different sense receptors are specialized for particular kinds of information, which may then be processed by an animal's brain. Animals are able to use their perceptions and memories to guide their actions. (4-LS1-2)	**LS2.C: Ecosystem dynamics, functioning, and resilience** • When the environment changes in ways that affect a place's physical characteristics, temperature, or availability of resources, some organisms survive and reproduce, others move to new locations, yet others move into the transformed environment, and some die. (3-LS4-4, extended from grade 3) **LS4.D: Biodiversity and humans** • Populations live in a variety of habitats, and change in those habitats affects the organisms living there. (3-LS4-4, extended from grade 3)	Systems and system models Energy and matter Structure and function Stability and change

Environments Module—FOSS Next Generation

ENVIRONMENTS — Framework and NGSS

Science and Engineering Practices	Connections to Common Core State Standards—ELA
Inv. 3: Brine Shrimp Hatching Developing and using models Planning and carrying out investigations Analyzing and interpreting data Using mathematics and computational thinking Constructing explanations Engaging in argument from evidence Obtaining, evaluating, and communicating information	RI 1: Refer to details/examples when explaining what the text says and when drawing inferences from text. RI 2: Determine the main idea of a text and explain how it is supported by key details; summarize the text. RI 3: Explain procedures or concepts in a scientific text. RI 4: Determine the meaning of general academic domain-specific words or phrases. RI 5: Describe the overall structure of information in a text. RI 6: Compare and contrast a firsthand and secondhand account of the same topic. RI 7: Interpret information presented visually; explain how information contributes to an understanding of the text. RI 9: Integrate information from two texts on a topic. W 5: Strengthen writing by revising. W 8: Gather relevant information from experiences and print, and categorize the information. W 9: Draw evidence from informational texts to support reflection. SL 1: Engage in collaborative discussions. SL 2: Paraphrase information presented orally. SL 4: Report on a text in an organized manner, using appropriate facts and relevant details.
Inv. 4: Range of Tolerance Planning and carrying out investigations Analyzing and interpreting data Constructing explanations Engaging in argument from evidence Obtaining, evaluating, and communicating information	RI 1: Refer to details/examples when explaining what the text says and when drawing inferences from text. RI 2: Determine the main idea of a text and explain how it is supported by key details; summarize the text. RI 4: Determine the meaning of general academic domain-specific words or phrases. RI 7: Interpret information presented visually; explain how information contributes to an understanding of the text. RI 8: Explain how an author uses reasons and evidence to support particular points in a text. RI 9: Integrate information from two texts on a topic. W 8: Gather relevant information from experiences and print, and categorize the information. SL 1: Engage in collaborative discussions. SL 3: Identify the reasons and evidence a speaker provides. SL 4: Report on a text in an organized manner, using appropriate facts and relevant details. L 4: Determine or clarify the meaning of unknown and multiple-meaning words and phrases. L 5: Demonstrate understanding of word relationships.

Connections to NGSS by Investigation

Disciplinary Core Ideas		Crosscutting Concepts
LS1.A: Structure and function • Plants and animals have both internal and external structures that serve various functions in growth, survival, behavior, and reproduction. (4-LS1-1) **LS2.C: Ecosystem dynamics, functioning, and resilience** • When the environment changes in ways that affect a place's physical characteristics, temperature, or availability of resources, some organisms survive and reproduce, others move to new locations, yet others move into the transformed environment, and some die. (3-LS4-4, extended from grade 3)	**LS4.B: Natural selection** • Sometimes the differences in characteristics between individuals of the same species provide advantages in surviving, finding mates, and reproducing. (3-LS4-2, extended from grade 3) **LS4.D: Biodiversity and humans** • Populations live in a variety of habitats, and change in those habitats affects the organisms living there. (3-LS4-4, extended from grade 3) **ESS3.C: Human impacts on Earth systems** • Human activities in agriculture, industry, and everyday life have had major effects on land, vegetation, streams, oceans, air, and even outer space. But individuals and communities are doing things to help protect Earth's resources and environments. (Foundational for grade 5)	Cause and effect Systems and system models Structure and function
LS1.A: Structure and function • Plants and animals have both internal and external structures that serve various functions in growth, survival, behavior, and reproduction. (4-LS1-1) **LS2.C: Ecosystem dynamics, functioning, and resilience** • When the environment changes in ways that affect a place's physical characteristics, temperature, or availability of resources, some organisms survive and reproduce, others move to new locations, yet others move into the transformed environment, and some die. (3-LS4-4, extended from grade 3)	**LS4.D: Biodiversity and humans** • Populations live in a variety of habitats, and change in those habitats affects the organisms living there. (3-LS4-4, extended from grade 3)	Patterns Cause and effect Structure and function

Environments Module—FOSS Next Generation

ENVIRONMENTS — Framework and NGSS

RECOMMENDED FOSS NEXT GENERATION K–8 SCOPE AND SEQUENCE

Grade	Integrated Middle Grades				
6–8	Heredity and Adaptation*	Electromagnetic Force*	Gravity and Kinetic Energy*	Waves*	Planetary Science
	Chemical Interactions		Earth History		Populations and Ecosystems
	Weather and Water		Diversity of Life		Human Systems Interactions*

*Half-length courses   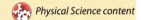 Physical Science content    Earth Science content    Life Science content   Engineering content

Grade	Physical Science	Earth Science	Life Science
5	Mixtures and Solutions	Earth and Sun	Living Systems
4	Energy	Soils, Rocks, and Landforms	Environments
3	Motion and Matter	Water and Climate	Structures of Life
2	Solids and Liquids	Pebbles, Sand, and Silt	Insects and Plants
1	Sound and Light	Air and Weather	Plants and Animals
K	Materials and Motion	Trees and Weather	Animals Two by Two

Materials

ENVIRONMENTS — *Materials*

Contents

Introduction	53
Kit Inventory List	54
Materials Supplied by the Teacher	56
Planning for Live Organisms	58
Preparing the Kit for Your Classroom	62
Care, Reuse, and Recycling	66

INTRODUCTION

The Environments kit contains

- *Teacher Toolkit: Environments*
  - 1 *Investigations Guide: Environments*
  - 1 *Teacher Resources: Environments*
  - 1 *FOSS Science Resources: Environments*
- *FOSS Science Resources: Environments* (class set of student books)
- Permanent equipment for one class of 32 students
- Consumable equipment for three classes of 32 students

FOSS modules use central materials distribution. You organize all the materials for an investigation on a single table called the materials station. As the investigation progresses, one member of each group gets materials as they are needed, and another returns the materials when the investigation is completed. You place items at the station—students do the rest.

Individual photos of each piece of FOSS equipment are available online for printing. For updates to information on materials used in this module and access to the Safety Data Sheets (SDS), go to www.FOSSweb.com. Links to replacement-part lists and customer service are also available on FOSSweb.

▶ **NOTE**
To see how all of the materials in the module are set up and used, view the teacher preparation video on FOSSweb.

▶ **NOTE**
Delta Education Customer Service can be reached at 1-800-258-1302.

Full Option Science System

53

ENVIRONMENTS — Materials

KIT INVENTORY List

Drawer 1 of 3

Equipment Condition

* The student books, if included in your purchase, are shipped separately.

Print Materials

1	*Teacher Toolkit: Environments* (1 *Investigations Guide*, 1 *Teacher Resources*, and 1 *FOSS Science Resources: Environments*)	
32	*FOSS Science Resources: Environments,* student books *	
1	Poster set, *Conservation,* 4/set	
2	Posters, *FOSS Science Safety* and *FOSS Outdoor Safety*	

Shared Items

8	Basins, 8 L	
10	Containers, 1/4 L	
40	Containers, 1/2 L	
50	Cup lids, plastic ✪	
50	Cups, plastic, 250 mL ✪	
8	FOSS trays, plastic	

▶ **NOTE**
The teacher toolkit is shipped separately. However, there is space in drawer 1 to store your toolkit.

Drawer 2 of 3

Shared Items (*continued*)

12	Basin covers, 6 L	
12	Basins, clear plastic, 6 L	
8	Beakers, 100 mL	
25	Boundary markers (vinyl stake wire flags)	
1	Collecting net (fish net)	
50	Craft sticks ✪	
32	Hand lenses	
8	Minispoons	
2	Pitchers	
8	Spoons, 5 mL	
1	Transparent tape, roll ✪	
16	Tray columns, plastic tube	
10	Vials, with caps, 12 dram	
100	Zip bags, 1 L ✪	

Drawer 3 of 3

Items for Investigation 1

10	Construction paper, black, sheets ✪	
12	Container lids, 1/4 L containers ✪	
8	Critter Replicators	

✪ These items might occasionally need replacement.

Full Option Science System

Equipment Condition

4	Posters, *Mealworm Larva, Mealworm Pupa, Mealworm Adult (Darkling Beetle), Mealworm Stages (Darkling Beetle Stages)*	
8	Screens, large mesh, 15 cm square	
10	Spoons, plastic	

Items for Investigation 2

1	Baster	
2	Fish tunnels	
8	Food Web cards, sets, Woods Ecosystem, 20/set	
1	Gravel, bag, 1 kg ✪	
4	Noisemakers, clackers	
4	Noisemakers, clapping hands	
4	Noisemakers, clickers	
4	Noisemakers, cow bells	
4	Noisemakers, finger cymbals	
4	Noisemakers, gear clicker	
4	Noisemakers, jingle bells	
32	Paper bags	
50	Rubber bands, #33 ✪	
2	Thermometers, Celsius	

Items for Investigations 3 and 4

1	Beaker, 1 L	
8	Container lids, 1 L containers	
8	Containers, 1 L	
8	Food Web cards, sets, Mono Lake, 12/set	
8	Metric rulers, clear plastic, 15 cm	

Consumable Items

1	Animal shapes set (7 sheets each brown, gray, green, red)	
1	Brine shrimp eggs, vial	
1	Fish food, flakes, package	
100	Index cards	
1	Kosher salt, box, 3 lb (1.5 kg)/box	
1	Seed package, barley	
1	Seed package, clover	
3	Seed packages, corn	
3	Seed packages, pea	
1	Seed package, radish	
4	Self-stick notes, pads, 100/pad	
1	Water conditioner, bottle	
40	Zip bags, 4 L	

▶ **NOTE**
This module includes access to FOSSweb, which includes the streaming videos, interactive simulations, virtual investigations, and tutorials used throughout the module.

✪ These items might occasionally need replacement.

Environments Module—FOSS Next Generation

ENVIRONMENTS — *Materials*

> **NOTE**
> Throughout the *Investigations Guide*, we refer to materials not provided in the kit as "teacher-supplied." These materials are generally common or consumable items that schools and/or classrooms already have, such as rulers, paper towels, and computers. If your school/classroom does not have these items, they can be provided by teachers, schools, districts, or materials centers (if applicable). You can also borrow the items from other departments or classrooms, or request these items as community donations.

MATERIALS *Supplied by the Teacher*

Each part of each investigation has a Materials section that describes the materials required for that part. It lists materials needed for each student or group of students and for the class.

Be aware that you must supply some items. These are indicated with an asterisk (★) in the materials list for each part of the investigation. Here is a summary list of those items by investigation.

For all investigations
- Chart paper and marking pen
- Drawing utensils (pencils, crayons, colored pencils, marking pens)
- Glue sticks
- Paper towels
- 1 Projection system
- Science notebooks (composition books)
- 32 Scissors
- Self-stick notes (if you need additional for review sessions)

For outdoor investigations
- 1 Bag for carrying materials
- 1 Camera (optional)
- 32 Clipboards
- 1 Marking pen, whiteboard
- 1 Timer, minutes
- Whistle
- 1 Whiteboard, portable, or chart paper and cardboard backing

Investigation 1: Environmental Factors
- Apple, carrot, or potato
- Bran, wheat
- Cold environment such as a refrigerator
- 150 Isopods
- Leaf and ground litter, moist
- Low-power microscopes (optional)
- 1 Paring knife
- 150 Mealworms
- Newspaper
- Potting soil
- 16 Recycled paper, 20 × 30 cm, sheets

56   Full Option Science System

1 Pushpin
    1 Sheet, white (optional)
- Water

Investigation 2: Aquatic Environments

   30 Amphipods (especially *Gammarus*)
    1 Bottle, plastic
  4–6 Elodea sprigs
    2 Goldfish
   10 Guppies
    1 Hole punch
  520 Food markers, such as raw elbow macaroni or short pieces of plastic straws
    1 Paper cutter
- Paper bags, large (optional)

10–12 Pond snails, small
- Rice or beans (optional for noisemakers)
- Rubber bands (optional for noisemakers)
- Tap water, aged or treated
- Recycled paper, white, 20 × 30 cm, sheets

Investigation 3: Brine Shrimp Hatching

    1 Marking pen, permanent, black
    1 Paper cutter
   32 Safety goggles
- Bottled water
- White paper
- Yeast (optional)

Investigation 4: Range of Tolerance

    2 Sheets of cardboard or poster board (about 76 × 60 cm)
- Chart paper

   64 Marking pens, pencils, or crayons (8 black, 8 brown, 8 red, 8 green, 8 blue, 8 yellow, 8 orange, 8 purple)
- Newspaper

    8 Paper, white, 20 × 30 cm, sheets
   32 Safety goggles
- Potting soil, 24 liters
- Water

Environments Module—FOSS Next Generation

ENVIRONMENTS — Materials

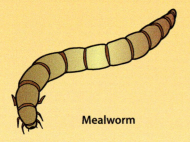
Mealworm

Pillbugs

Sowbug

▶ **SAFETY NOTE**
Find out if any students are allergic to wheat. The mealworms are often shipped to the school in wheat bran.

TEACHING NOTE

For the latest information about aquatic-organism care and regional regulations, go to FOSSweb.

PLANNING *for Live Organisms*

Some organisms come in the kit (seeds of barley, corn, clover, radish, and pea, and brine shrimp eggs). Other organisms you will need to provide. Plan ahead so that you have healthy organisms in your classroom when students are ready to start the investigations.

1. Check seeds (Investigations 1 and 4)

There are enough seeds in a new kit for three classes of 32 students to conduct all the investigations, with seeds left over. Check the package date on the seeds, and if the seeds are more than 2 years old, consider ordering fresh seeds. This will increase the chances of successful seed germination.

We recommend that you order the seeds from Delta Education to make sure that you have the best variety for classroom use and student safety. If you need the seeds immediately and want to buy them at a local store, get untreated seeds. Some seeds, particularly corn, are sometimes powdered with fungicide (you can tell because they will be pink instead of yellow). Provide only untreated seeds for classroom use. The barley seeds used in this module are special salt-tolerant seeds that will not be available in local stores.

2. Acquire mealworms and isopods (Investigation 1)

You will need about 150 mealworms and 150 isopods for Investigation 1. Purchase mealworm larvae from a local pet store or biological supply company and put the larvae, some bran, and a piece of carrot or apple in a plastic container. Depending on the age of the larvae, you can get adult beetles in a few weeks. If you plan ahead, you can have a ready supply of adult beetles (*Tenebrio*) by maintaining a culture in the classroom.

Isopods are best collected by you and your students from local outdoor areas. You can order isopods, as well as beetles, from Delta Education.

3. Acquire aquatic organisms (Investigation 2)

Goldfish, guppies, pond snails, and elodea can be purchased at local pet stores. We encourage you to use local sources for these, especially the goldfish and guppies, as they do not always fare well during shipment. Work with your local pet shop to get healthy, inexpensive goldfish, but not feeder goldfish. It has been our experience that feeder goldfish, although they are the cheapest, don't always survive in the classroom aquarium. Freshwater aquatic amphipods such as *Gammarus* (scuds) need to be ordered from Delta Education.

4. Caring for fish

Common goldfish are usually bright orange, but they can be white, black, or multicolored. The stock from which they are derived is a flat tan gray, and occasionally you will find a native-colored fish among the feeders. It is impossible to tell males and females apart, unless, of course, you are a goldfish.

Guppies are small fish that bear live young. The feeder-guppy females are larger and usually a uniform beige or silver-gray. Their abdomens become quite large when they are gravid (carrying young). The males are smaller and have longer, flowing tails. Males are the ones with spots of multiple colors. Fancy guppies that have been bred for showy colors can be dazzling.

Goldfish and guppies are among the hardiest aquarium fish and the easiest to care for. Students will enjoy observing their behavior and caring for them as they become part of your classroom. Although they are easy to care for, there are some things that you should keep in mind to ensure success with your new aquarium. Suitable water, sufficient oxygen, correct temperature (within a fairly large range for these fish), and correct feeding are the most important things to consider.

Chlorine used in water treatment is toxic to fish. Chlorine will diffuse out of water when it is exposed to air for a day. This is called aging. Plan to keep some aged tap water on hand at all times for the goldfish. For specific information on the water needs of your fish, refer to Investigation 2, Getting Ready for Part 1.

Goldfish and guppies don't place many demands on the aquarist. They need unpolluted water, but it is not necessary to provide extra oxygen with an air pump. You might see some mortality when you introduce new fish into your aquarium, but this is often due to transportation stress. As long as the fish are not crowded, they will be able to get enough oxygen just from what is dissolved at the surface of the water. (You might have heard that the rule of thumb is 3–4 liters (L) of water per goldfish.)

Goldfish and guppies will eat a wide variety of food, but the most convenient is a commercial flake food, provided in the kit. This kind of food floats, and the fish will quickly learn to come to the surface to eat. They should be fed once a day, but will do fine without food over the weekend. The most important thing about feeding is not to overfeed! Overfeeding can be lethal. To determine the proper amount of food, watch to make sure that the fish eat their food completely in a few minutes. If you keep plants such as elodea in the aquarium, the fish might nibble on the

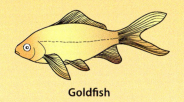
Goldfish

Female guppy

Male guppy

Environments Module—FOSS Next Generation

ENVIRONMENTS — Materials

greenery for additional food from time to time. Other things fish will eat are insect larvae, *Tubifex* worms, aquatic plants, and snail eggs.

Goldfish and guppies generally do not require an aquarium heater, although you might want to purchase one if your classroom gets below 15°C (65°F) on weekends. Fish cannot tolerate extremely warm water, either. Check the water temperature in hot weather. If the water is above 29.5°C (85°F), add cool, aged water to lower the temperature.

The aquarium requires some light for the plants to grow, but should not be in direct sunlight. Direct sunlight encourages algae growth, which turns the water green. Nutrients added to the water in the form of fish waste also encourage the algae growth.

Guppies are quite prolific and will probably give birth during their stay in your classroom. In fact, you might observe the arrival of baby guppies a day or two after the adults are put in their basin aquarium. The stress of transportation might induce a gravid female to release the babies. Adult guppies will eat the young, so you should supply the aquarium with plenty of *Elodea* in which the babies can hide, or move the adults to a separate tank. Students will enjoy watching the baby guppies grow. Goldfish lay eggs; they do not give live birth. Although they are prolific in nature, they usually will not breed in a small aquarium.

If a fish dies, you might want to leave it in the aquarium until students come to school, but dead fish foul the water, so they should be removed from the aquarium as soon as possible. Some students might be upset that a fish has died. Rather than simply throwing a dead fish away, take it outside and bury it in the ground near a plant. Discuss how this will help fertilize the plant, giving it additional nutrients to help it grow as the body of the fish decays.

Live guppies and goldfish should never be released in ponds or creeks. Allow them to live out their short lives in your classroom.

5. Provide aquatic plants (Investigation 2)

Buy or collect from a local pond some small aquatic plants for the aquarium. We recommend getting 4–6 sprigs of elodea, also known as *Anacharis*. (It looks like a little green feather boa.) You can order it from Delta Education, or you can pick it up locally at a pet store that deals with fish. If elodea is not available, try another inexpensive aquatic plant.

NOTE: There are several different species of elodea, and some of them are restricted in some states. Ask your supplier for the species that can safely be distributed in your state.

Elodea

60   Full Option Science System

6. **Check brine shrimp (Investigation 3)**
   Brine shrimp eggs don't last forever in nature or in a FOSS kit. Check the package date on the eggs; if it is more than 2 years old, order new eggs. Brine shrimp eggs are usually available at local pet shops. See Step 5 of Getting Ready for Investigation 3, Part 1, for details on how to test the eggs with your local bottled water.

7. **Dealing with organisms at the end of the module**
   You have a number of options for dealing with the organisms at the end of the module. The organisms might find a permanent home in your classroom. You will need to provide containers for permanent habitats if the kit will be used by another teacher. Or check with your district to see if there is a plan for reusing FOSS living organisms.

   Aquatic organisms obtained from pet stores or Delta Education should never be released into the local environment. There is always a chance that an introduced species might displace a native species in the environment, so releasing such organisms is never an option. This applies to any plants such as elodea as well.

   As a last resort, you can put the organisms in a bag and place them in a freezer overnight to euthanize them prior to disposal.

8. **Respect living organisms**
   FOSS believes that studying live organisms is a critical part of any life science curriculum—it is especially important to support the philosophy that children learn best through direct experiences. FOSS is committed to including the study of live organisms in the program.

   We know that the use of organisms comes with a unique set of challenges, but we think it's well worth the effort. We continue to support and abide by federal and state regulations and NSTA (National Science Teachers Association) guidelines for the responsible treatment of animals in the classroom, while taking steps to ensure that children have hands-on life science experiences, and teachers and school districts have a variety of options to obtain organisms.

▶ **SAFETY NOTE**
Find out if any students are allergic to seafood and/or shellfish as they are likely to also be allergic to brine shrimp.

TEACHING NOTE
It is important for students to understand the reasons for not releasing nonnative plants and animals into local environments. For the latest information about aquatic-organism care and regional regulations, go to FOSSweb.

Environments Module—FOSS Next Generation

ENVIRONMENTS — *Materials*

PREPARING *the Kit for Your Classroom*

Some preparation is required each time you use the kit. Doing these things before beginning the module will make daily setup quicker and easier.

1. Check consumable materials

A number of items in the kit are listed as consumable. Some of these items may be used up during the investigations (craft sticks, seeds, string, brine shrimp eggs, and fish food), and others will wear out and need replacement (zip bags and plastic cups). Before throwing items out, consider ways to recycle them, and get your students involved in this process.

2. Care and reuse of containers

All the containers in the module are to be rinsed and reused. Do not use soap to clean containers that will become environments for organisms later. It is a good idea to rinse containers before using them in your class.

3. Plan for use of the equipment

Many of these investigations require using the equipment continuously for an extended period of time. The 6 liter (L) basins are used as terrariums for Investigation 1 and for two class aquariums in Investigation 3. The 1/2 L containers are used in the last investigation as planters.

4. Locate the FOSS trays and tubes

The clear plastic FOSS tray was designed specially for this module. Each group of four students uses one tray. It holds four 1/2 L planters or brine shrimp hatcheries. The trays can be stacked two high to conserve space when storing experiments in progress. Use four plastic tubes to provide the support columns between two trays. Check that all the trays and tubes are in the kit.

5. Familiarize yourself with the basins

Familiarize yourself with the different basins used in this module.

- Twelve clear, 6 L basins are provided. They are used for a beetle habitat (one), isopod habitats (eight), class aquaria (two), and temporary environment for aquatic plants, snails, and scuds (one or two).

- Eight 8 L basins are used for collecting leaf-litter critters outdoors and for distributing soil.

6. Familiarize yourself with the spoons

The kit has three sizes of spoons. The smallest is the minispoon. It is used to measure clover seeds and brine shrimp eggs. The

Full Option Science System

5 milliliter (mL) spoon is used to measure salt. The plastic spoon that you might take on a picnic is used to move little animals around in their environments.

7. Provide soil
You must provide potting soil. The salty soil from Investigation 4 should be discarded; new soil should be purchased when a new class starts the module. You will need a total of 27 L.

- Investigation 1: 3 L of soil
- Investigation 4: 24 L of soil

8. Check kosher salt
Kosher salt is used in Investigations 3 and 4. In each of these investigations, students prepare salt concentrations according to a recipe. These recipes are based on the brand of kosher salt packaged in the equipment kit—one 5 mL spoon of the salt has a mass of about 4 grams (g). Other brands of kosher salt have different grain size (and density). One 5 mL spoon of another brand might weigh 7 g. Be aware that not all kosher salt is exactly the same, and, even if you follow the recipes in the investigation perfectly, the resulting concentrations might be different, and you will get different outcomes in the experiments.

9. Check zip bags
In Investigation 4, students put the planters in large zip bags to investigate water tolerance of young plants. These zip bags should be used by three classes before they are replaced. Check the supply of 4 L zip bags in the kit.

10. Check Critter Replicators
Check the condition of the 8 Critter Replicators used in Investigation 1, Part 3. Make repairs as needed. Teacher masters for the covers and wheel transparencies are included in *Teacher Resources*.

11. Check Food Web Cards
The Food Web Cards contain two different sets of organism cards—Woods Ecosystem and Mono Lake. There are eight copies of each set of cards. Check that all the sets are complete.

12. Print or photocopy notebook sheets
You will need to print or make copies of science notebook sheets before each investigation. See Getting Ready for Investigation 1, Part 1, for ways to organize the science notebook sheets for this module. If you use a projection system, you can download electronic copies of the sheets from FOSSweb for projection.

Woods Ecosystem organisms, 20/set
- American robin
- Aquatic snail
- Bacteria
- Black bear
- Brook trout
- Chipmunk
- Coyote
- Dead plants and animals
- Earthworm
- Grama grass
- Great blue heron
- Green algae
- Grouse
- Hare
- Mayfly
- Pine trees
- Red-tailed hawk
- Scuds
- *Tubifex* worm
- Wild blueberry

Mono Lake organisms, 12/set
- Bottom algae
- Brine fly
- Brine shrimp
- California gull
- Caspian tern
- Coyote
- Eared grebe
- Floating algae
- Halobacteria
- Red-necked phalarope
- Snowy plover
- Wilson's phalarope

Environments Module—FOSS Next Generation 63

ENVIRONMENTS — *Materials*

EL NOTE

You might want to print out the FOSS equipment photo cards (from FOSSweb) to add to the word wall to help students with vocabulary.

WARNING — This set contains chemicals that may be harmful if misused. Read cautions on individual containers carefully. Not to be used by children except under adult supervision.

TEACHING NOTE

Families can get more information on Home/School Connections from FOSSweb.

13. Plan for the word wall

As the module progresses, you will add new vocabulary words to a word wall or pocket chart and model writing and responding to focus questions. Plan how you will do this in your classroom.

You might also find it beneficial to use a pocket chart to display the equipment photo cards as reference for students as they gather needed items from the materials station for each part. Print the photo cards from FOSSweb.

14. Consider safety issues indoors and outdoors

Two safety posters are included in the kit, *FOSS Science Safety* and *FOSS Outdoor Safety.* You should review the guidelines with students and post the posters in the room as a reminder. Getting Ready for Investigation 1, Part 1, offers suggestions for this discussion. Also be aware of any allergies that students in your class might have to plants (peas and clover seeds are legumes), including wheat bran used to feed mealworms. The mealworms shipped from Delta Education might include wheat bran.

Use the four *Conservation* posters to discuss the importance of conserving natural resources.

You will see a warning label on some science notebook sheets. The label is required by the US Consumer Product Safety Commission (CPSC) whenever students work with chemicals. The label should act as a reminder to you and students to exercise particular safety precautions when working with materials in the investigations where the sheets are used. In this module, kosher salt is the chemical requiring this safety labeling.

CPSC also requires that chemicals be stored in a location away from the kit in the classroom.

15. Plan for letter home and home/school connections

You will need to print or make copies of teacher master 1, *Letter to Family*, for the module and of the Home/School Connection teacher masters for each investigation. The *Letter to Family* and *Home/School Connections* are also available electronically on FOSSweb.

16. Check FOSSweb for resources

Go to FOSSweb, register as a FOSS teacher, and review the print and digital resources available for this module, including the eGuide, eBook, Resources by Investigation, and *Teacher Resources*. Be sure to check FOSSweb often for updates and new resources.

ENVIRONMENTS — Materials

CARE, Reuse, and Recycling

When you finish teaching the module, inventory the kit carefully. Note the items that were used up, lost, or broken, and immediately arrange to replace the items. Use a photocopy of the materials list (the Kit Inventory List), and put your marks in the "Equipment Condition" column. Refill packages and replacement parts are available for FOSS by calling Delta Education at 1-800-258-1302 or by using the online replacement-part catalog (www.DeltaEducation.com).

Standard refill packages of consumable items are available from Delta Education. A refill package for a module includes sufficient quantities of all consumable materials (except those provided by the teacher) to use the kit with three classes of 32 students.

Here are a few tips on storing the equipment after use.

- Clean all the containers, lids, cups, vials, and basins with water only (no soap), and dry them thoroughly.
- Make sure the gravel is dry before returning it to the kit.
- Recycle the soil that did not have salt added to it. Put it in a secure bag.
- Repackage any remaining seeds, and label the package with the date they were resealed.
- Make sure this kit is stored in a location free from critters that might enjoy a meal of seeds.
- Sort the organism cards, and place each set in its plastic bag.
- Make sure all the posters are stored flat on the bottom of Drawer 1.
- Check quantity of consumables, and order more if necessary.

The items in the kit have been selected for their ease of use and durability. Small items should be inventoried (a good job for students under your supervision) and put into zip bags for storage. Any items that are no longer useful for science should be properly recycled. This is a good opportunity to get students involved in making decisions about what items can be recycled.

Technology

ENVIRONMENTS — *Technology*

Contents

Introduction	67
Technology for Students	68
Technology for Teachers	71
Requirements for Accessing FOSSweb	76
Troubleshooting and Technical Support	78

INTRODUCTION

Technology is an integral part of the teaching and learning with FOSS Next Generation. FOSSweb is the Internet access to FOSS digital resources. FOSSweb gives students the opportunity to interact with simulations, virtual investigations, tutorials, images, and text—activities that enhance understanding of core ideas. It provides support for teachers, administrators, and families who are actively involved in implementing FOSS.

Different types of online activities are incorporated into investigations where appropriate. Each multimedia use is marked with the technology icon in the *Investigations Guide*. You will sometimes show videos to the class. At other times, individuals or small groups of students will work online to review concepts or reinforce their understanding. Tutorials on specific elements of concepts provide opportunities for review and differentiated instruction.

▶ **NOTE**
To get the most current information, download the latest Technology chapter on FOSSweb.

To use these digital resources, you should have at least one computer with Internet access that can be displayed to the class by an LCD projector with an interactive whiteboard or a large screen. Access to enough devices for students to work in small groups or one on one is recommended for other parts.

All FOSS online activities are available at www.FOSSweb.com for teachers, students, and families. We recommend you access FOSSweb well before starting the module, to set up your teacher-user account and to become familiar with the resources.

Full Option Science System

ENVIRONMENTS — *Technology*

TECHNOLOGY *for Students*

FOSS is committed to providing a rich, accessible technology experience for all FOSS students. Here are brief descriptions of selected resources for students on FOSSweb.

Online activities. The online simulations and activities were designed to support students' learning at all grades. They include virtual investigations and tutorials, grades 3–5, that review the active investigations and support students who have difficulties with the materials or who have been absent. Summaries of some of the online activities are on the next page.

FOSS Science Resources—eBooks. The student book is available as an audio book on FOSSweb, accessible at school or at home. In addition, as premium content, *FOSS Science Resources* is available as an eBook on computer or tablet, either as a read-only PDF or in an interactive format that allows text to be read and provides points of interactivity. The eBook can also be projected for guided reading with the whole class.

Media library. A variety of media enhances students' learning and provides them with opportunities to obtain, evaluate, and communicate information. FOSS has reviewed print books and digital resources that are appropriate for students and prepared a list of these resources with links to content websites. There is also a list of regional resources for virtual and actual field trips for students to use in gathering information for projects, and a database of science and engineering careers. Other resources include vocabulary lists to promote use of academic language.

Home/school connections. Each module includes a letter to families, providing an overview of the goals and objectives of the module. There is also a Module Summary available for families to download. Most investigations have a home/school science activity that connects the classroom experiences with students' lives outside of school. These connections are available as PDFs on FOSSweb.

Class pages. Teachers with a FOSSweb account can easily set up class pages with notes and assignments for each class. Students and families can then access this class information online, using the teacher-assigned class login.

> **NOTE**
> The following student-facing resources are available in Spanish on FOSSweb using a teacher's class page.
> - Vocabulary
> - Equipment photo cards
> - eBooks
> - Select streaming videos
> - Home/school connections
> - Audio books

Technology for Students

Environments Online Activities

Here is a sampling of the online activities used in the **Environments Module** investigations.

Investigation 2, Part 3: Population Simulation

- **"Virtual Terrarium"**
  Students set up a virtual terrarium, selecting the number of isopods, number of plants (providing oxygen), amount of food and moisture (twigs and decomposing leaves), and temperature from a heat lamp. Students observe how many isopods survive after 30 days.

- **"Virtual Aquarium"**
  Students set up a virtual aquarium, selecting the number of fish, number of plants, amount of food, temperature, and whether the bubbler is on or off. Students observe how many fish survive after 30 days.

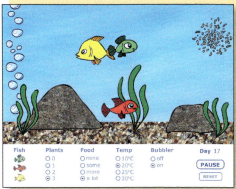

Investigation 3, Part 2: Determining Range of Tolerance

- **"Food Webs"**
  After reading about Mono Lake and working with the Mono Lake organisms cards to study the ecosystem's food web, students can use the "Food Webs" online activity to design possible food webs.

  At the end of this part, students can study food webs in three other ecosystems.

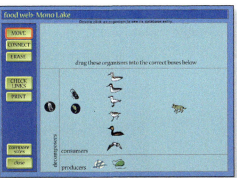

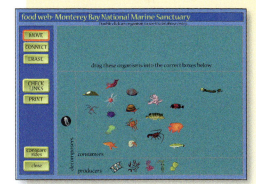

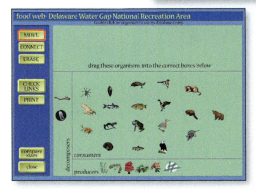

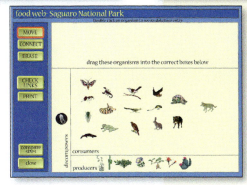

Environments Module—FOSS Next Generation

69

ENVIRONMENTS — *Technology*

Investigation 3, Part 3: Determining Viability

- **"Trout Range of Tolerance" (virtual investigation)**
  In this virtual investigation, students are presented with a scenario about a water company that wants to build a dam on a river. The dam will increase the water temperature and the students are asked to analyze experimental data dealing with trout egg hatching in different temperatures. Based on the analysis of trout egg hatching at different temperatures, students determine the range of tolerance for water temperature for trout eggs and then write a letter to the water company with a recommendation about the environmental impact of the dam on the trout survival.

Investigation 4, Part 1: Water or Salt Tolerance and Plants

- **"Analyzing Environmental Experiments" (tutorial)**
  In this tutorial, students analyze the design of different experiments to determine if they are testing for preferred environmental condition, range of tolerance, or optimum conditions for growth and survival.

70 Full Option Science System

Technology for Teachers

TECHNOLOGY *for Teachers*

The teacher side of FOSSweb provides access to all the student resources plus those designed for teaching FOSS. By creating a FOSSweb user account and activating your modules, you can personalize FOSSweb for easy access to your instructional materials. You can also set up a class login for students and their families.

Creating a FOSSweb Teacher Account

Setting up an account. Set up a teacher account on FOSSweb before you begin teaching a module. Go to FOSSweb and register for an account with your school e-mail address. Complete registration instructions are available online. If you have a problem, go to the Connecting with FOSS pull-down menu, and look at Technical Help and Access Codes. You can also access online tutorials for getting started with FOSSweb at www.FOSSweb.com/fossweb-walkthrough-videos.

Entering your access code. Once your account is set up, go to FOSSweb and log in. To gain access to all the teacher resources for your module, you will need to enter your access code. Your access code should be printed on the inside cover of your *Investigations Guide*. If you cannot find your FOSSweb access code, contact your school administrator, your district science coordinator, or the purchasing agent for your school or district.

Familiarize yourself with the layout of the site and the additional resources available when you log in to your account. From the module detail page, you will be able to access teacher masters, science notebook masters, assessment masters, the FOSSmap online assessment component, and other digital resources not available to "guests."

Explore the Resources by Investigation, as this will help you plan. This page makes it simple to select the investigation you are teaching, and view all the digital resources organized by part. Resources by Investigation provides immediate access to the streaming videos, online activities, science notebook masters, teacher masters, and other digital resources for each investigation part.

Setting up class pages and student accounts. To enable your students to log in to FOSSweb to see class assignments and student-facing digital resources, set up a class page and generate a username and password for the class. To do this, log in to FOSSweb and go to your teacher page. Under "My Class Pages," follow the instructions to create a new class page and to leave notes for students. Note: student access to the student eBook from your class page requires premium content.

▶ **NOTE**
For more information about FOSS premium content, including pricing and ordering, contact your local Delta sales representative by visiting www.DeltaEducation.com or by calling 1-800-258-1302.

Environments Module—FOSS Next Generation

ENVIRONMENTS — *Technology*

Support for Teaching FOSS

FOSSweb is designed to support teachers using FOSS. FOSSweb is your portal to instructional tools to make teaching efficient and effective. Here are some of the tools available to teachers.

- **Grade-Level Planning Guide.** The Planning Guide provides strategies for three-dimensional teaching and learning.

- **Resources by Investigation.** The Resources by Investigation organizes in one place all the print and online instructional materials you need for each part of each investigation.

- **Investigations eGuide.** The eGuide is the complete *Investigations Guide* component of the *Teacher Toolkit*, in an electronic web-based format for computers or tablets. If your district rotates modules among several teachers, this option allows all teachers easy access to *Investigations Guide* at all times.

- **Teacher preparation videos.** Videos present information to help you prepare for a module, including detailed investigation information, equipment setup and use, safety, and what students do and learn in each part of the investigation.

- **Interactive whiteboard resources for grades K–5.** You can use these interactive files with or without an interactive whiteboard to facilitate each part of each investigation. You'll need to download the appropriate software to access the files. Links for software downloads are on FOSSweb.

- **Focus questions.** The focus questions address the phenomenon for each part of each investigation, and are formatted for classroom projection and for printing, so that students can glue each focus question into their science notebooks.

- **Module updates.** Important updates cover teacher materials, student equipment, and safety considerations.

- **Module teaching notes.** These notes include teaching suggestions and enhancements to the module, sent in by experienced FOSS users.

- **Home/school connections.** These masters include an introductory letter home (with ideas to reinforce the concepts being taught) and the home/school connection sheets.

- **Regional resources.** Listings of resources for your geographic region are provided for virtual and actual field trips and for students to use as individual or class projects.

- **Access to FOSS developers.** Through FOSSweb, teachers have a connection to the FOSS developers and expert FOSS teachers.

▶ **NOTE**
There are two versions of the eGuide, a PDF-based eGuide that mimics the hard copy guide, and an HTML interactive eGuide that allows you to write instructional notes and to interface with online resources from the guide.

▶ **NOTE**
The following resources are available on FOSSweb in Spanish.

Teacher-facing resources:
- Notebook masters
- Teacher masters
- Assessment masters
- Focus questions
- Interactive whiteboard files

Student-facing resources:
- Vocabulary
- Equipment photo cards
- eBooks
- Select streaming videos
- Home/school connections
- Audio books

Technology for Teachers

Technology Components of FOSS Assessment System

FOSSmap for teachers and online assessments for students are the technology components of the FOSS assessment system.

For teachers. FOSSmap is where you set up your class, schedule online assessments, review/record codes, and run reports. The reports are diagnostic and will help you to know what students understand and what they still need help with individually and as a class.

The teacher page of FOSSweb has a direct link to FOSSmap. Once you have a login and password for FOSSweb, use the same login and password to access FOSSmap. FOSSmap is a secure site that only you can see. FOSSmap tutorials will get you started with these technology components.

For students. FOSSmap.com/icheck is the URL for students in grades 3–5 who take the assessments online (*Survey/Posttest*, I-Checks, and interim assessments). Students can access this site only when you have scheduled an assessment for them to take. Access codes are generated for each student in the FOSSmap program and can be printed out on mailing labels. Each access code is good for all the assessments taken in one module. When you change modules, students get new access codes.

For more information about the FOSS assessment system and the technology components, see the Assessment chapter.

▶ **NOTE**
FOSSmap has a number of short online tutorials to get you started. Titles include:
- Overview
- Getting Started
- Module Homepage
- Embedded Assessment
- Scheduling Online Assessments
- Taking an Online Assessment (Teacher Edition)
- How to Code an Assessment
- Creating Reports

Environments Module—FOSS Next Generation

ENVIRONMENTS — *Technology*

Technology for Differentiated Instruction

Some resources are for differentiated instruction. They can be used by students at home or by you as part of classroom instruction.

- **Online activities.** The online simulations and activities described earlier in this chapter are designed to support student learning and are often used during instruction. They include virtual investigations and student tutorials for grades 3–5 that you can use to support students who have difficulties with the materials or who have been absent. Virtual investigations require students to record data and answer concluding questions in their notebooks. In some cases, the notebook sheet used in the classroom investigation is also used for the virtual investigation.

- **Vocabulary.** The online word list has science-related vocabulary and definitions used in the module (in both English and Spanish).

- **Equipment photo cards.** Equipment cards provide labeled photos of equipment that students use in the investigations. Cards can be printed and posted on the word wall as part of instruction.

- **Student eBooks.** Student access to audio-only *FOSS Science Resources* books requires basic access. With premium content, students can access the books from any Internet-enabled device. The eBooks are available in PDF and interactive versions. The PDF version mimics the hard copy book. The interactive eBook reads the text to students—highlighting the text as it is read—and provides students with video clips and online activities.

- **Streaming videos used for extensions.** Some videos are part of the instruction in the investigation and are in Resources by Investigation for each part. Those videos also appear again in the digital resources under "Streaming Videos" along with other videos that extend concepts presented in a module.

- **Recommended books, websites, and careers database.** FOSS-recommended books, websites, and a Science and Engineering Careers Database that introduces students to a variety of career options and diversity of individuals engaged in those careers are provided.

- **Regional resources.** This list provides local resources that can be used to enhance instruction. The list includes website links and PDF documents from local sources.

▶ **NOTE**
The eBook is premium content.

Technology for Teachers

Support for Classroom Materials Management

- **Materials chapter.** A PDF of the Materials chapter in *Investigations Guide* is available to help you prepare for teaching. A list, organized by drawer, shows the materials included in the FOSS kit for a given module. You can print and use this list for inventory and to monitor equipment condition.
- **Safety Data Sheets (SDS).** A link takes you to the latest safety sheets with information from materials manufacturers on the safe handling and disposal of materials.
- **Plant and animal care.** This section includes information on caring for organisms used in the investigations.

Professional Learning Connections

FOSSweb provides PDF files of professional development chapters, mostly from *Teacher Resources*, that explain how to integrate instruction to improve learning. Some of them are

- Sense-Making Discussions for Three-Dimensional Learning
- Science-Centered Language Development
- FOSS and Common Core English Language Arts and Math
- Access and Equity
- Taking FOSS Outdoors

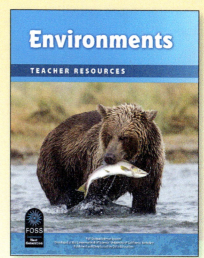

Environments Module—FOSS Next Generation

ENVIRONMENTS — Technology

REQUIREMENTS for Accessing FOSSweb

FOSSweb Technical Requirements

To use FOSSweb, your computer must meet minimum system requirements and have a compatible browser and recent versions of Flash Player, QuickTime, and Adobe Reader. Many online activities have been updated to an HTML5 version compatible with all devices. (Those designated with "Flash" after the title require Flash Player.) The system requirements are subject to change. It is strongly recommended that you visit FOSSweb to review the most recent minimum system requirements and any plug-in requirements. There, you can access the "Tech Specs and Info" page to confirm that your browser has the minimum requirements to support the online activities.

Preparing your browser. FOSSweb requires a supported browser for Windows or Mac OS with a current version of the Flash Player plug-in, the QuickTime plug-in, and Adobe Reader or an equivalent PDF reader program. You may need administrator privileges on your computer in order to install the required programs and/or help from your school's technology coordinator.

By accessing the "Tech Specs and Info" page on FOSSweb, you can check compatibility for each computer you will use to access FOSSweb, including your classroom computer, computers in a school computer lab, and a home computer. The information on FOSSweb contains the most up-to-date technical requirements for all devices, including tablets and mobile devices.

Support for plug-ins and reader. Flash Player and Adobe Reader are available on www.adobe.com as free downloads. QuickTime is available for free from www.apple.com. FOSS does not support these programs. Please go to the program's website for troubleshooting information.

> **NOTE**
> It is strongly recommended that you visit FOSSweb to review the most recent minimum system requirements.

Full Option Science System

Requirements for Accessing FOSSweb

Other FOSSweb Considerations

Firewall or proxy settings. If your school has a firewall or proxy server, contact your IT administrator to add explicit exceptions in your proxy server and firewall for FOSSweb Akamai video servers. For more specific information on servers for firewalls, refer to "Tech Specs and Info" on FOSSweb.

Classroom technology setup. FOSS has a number of digital resources and makes every effort to accommodate users with different levels of access to technology. The digital resources can be used in a variety of ways and can be adapted to a number of classroom setups.

Teachers with classroom computers and an LCD projector, interactive whiteboard, or a large screen will be able to show online materials to the class. If you have access to a computer lab, or enough computers in your classroom for students to work in small groups, you can set up time for students to use the FOSSweb digital resources during the school day. Teachers who have access to only a single computer will find a variety of resources on FOSSweb that can be used to assist with teacher preparation and materials management.

Teachers who have tablets available for student use and have premium content can download the FOSS eBook app onto devices for easy student access to the FOSS eBooks. Instructions for downloading the app can be found on FOSSweb on the Module Detail Page for any module. You'll find them under the Digital-Only Resources section and then under the tab for Student eBooks.

Displaying online content. Throughout each module, you may occasionally want to project online components for instruction through your computer. To do this, you will need a computer with Internet access and either an LCD projector and a large screen, an interactive whiteboard, or a document camera arranged for the class to see.

You might want to display the notebook and teacher masters to the class. In Resources by Investigation, you'll have the option of downloading the masters to project or to copy. Choose "for Display" if you plan on projecting to the class. These masters are optimized for a projection system and allow text entry directly onto the sheet from the computer. The "for Print" versions are sized to minimize paper use when photocopying for the class.

▶ **NOTE**
FOSSweb activities are designed for a minimum screen size of 1024 × 768. It is recommended that you adjust your screen resolution to 1024 × 768 or higher.

Environments Module—FOSS Next Generation

ENVIRONMENTS – Technology

> **NOTE**
> The FOSS digital resources are available online on FOSSweb. You can always access the most up-to-date technology information, including help and troubleshooting, on FOSSweb.

TROUBLESHOOTING and Technical Support

If you experience trouble with FOSSweb, you can troubleshoot in a variety of ways.

1. First, test your browser to make sure you have the correct plug-in and browser versions. Even if you have the necessary plug-ins installed on your computer, they may not be recent enough to run FOSSweb correctly. Go to FOSSweb, and select the "Tech Specs and Info" page to review the most recent system requirements and check your browser.
2. Check the FAQs on FOSSweb for additional information that may help resolve the problem.
3. Try emptying the cache from your browser and/or quitting and relaunching.
4. Restart your computer, and make sure all computer hardware turns on and is connected correctly.

If you are still experiencing problems after taking these steps, send FOSS Technical Support an e-mail to support@FOSSweb.com. In addition to describing the problem you are experiencing, include the following information about your computer: Mac or PC, operating system, browser name and version, plug-in names and versions. This will help us troubleshoot the problem.

Where to Get Help

For further questions about FOSSweb, please don't hesitate to contact our technical support team.

Account questions/help logging in

School Specialty Online Support

loginhelp@schoolspecialty.com

Phone: 1-800-513-2465, 8:30 a.m.–6:00 p.m. ET

5:30 a.m.–3:00 p.m. PT

General FOSSweb technical questions

FOSSweb Tech Support

support@fossweb.com

Investigation 1: Environmental Factors

Life of a Mealworm

	Observation 1	Observation 2	Comments
Week 1			
Week 2			
Week 3			
Week 4			

Draw a picture of a mealworm.

Life of a Mealworm

	Observation 1	Observation 2	Comments
Week 5			
Week 6			
Week 7			
Week 8			

INVESTIGATION 1 – Environmental Factors

Part 1
Observing Mealworms 92

Part 2
Designing an Isopod Environment 117

Part 3
Leaf-Litter Critters 130

Guiding question for phenomenon:
How do the structures of terrestrial organisms function to support the survival of the organisms in that environment?

Science and Engineering Practices
- Asking questions
- Developing and using models
- Planning and carrying out investigations
- Analyzing and interpreting data
- Constructing explanations
- Obtaining, evaluating, and communicating information

PURPOSE

Students investigate the phenomena of mealworms and beetles responding to their environment by carrying out a series of investigations with variables of water, light, and temperature.

Content

- An environment is everything living and nonliving that surrounds and influences an organism.
- A relationship exists between environmental factors and how well organisms grow.
- Animals have structures and behaviors that function to support survival, growth, and reproduction.
- Every organism has a set of preferred environmental conditions.

Practices

- Design an investigation so that the effect of one environmental factor can be observed. Record changes in animals and their environments over time.
- Determine an organism's environmental preferences for various nonliving environmental factors to better understand the environment in which it will survive.

Disciplinary Core Ideas

LS1: How do organisms live, grow, respond to their environment, and reproduce?
LS1.A: Structure and function
LS1.D: Information processing
LS2: How and why do organisms interact with their environment and what are the effects of these interactions?
LS2.C: Ecosystem dynamics, functioning, and resilience
LS4: How can there be so many similarities among organisms yet so many different kinds of plants, animals, and microorganisms?
LS4.D: Biodiversity and humans

Crosscutting Concepts
- Cause and effect
- Systems and system models
- Structure and function

FOSS Full Option Science System

INVESTIGATION 1 – Environmental Factors

Investigation Summary	Time	Focus Question for Phenomenon, Practices
PART 1 — **Observing Mealworms** Students observe mealworms as a phenomenon and describe their structures and behaviors. They ask questions to determine what they need to do to provide a proper environment for the mealworms to thrive. Each group sets up a mealworm environment and keeps it at room temperature. The class keeps additional mealworm environments at a colder temperature.	**Assessment** 1 Session **Active Inv.** 1 Session* **Reading** 1 Session	**How do mealworm structures and behaviors help them grow and survive?** **Practices** Planning and carrying out investigations Analyzing and interpreting data Constructing explanations Obtaining, evaluating, and communicating information
PART 2 — **Designing an Isopod Environment** Students observe isopods as a phenomenon. The class conducts two different investigations to find out how isopods respond to the environmental factors of water and light. Based on their findings, students design an isopod environment in a terrarium. Students obtain information about isopod structures and their functions in survival first-hand and through readings.	**Active Inv.** 4 Sessions	**What moisture conditions do isopods prefer?** **What light conditions do isopods prefer?** **Practices** Asking questions Developing and using models Planning and carrying out investigations Analyzing and interpreting data Constructing explanations Engaging in argument from evidence Obtaining, evaluating, and communicating information
PART 3 — **Leaf-Litter Critters** Students go to the schoolyard to collect, observe, and sort small animals living in natural ground litter. They use a Critter Replicator to become familiar with the anatomical parts of animals they find in the leaf litter. Students consider adding found organisms to their group's isopod environment. Students use a concept grid to organize the information they have gathered firsthand in class and outdoors, and through readings and videos about structures and functions of terrestrial organisms and how they meet their needs for survival.	**Active Inv.** 1 Session **Reading/ Video** 3 Sessions **Assessment** 2 Sessions	**What are the characteristics of animals living in the leaf-litter environment?** **Practices** Asking questions Planning and carrying out investigations Analyzing and interpreting data Obtaining, evaluating, and communicating information

*A class session is 45–50 minutes.

At a Glance

Content Related to DCIs	Writing/Reading	Assessment
• An environment is everything living and nonliving that surrounds and influences an organism. • A relationship exists between environmental factors and how well organisms grow. • Animals have structures and behaviors that function to support survival, growth, and reproduction. These include sensory system structures.	**Science Notebook Entry** *Mealworm Observations* *Life of a Mealworm* **Science Resources Book** "Two Terrestrial Environments" "Darkling Beetles" **Video** *Deserts*	**Benchmark Assessment** *Survey* **Embedded Assessment** Science notebook entry
• Designing an investigation involves controlling the factors so that the effect of one factor can be observed. • Every organism has a set of preferred environmental conditions. • Isopods prefer moist environments; isopods prefer dark environments.	**Science Notebook Entry** *Isopod Investigation* *Isopod Environment Map* **Science Resources Book** "Setting Up a Terrarium" "Isopods"	**Embedded Assessment** Performance assessment Response sheet
• Every organism has a set of preferred environmental conditions. • An environment is everything living and nonliving that surrounds and influences an organism. • A relationship exists between environmental factors and how well organisms grow.	**Science Notebook Entry** *Critter Record* *Critter Body Parts* (optional) **Science Resources Book** "Amazon Rain Forest Journal" **Videos** *Animals of the Rain Forest* *Animal Needs*	**Benchmark Assessment** *Investigation 1 I-Check* **NGSS Performance Expectations addressed in this investigation** 4-LS1-1 4-LS1-2

Environments Module—FOSS Next Generation   **81**

INVESTIGATION 1 – Environmental Factors

BACKGROUND *for the Teacher*

TEACHING NOTE

Refer to the grade-level Planning Guide chapter in Teacher Resources for a summary explanation of the phenomena students investigate in this module using a three-dimensional learning approach.

The anchor phenomena investigated in this module are the ways that animals and plants interact with their environment and the structures involved in the process. The driving question for the module is how do the structures of an organism allow it to survive in its environment? The guiding question for this investigation is how do the structures of terrestrial organisms function to support the survival of the organisms in that environment?

All **organisms**, including plants and animals, have specific requirements for successful growth, development, and reproduction. One of the requirements is a supportive environment. The sum of the external conditions that affect reproduction and survival is an organism's **environment**. The external conditions, called **environmental factors**, include **nonliving**, or abiotic, factors such as water, light, air, chemicals, and temperature, as well as **living**, or biotic, factors, which are the influences of other organisms.

You would not expect to find a polar bear living in the deserts of Southern California, nor would you expect to find a barrel cactus growing in the arctic tundra. Organisms are found in environments that provide them with suitable living **conditions**. The distribution and abundance of an organism depend in part on the quality and quantity of each environmental factor that affects the organism throughout its **life cycle**. If a particular factor is deficient, such as food for an animal or an essential nutrient for a plant, or if a factor is in excess, such as extremes in temperature or chemical concentration, the organism might not be able to survive.

In these first investigations, students focus on organisms in their environments. They look at how the environmental factors, both nonliving and living, support the organisms. The term *environment* encompasses both the community of organisms (living components) and the physical surroundings (nonliving components) at a given time.

Later in the module, students will learn another term, *ecosystem*. An ecosystem is a group (or community) of living organisms, the nonliving factors that surround it, and the interactions that take place between them and their environment. This term is important when we study feeding relationships in food chains and food webs. Students may use the two terms, *environment* and *ecosystem,* interchangeably, which is a fine first step in their understanding.

Eventually students should understand that environmental biologists, or ecologists, observe organisms in natural environments to better understand the interactions that influence their success. Ecologists use the information gathered in nature to design controlled laboratory investigations, which in turn provide even more information about the organisms and their interactions with the environment.

> *"All organisms, including plants and animals, have specific requirements for successful growth, development, and reproduction."*

Full Option Science System

Background for the Teacher

Students, as well as scientists, need to have experiences with organisms in environments before designing experiments. This investigation brings a bit of the living world, in the form of a terrarium, into the classroom to provide introductory experiences with organisms and their environments.

We are all creatures of comfort to some degree. On a hot, humid day we're likely to seek out a place where it's possible to relax in the shade of a tree down by the shore, where a breeze is moving the air a little. All the better if there are pleasant sights, smells, and sounds in the environment to enhance the enjoyment of the moment. In choosing a comfortable, secure place to repose, we are demonstrating environmental preference.

Environmental preference refers to the mix of environmental factors that you choose to have around you. You can survive outside your **preferred environment**, but it won't be as pleasant. For instance, on that stifling day, you might very well walk across 200 meters (m) of hot asphalt and stand in line for 10 minutes in direct sun to get a large soft drink. You can survive in these conditions for a short time but it is not very comfortable.

Darkling beetles of the genus *Tenebrio* are cut from a different fabric. They have no need for moisture in their environment, being able to complete their whole life cycle without ever touching the minutest drop of water. Darkling beetles extract all the water they need to survive from the meager supply stored in the grains they eat. Temperatures in the grain storage areas where they live can soar, and the beetles thrive. In fact, darkling beetles prefer a warm, dry environment in which to live and reproduce.

Isopods, also known as **pill bugs** and **sow bugs**, are related to crayfish and crabs. Like their aquatic kin, isopods breathe with gill-like organs instead of lungs. For this reason, isopods require moist environments. Because they eat primarily dead and decaying plant material, they prefer to live in leaf litter and rotting wood. Because they are fairly defenseless, having no tooth, claw, stinger, or bad odor to deter predators, they prefer to be under or wedged into some form of protective cover. All in all, isopods prefer an environment under a cool, damp rock, buried in moldering leaves and twigs.

"Animals can move somewhat freely in their environments, so it is possible for them to vote with their feet to indicate their preferred environment."

We have a pretty good idea of human environmental preference for a large number of environmental factors, but for other kinds of organisms we need to conduct careful investigations. Animals can move somewhat freely in their environments, so it is possible for them to vote with their feet to indicate their preferred environment. In this way, we can discover through laboratory investigations the environmental preferences of a great many animals. This provides insight into the environment in which these organisms can survive.

Environments Module—FOSS Next Generation

83

INVESTIGATION 1 – Environmental Factors

How Do Mealworm Structures Help Them Grow and Survive?

The only insect used in the **Environments Module** is the darkling beetle (*Tenebrio molitor*). It is called darkling perhaps because it leads a cryptic life in dim recesses, coming out after dark, or because it is dark brown. Like all insects, darkling beetles have three body parts (head, thorax, abdomen) and three pairs of legs. Like many insects, darkling beetles have wings, but they are not completely formed, so the beetles rarely attempt to fly. Mostly they scurry about on their six long legs.

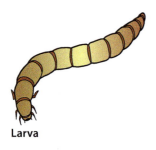
Larva

Darkling beetles follow a life history known as complete metamorphosis. Like butterflies and moths, they go through four distinct **stages** during their life cycle. A female beetle lays eggs, as many as 500 in her brief adult lifetime of a month or two. The eggs are about the size of the period at the end of this sentence. After a couple of weeks, the equally tiny **larvae** emerge from the eggs. The larvae are known as **mealworms**, but of course they are not true worms. The larvae are golden yellow and have 12 body segments. They are the counterpart of the familiar caterpillar in the butterfly life cycle. Mealworms pull themselves around on six stubby legs that are all crowded at the front.

Pupa, ventral view

Pupa, dorsal view

The larvae seem to have two purposes in life: eat and grow. These beetle larvae are arthropods, and like all members of their phylum, they wear their skeleton on the outside like a suit of armor. This is very practical when they are being attacked, but very inconvenient when they are trying to grow. The arthropods have solved this problem by **molting**, shedding their skinlike cuticle, periodically. Immediately after molting, the soft, white larvae expand before the new larger skin hardens. This process may repeat half a dozen or more times over a 3-month period, until the larvae are about 2 centimeters (cm) long and they **pupate**. The final larval molt reveals the next stage, the **pupa**.

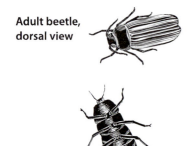
Adult beetle, dorsal view

Adult beetle, ventral view

The pupae don't eat, and they don't move except for a twitch or two. Inside, however, the mealworm is turning into a beetle, much the same as a caterpillar turns into a butterfly while sequestered inside the chrysalis. In 2 or 3 weeks, the pupa splits open and out walks an **adult** beetle, white at first, but soon turning brown and finally black after a day. The adult beetles mate and lay eggs, and the cycle repeats.

Mealworms and darkling beetles are rarely seen in the wild, but when they are, it is likely to be in a field where wild grasses flourish and seeds are plentiful. They are most often found in barns, grain-storage facilities, and food-preparation areas. This organism has benefited by living close to human enterprises, because we unwittingly provide a much better environment for the success of *Tenebrio* than could be found in the natural world. For this reason, they have become a minor pest in grain-storage areas.

Background for the Teacher

Mealworms and darkling beetles are excellent classroom animals for making **observations**—they exhibit interesting **structures** and **behaviors**; they are small but not tiny; they don't bite, smell, fly, or jump; and they are extremely easy to care for. Mealworms live right in a container of their food source: bran, cornmeal, rolled oats, breakfast flakes, or chick starter mash. All are excellent foods, but bran and chick starter are recommended. The food must be kept dry. Mealworms can go through their complete life cycle without any added water (they are very efficient at extracting water from their food), but it is recommended that small bits of apple, potato, or carrot be added from time to time.

Mealworms should be kept in large, relatively flat containers. They seem to thrive best when the colony has a large surface area. Keep the bran about 2 or 3 cm deep in a basin, bus tray, aquarium, or plastic shoe box. If the container sides are steep and smooth, it is not necessary to keep the container covered. Adults and larvae seem to prefer hiding under bits of paper or light cardboard; the pupae give no indication that they care.

The mealworm's preferred environment is very dry, moderately warm, and dark. A bit of apple provides extra moisture for the mealworms and seems to stimulate rapid growth. As the temperature increases, so does the rate at which mealworms advance through their life cycle. Under ideal conditions in a classroom, the complete life cycle can take as little as 3 months, but more likely it will take 4 months. Cold slows the process almost to the point of suspended animation. Mealworms can be put into the refrigerator (not the freezer) for a time to stop metamorphosis.

What Do Isopods Need from Their Environment?

Iso is Greek for "similar or equal." *Pod* means "foot." Put them together, and you have the isopod, an organism that has an equal number of similar feet or legs on both sides. Isopods have 14 legs that all **function** the same. This distinguishes them from closely related organisms that have legs that are modified to perform different functions, such as walking, feeding, feeling, and grasping.

Sow bug

The many species of isopods around the world share certain characteristics. Isopods are crustaceans, distant kin of shrimps, crabs, and crayfish. Like all crustaceans, isopods have a segmented outer shell (seven overlapping plates) that provides a measure of protection from the environment and predators. Like their aquatic relatives, isopods get the oxygen they need to survive through gill-like structures at the bases of their legs, rather than through lungs like most terrestrial organisms. That is why isopods must keep moist at all times—if they dry, they die.

Pill bug

Environments Module—FOSS Next Generation

INVESTIGATION 1 – *Environmental Factors*

Pill bugs

Sow bugs

Two kinds of isopods are of interest as classroom organisms. The genus *Armadillidium* (arm•uh•duh•LID•e•um) is known casually as the pill bug or roly-poly. It gets these names from its habit of rolling into a tight sphere when threatened or stressed. The pill bug has a highly domed shape, short legs, and inconspicuous antennae. Its defensive rolled posture makes the bug hard for a predator to grip and helps it resist drying out.

Pill bugs move slowly and have a difficult time righting themselves if they roll onto their backs on a smooth surface. They range from light brown to dark gray or black. Often they have white, cream, or yellowish spots on their backs. The largest individuals of this kind of isopod can be 1 cm long, but most are 7 or 8 millimeters (mm).

The second isopod used extensively in classrooms, genus *Porcellio* (por•sel•EE•oh), is commonly called the sow bug or wood louse. These names are potentially confusing because *Porcellio* don't show a particular affinity for swine, nor are they lice. They are relatively flat with legs that extend a little bit beyond the edge of the shell, and they have powerful antennae to sense their environment. They move rather quickly and will use their long **antennae** and little spikelike tail projections to right themselves if they happen to roll onto their backs. Sow bugs come in a surprising array of colors, including tan, orange, purple, and blue, as well as the usual battleship gray. Their size is similar to that of the pill bug.

In the wild, isopods are not usually seen out and about. They are members of that large category of animals known descriptively (not taxonomically) as cryptozoa, or hidden animals. They are most often found in layers of duff and leaf litter, under rocks or logs, or burrowed a short distance under the surface of the soil. The environment they seek is moist and dark, in or near dead and decomposing wood and other plant material. Wood is their main source of food, accounting, perhaps, for their common name of wood lice. Isopods are not above eating fresh strawberries and carrots, however, making them a minor pest in the garden.

There are both male and female isopods, but only another isopod can reliably tell them apart. After mating, the female lays several dozen eggs, which she carries in a compact white package on her underside between her legs. This package is a specialized brood pouch, the marsupium, in which the eggs develop for 3 or 4 weeks before hatching. A few days after hatching, a swarm of fully formed, minute isopods strike out into the world. They are nearly invisible at first but soon grow to a size that can be seen by the unaided eye. Like all crustaceans that carry a hard outer shell, isopods must shed their shells in order to grow. In the molting

Background for the Teacher

process, the shell is cast off, and the new soft shell underneath expands before hardening. Interestingly, the whole shell is not shed at once; first the rear (posterior) shell segments are shed, and 2 or 3 days later the front (anterior) ones fall off.

Isopods are also excellent classroom animals—they, too, exhibit interesting structures and behaviors; they are small but not tiny; they don't bite, smell, fly, or jump; and they are easy to care for. Isopods can live in just about any vessel, from a recycled margarine tub to a 50 liter (L) **terrarium**. If the container is smooth-sided, it doesn't even have to be covered, because isopods can't climb smooth surfaces at all. A layer of soil covered with some dead leaves, twigs, and bark is great, but isopods will do fine with some paper towels or newspaper laid on the soil. They do like to have some structure to crawl under.

The most important thing to remember is that the soil must be kept moist at all times—not wet, but moist—so that the isopods don't dry out. A chunk of raw potato in the container with the isopods serves as a source of both food and moisture. Otherwise they will eat the decomposing leaves and twigs or the paper towels and newspaper.

What Are the Characteristics of Animals Living in the Leaf-Litter Environment?

In the leaves under our feet is a world teeming with animals. As plants die and drop to the ground, they form a layer of decaying leaves, sticks, and bark called leaf litter. This natural litter layer forms a constantly changing habitat in which many organisms live.

Rain washes litter away, wind blows it around, and the Sun dries it out. Where the litter is thick, however, only the upper layers dry out. The layers next to the soil provide a continuously moist environment. In this moist area, tiny organisms such as fungi and bacteria feed on bark, leaves, and twigs, and break them down (decompose them) into smaller and smaller pieces. Decomposers also release minerals back to the soil. At the same time that the lower layers are being broken down, new plant material is being deposited on top of the existing layer, ensuring a continuous litter habitat.

Animals that live in natural ground litter (for example, insects, slugs, spiders, and salamanders) are generally small. Their small size allows these animals to crawl into tiny crevices between pieces of decomposing plant and animal matter. Their size also makes these animals easy to overlook.

"Isopods are also excellent classroom animals—they, too, exhibit interesting behaviors; they are small but not tiny; they don't bite, smell, fly, or jump; and they are easy to care for."

Environments Module—FOSS Next Generation

87

INVESTIGATION 1 – Environmental Factors

TEACHING CHILDREN about Environmental Factors

Developing Disciplinary Core Ideas (DCI)

An environment is generally thought of as a place where plants and animals live. To fully understand this complex biological concept, however, one must know about the conditions and factors that affect an individual's and a population's quality of life and their ability to survive within an environment.

Research indicates that upper-elementary students have a vague idea of what an environment is. They think of it as a place where something lives. They are unlikely to think of an environment in terms of the specific factors that it provides for an organism to live and survive. They are unlikely to consider the uniqueness of an environment or the differences between one environment and another.

It takes many experiences with environments, both formal and informal, to understand the biological aspects of the concept. These aspects are not learned through one activity or in one course. They are learned over many years of experience. If the experiences are thoughtfully organized, as in the FOSS Program, students' understanding will grow progressively throughout their school years.

The animals in our world are so numerous that they cannot all be studied in detail. The animals in this investigation were chosen because they are sturdy. They are not injured with general handling by students. And their behaviors in response to changes in environmental conditions are easily recognized.

When introducing animals to students, capitalize on what they already know and be sensitive to students' feelings. Some students are reluctant to touch certain types of animals. This personal feeling should be respected, and a student should never be made to feel bad or left out of an experience because of this reluctance.

Teachers report that many students lose their reluctance as they see other students handling and exploring the animals. Reluctance to touch an animal is learned from others who are reluctant. Thus, observing others who handle animals freely provides an alternative model. When they are ready, many will explore with their classmates, gaining confidence as they do so.

Environmental factors are best studied one at a time. Tests that isolate one factor to study its effect on an animal are useful for learning about the relationships between factors and animals. Such tests predominate throughout this investigation.

NGSS Foundation Box for DCI

LS1.A: Structure and function
- Plants and animals have both internal and external structures that serve various functions in growth, survival, behavior, and reproduction. (4-LS1-1)

LS1.D: Information processing
- Different sense receptors are specialized for particular kinds of information, which may then be processed by an animal's brain. Animals are able to use their perceptions and memories to guide their actions. (4-LS1-2)

LS2.C: Ecosystem dynamics, functioning, and resilience
- When the environment changes in ways that affect a place's physical characteristics, temperature, or availability of resources, some organisms survive and reproduce, others move to new locations, yet others move into the transformed environment, and some die. (3–LS4-4, extended from grade 3)

LS4.D: Biodiversity and humans
- Populations live in a variety of habitats, and change in those habitats affects the organisms living there. (3–LS4-4, extended from grade 3)

Teaching Children about Environmental Factors

While we can provide laboratory investigations to study the environmental preferences of some common animals, it is harder to have students study interdependence of organisms in the classroom. We hope that students will observe interactions between plants and animals as they tend their terrariums and aquariums throughout the module.

The experiences students have in this investigation contribute to the disciplinary core ideas **LS1.A: Structure and function; LS1.D: Information processing; LS2.C: Ecosystem dynamics, functioning, and resilience;** and **LS4.D: Biodiversity and humans.** In addition to teaching children about the core idea of organisms interacting with their environment, each investigation in the **Environments Module** provides students with opportunities to experience two other dimensions of the scientific enterprise—the dimension of science and engineering practices and the dimension of crosscutting concepts.

Engaging in Science and Engineering Practices (SEP)

Engaging in the practices of science helps students understand how scientific knowledge develops; such direct involvement gives them an appreciation of the wide range of approaches that are used to investigate, model, and explain the world. Engaging in the practices of engineering likewise helps students understand the work of engineers, as well as the links between engineering and science. Participation in these practices also helps students form an understanding of the crosscutting concepts and disciplinary ideas of science and engineering; moreover, it makes students' knowledge more meaningful and embeds it more deeply into their worldview. (National Research Council, *A Framework for K–12 Science Education*, 2012, page 42).

The focus questions and notebook sheets provide scaffolds for students in this investigation, scaffolds that can be carefully removed in later investigations, as students become more experienced engaging in practices. In this first investigation, students engage in these practices while using others as well.

- **Asking questions** about the preferred environmental conditions of isopods and design experiments based on cause-and-effect relationships.
- **Developing and using models** to test environmental preferences of isopods to understand a natural system.
- **Planning and carrying out investigations** dealing with the effect of different environmental conditions on mealworms and isopods. Collecting data to serve as the basis for evidence for a claim.

NGSS Foundation Box for SEP

- **Ask questions** that can be investigated based on patterns such as cause-and-effect relationships.
- **Use a model** to test cause-and-effect relationships to interactions concerning the functioning of a natural system.
- **Plan and conduct an investigation** collaboratively to produce data to serve as the basis for evidence.
- **Make observations and/or measurements to produce data** to serve as the basis for evidence for an explanation of a phenomenon or test a design solution.
- **Analyze and interpret data** to make sense of phenomena using logical reasoning.
- **Compare and contrast data** collected by different groups in order to discuss similarities and differences in their findings.
- **Organize simple data sets** to reveal patterns that suggest relationships.
- **Use evidence** (e.g., observations, patterns) to support an explanation.
- **Identify the evidence** that supports particular points in an explanation.
- **Respectfully provide and receive critiques from peers** about a proposed procedure, explanation, or model by citing relevant evidence and posing specific questions.
- **Obtain and combine information** from books and other reliable media to explain phenomena.

Environments Module—FOSS Next Generation

89

INVESTIGATION 1 – Environmental Factors

NGSS Foundation Box for CC

- **Cause and effect:** Cause-and-effect relationships are routinely identified and used to explain change.
- **Systems and system models:** A system can be described in terms of its components and their interactions.
- **Structure and function:** Substructures have shapes and parts that serve functions.

Adult
Antennae
Behavior
Condition
Darkling beetle
Environment
Environmental factor
Function
Inference
Isopod
Larva
Life cycle
Living
Mealworm
Molting
Nonliving
Observation
Organism
Pill bug
Preferred environment
Pupa
Pupate
Sow bug
Stage
Structure

- **Analyzing and interpreting data** to find preferences of organisms for different environmental conditions in order to reveal patterns that can be used to make predictions. Compare data observed by different groups to look for similarities and differences. Organize simple data sets about where organisms were collected outdoors to find relationships between the organisms' structures and the environment in which they are found.

- **Constructing explanations** using evidence, such as the relationships between environmental conditions and the structures and behaviors of isopods, beetles, and other small animals.

- **Engaging in argument from evidence** about the preferred environmental conditions for organisms. Critique the investigations of others and offer suggestions for improvement.

- **Obtaining, evaluating, and communicating information** from books and media and integrating that with their firsthand experiences to construct explanations about how organisms interact with their environment.

Exposing Crosscutting Concepts (CC)

The third dimension of instruction involves the crosscutting concepts, sometimes referred to as the unifying principles, themes, or big ideas, that are fundamental to the understanding of science and engineering.

These concepts should become common and familiar touchstones across the disciplines and grade levels. Explicit reference to the concepts, as well as their emergence in multiple disciplinary contexts, can help students develop a cumulative, coherent, and usable understanding of science and engineering. (National Research Council 2012, page 83)

In this first investigation, the focus is on these crosscutting concepts.

- **Cause and effect.** Cause-and-effect relationships between environmental conditions and the structures and behaviors of organisms can be identified and used to predict future events.

- **Systems and system models.** Environments can be described in terms of their living and nonliving components and their interactions. The system (the environment) provides the basic needs for the organisms living there.

- **Structure and function.** The form of an organism's structures relate to their functions.

Teaching Children about Environmental Factors

Conceptual Flow

The anchor phenomena investigated in this module are the ways that animals and plants interact with their environment and the structures involved in the process. The driving question for the module is how do the structures of an organism allow it to survive in its environment? The first investigation engages students with terrestrial organisms. The guiding question for this investigation is how do the structures of terrestrial organisms function to support the survival of the organisms in that environment?

The **conceptual flow** for this investigation starts in Part 1 with students observing **mealworms** as a phenomenon and describing the organisms' structures and behaviors. After discussing the needs of organisms, each group of students places mealworms in a container with air, food, and a tiny bit of apple or carrot. Students describe the environment they provided for the mealworms and identify the **environmental factors that describe the conditions**. The **environment is the living and nonliving things that surround and influence an organism**. Students keep the containers at room temperature, and the class places additional mealworm environments in a refrigerator to see how the factor of air temperature affects the growth of mealworms. Students observe the mealworms over 6–8 weeks and monitor changes as the mealworms go through **life-cycle stages**: **egg** to **larva** to **pupa** to **adult** beetle. Students compare the growth of the organism in warm and cold environments. They have opportunities to observe variation in a population of organisms and to see inherited characteristics.

In Part 2, students observe related organisms as a phenomenon, and investigate the structures and behaviors of two different kinds of terrestrial **isopods**, pill bugs and sow bugs. They design investigations to test the isopods' **preferred condition** for two separate environmental factors—**moisture and light**. Based on the results of the tests, students **design an isopod environment**—a terrarium—and monitor the organisms over time.

In Part 3, students go outdoors to look for organisms that **live in natural leaf litter**. Students observe and describe the characteristics of the critters that live in environments similar to the isopod environment.

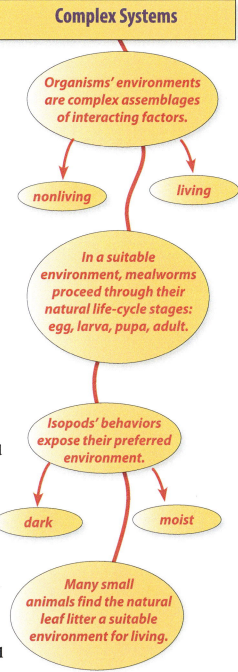

Environments Module—FOSS Next Generation

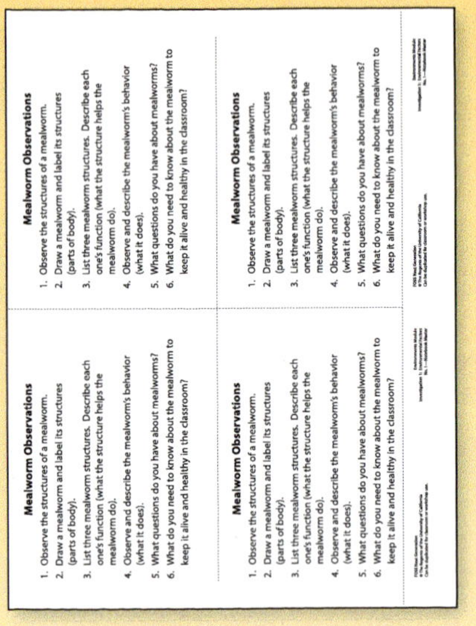

INVESTIGATION 1 – Environmental Factors

No. 1—Notebook Master

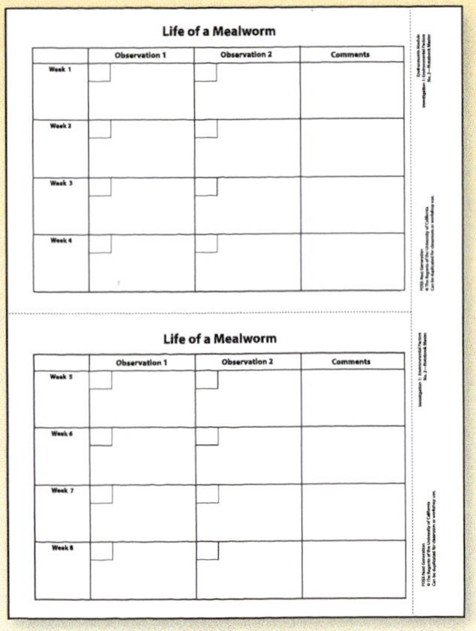

No. 2—Notebook Master

MATERIALS for
Part 1: Observing Mealworms

For each student
- 1 *Letter to Family* ★
- 1 Science notebook (composition book) ★
- ☐ 1 Notebook sheet 1, *Mealworm Observations*
- ☐ 1 Notebook sheet 2, *Life of a Mealworm*
- 1 FOSS Science Resources: Environments
  - "Two Terrestrial Environments"
  - "Darkling Beetles"

For each group
- 10 Mealworms ★
- 1 Container with lid, 1/4 L
- 4 Plastic cups
- 1 Plastic spoon
- 1 Sheet of recycled paper ★
- 1 Self-stick note
- 1 Critter Replicator
- 4 Hand lenses
- • Low-power microscopes (optional) ★

For the class
- 100 Mealworms ★
- 2 Containers with lid, 1/4 L
- 2 Self-stick notes
- 1 Basin, clear plastic, 6 L
- 2 Spoons, 5 mL
- • Wheat bran, 0.5 kg (1 lb.) (See Step 4 of Getting Ready.) ★
- • Apple, carrot, or potato ★
- 1 Paring knife ★
- 1 Push pin ★
- 5 Paper towel sheets (See Step 6 of Getting Ready.) ★
- • Cold environment such as a refrigerator ★

Full Option Science System

Part 1: Observing Mealworms

- 4 Posters—*Mealworm Larva, Mealworm Pupa, Mealworm Adult (Darkling Beetle), Mealworm Stages (Darkling Beetle Stages)*
- *FOSS Science Safety* and *FOSS Outdoor Safety* posters
- *Conservation* posters
- ❑ 1 Teacher master 1, *Letter to Family*
- ❑ • Teacher masters 2–5 (optional)

For embedded assessment
❑ • *Embedded Assessment Notes*

For benchmark assessment
❑ • *Survey*
❑ • *Assessment Record*

* Supplied by the teacher.   ❑ Use the duplication master to make copies.

No. 1—Teacher Master

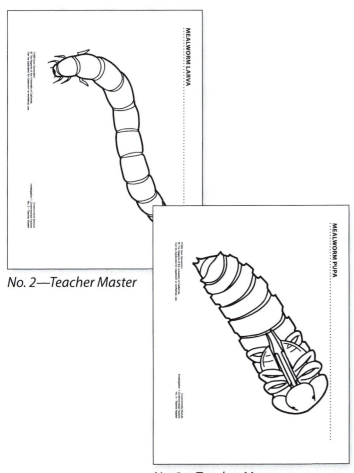

No. 2—Teacher Master

No. 3—Teacher Master

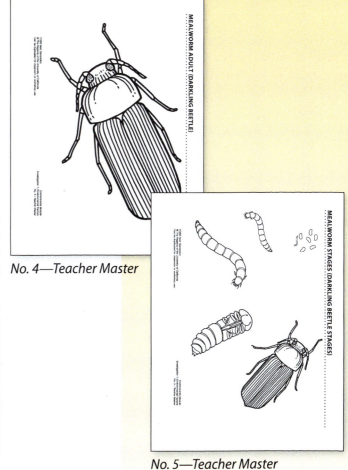

No. 4—Teacher Master

No. 5—Teacher Master

Environments Module—FOSS Next Generation

93

INVESTIGATION 1 – Environmental Factors

GETTING READY for
Part 1: Observing Mealworms

TEACHING NOTE

The Getting Ready section for Part 1 of Investigation 1 is longer than the corresponding section in the rest of the module. Several of the numbered items appear only here, but might apply to other parts as well, such as the discussion about setting up a materials station (3), planning a word wall (9), preparing student science notebooks (10), and printing or photocopying duplication masters (12).

▶ **NOTE**
To prepare for this investigation, view the teacher preparation video on FOSSweb.

▶ **SAFETY NOTE**
Find out if any students are allergic to wheat. The mealworms are often shipped to the school in wheat bran.

1. **Schedule the investigation**
   This part will take one session for the survey, one session for the active investigation, and one or two sessions for the reading and video. When you get to Step 20 of Guiding the Investigation, students begin to make observations of the mealworms two times a week for 6 weeks. After students view and discuss the video (Steps 21-22), you should go on to Part 2.

   Return to Part 1, Steps 23–37, over the next 6-8 weeks as students monitor changes to mealworms in the different environments. Plan on short observation sessions throughout the module so students can observe all the stages in the life cycle of this organism.

2. **Preview Part 1**
   Students observe mealworms and describe their structures and behaviors. They ask questions to determine what they need to do to provide a proper environment for the mealworms to thrive. Each group sets up a mealworm environment and keeps it at room temperature. The class keeps additional mealworm environments at a colder temperature. The focus question is **How do mealworm structures and behaviors help them grow and survive?**

3. **Plan student organization**
   This module is designed for groups of four students. Sometimes students work in pairs. Working in groups allows students to observe others, compare, share, and cooperate. Students can assume roles of Getters (often two students to get and return equipment at the materials station), Starters (to make sure that everyone has a turn to work with materials), or Reporters (to report group findings to the class).

   Plan to organize the materials for all the investigations in a convenient location where Getters can get the materials efficiently.

4. **Acquire wheat bran, potatoes, and a knife**
   Wheat bran is suggested as food for the mealworms. It is often available in bulk at large grocery stores. Otherwise, any grain-based product will suffice: wheat germ, rolled oats, oat bran, or bran breakfast cereals. About 0.5 kilogram (kg) of bran should be sufficient for a start. Put bulk wheat bran in the microwave for 1 minute to kill stray organisms. Students will use a 5 milliliter (mL) spoon to measure the bran into the mealworm 1/4 L containers.

94   Full Option Science System

Part 1: Observing Mealworms

Bring a paring knife to cut the fruit or vegetables. Put large chunks of white potato, apple, or carrot into the class mealworm culture once a week. A small cube of the same fruit or vegetable (the size of a pea) should be put into each 1/4 L container, too, but students should be cautioned that too much moisture in the containers will cause the bran to mold.

5. **Obtain mealworms**
   You will need about 180 regular mealworms (*Tenebrio molitor* larvae). Mealworms are available at pet stores, bait shops, and biological supply houses and from many online suppliers. They can also be ordered from Delta Education. Do not get "king" or "giant" mealworms, as they are a larger species (*Zophobas*) that takes much longer to go through a life cycle.

   Keep the mealworms in bran in the 6 L clear basin. Put half the bran, a vegetable piece, and the mealworms in a basin with no lid, and you have officially started your mealworm environment. Plan to keep it out of direct sun and accessible to students. Keep mealworms hidden until you introduce them to the class.

6. **Care for animals**
   Mealworms should be kept in the environmental container in the classroom until they are needed for the activity. On the day of the investigation, prepare a container of mealworms for each group. Cut a paper-towel liner for the bottom of each 1/4 L container. Trace the bottom of the container on one paper towel, then stack the toweling and cut enough liners for each container. Place one liner and 10 mealworms in a 1/4 L container with no bran for each group. You can use your fingers or plastic spoon to pick up the larvae.

   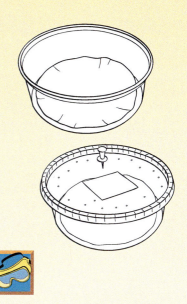

   Place a lid on each container. Place a self-stick note on the container lid. Using a push pin, make about 10-15 holes in the lid around the note. The holes will provide air for the mealworms and the note will be used later to identify each group's container.

7. **Plan for safety**
   Upper-elementary students must be allowed to demonstrate that they can act responsibly with materials, but they must be given guidelines for safe and appropriate use of materials. Work with students to develop those guidelines so that they participate in making behavior rules and understand the rationale for the rules. Encourage responsible actions toward other students. Display and discuss the *FOSS Science Safety* and *FOSS Outdoor Safety* posters in class. The posters are included in the kit.

Environments Module—FOSS Next Generation

INVESTIGATION 1 – *Environmental Factors*

Look for the safety-note icon in the Getting Ready and Guiding the Investigation sections, which will point out specific safety alerts throughout the module. Be aware of allergies your students have, including wheat allergies (wheat bran is the mealworm food). Mealworms are often shipped to schools in wheat bran.

This is a good time to introduce the set of four *Conservation* posters and discuss the importance of natural resources with students.

8. Send a letter home to families

Teacher master 1, *Letter to Family*, is a letter you can use to inform families about this module. The letter states the goals of the module and suggests some home experiences that can contribute to students' learning. Space is left at the top so that you can copy the letter onto your school letterhead.

9. Plan a word wall

As the module progresses, you will add new vocabulary words to chart paper for posting on a wall and/or to vocabulary cards or sentence strips for use in a pocket chart. For additional information, see the Science-Centered Language Development chapter.

10. Plan for student notebooks

Students will keep a record of their science investigations in individual science notebooks. They will record observations, responses to focus questions, and connections between their classroom learning and the world beyond. This record will be a useful reference document for students and a revealing testament for adults of each student's learning progress.

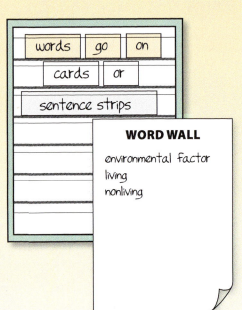

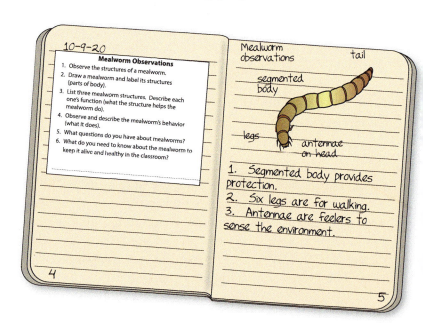

96

Full Option Science System

Part 1: Observing Mealworms

We recommend that students use bound composition books for their science notebooks. This ensures that student work becomes an organized, sequential record of student learning. Students' notebooks will be a combination of structured sheets photocopied from notebook masters provided in this teacher guide (and glued into the notebook) and unstructured student-generated drawings and writings on blank pages.

11. Plan to project images of notebook sheets

Throughout the module, you will see suggestions for projecting images of notebook sheets to orient students to their use or to share data that have been gathered. Use whatever technology is available to you. This may include a document camera, LCD projector and computer, interactive whiteboard, or overhead projector. Digital versions of all duplication masters for duplication or large format for projecting are available at www.FOSSweb.com.

12. Print or photocopy duplication masters

Two sets of duplication masters are used throughout this module: notebook masters and teacher masters. Digital versions of these duplication masters are also available on FOSSweb.

Notebook masters are sized to be cut apart and glued into a composition book. Each notebook master has two or four copies of the sheet, which should be cut apart.

Teacher masters serve various functions such as letter to family, home/school connections, and math extensions.

A notebook sheet or teacher master that requires printing or duplication is flagged with this icon ❏ in the materials list for the part. It is referred to as a notebook sheet in the materials list and a notebook master in the sidebar snapshot.

13. Plan for working with English learners

At important junctures in an investigation, you'll see an EL note in the sidebar. These notes suggest additional strategies for enhancing access to the science concepts for English learners. You'll also see references to the Science-Centered Language Development chapter in the teacher guide. Refer to this chapter for resources and examples to use when working with science vocabulary and readings in each investigation.

Each time new science vocabulary is introduced, you'll see the new-word icon in the sidebar. This icon lets you know not only that you'll be introducing important vocabulary, but also that you might want to spend more time with those students who need extra help.

> **TEACHING NOTE**
>
> Refer to the Science Notebooks in Grades 3–5 chapter for more details on how to integrate the effective use of science notebooks into all FOSS investigations.

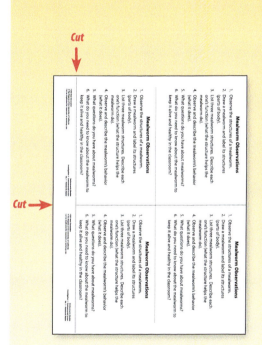

> **EL NOTE**
>
> Project the equipment photo card for each object and write the object's name on the word wall. Equipment photo cards are available on FOSSweb and can be downloaded and printed.

Environments Module—FOSS Next Generation 97

INVESTIGATION 1 – *Environmental Factors*

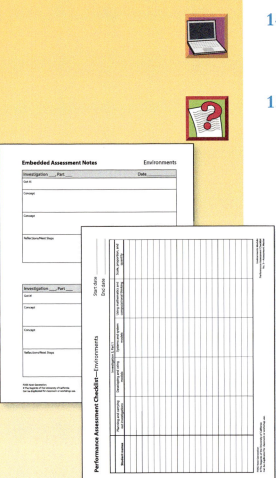

Assessment Masters

14. Plan for online activities

During instruction, you will need to project online activities through a computer for the class to see. The Getting Ready section for each part will indicate what to prepare.

15. Assess progress throughout the module

Research shows that frequent formative assessment leads to greater student understanding. Embedded (formative) assessments provide a variety of ways to gather information about students' thinking while their ideas are developing. These assessments are meant to be diagnostic. They provide you with information about students' learning so that you know if you need a next step to clarify understanding before going on to the next part of the investigation. Most Getting Ready sections describe an embedded-assessment strategy that you might find useful in that part. Assessment master 1, *Embedded Assessment Notes*, is a half sheet provided to help you analyze students' embedded-assessment data. (See the Assessment chapter for more on how to use this sheet.) In some parts, the embedded assessment involves a performance assessment. The *Performance Assessment Checklist* is used for recording these observations.

At the end of each investigation (except Investigation 4), there is an I-Check benchmark assessment. The questions on these assessments are summative—they examine the disciplinary core ideas, science and engineering practices, and crosscutting concepts students have learned up to that point in the curriculum. You can find out more about I-Check assessments in the Assessment chapter.

16. Plan benchmark assessment: *Survey*

Before beginning the module, have students take the *Survey* benchmark assessment. This assessment will inform you about students' prior knowledge and give you some insight into which concepts may need more focused instruction. Save the *Surveys* to compare to the identical *Posttest*, given after the module instruction has been completed. Masters for the *Survey* can be found in the assessment masters on FOSSweb. Record student performance on *Assessment Record* (assessment master no. 5). Or plan to administer the *Survey* using FOSSmap. For more information on the assessment system and online tools, see the Assessment chapter.

Full Option Science System

Part 1: Observing Mealworms

17. **Preview the video**

    Preview the video, *Deserts* (duration 25 minutes), and plan how best to have students view it. The video introduces biomes of the world, gives an overview of the desert biome characteristics, and then focuses on animals and plant structures that help them survive in the desert biome. In Step 21 of Guiding the Investigation, there is a detailed outline of the content of each video chapter, which may be helpful. The link to this video is in Resources by Investigation on FOSSweb.

18. **Plan to read *Science Resources*: "Two Terrestrial Environments" and "Darkling Beetles"**

    Plan to read the article "Two Terrestrial Environments" during a reading period once the mealworm environments are set up (Step 17 of Guiding the Investigation).

    The last article in the book, "Darkling Beetles," should be read about 8 weeks later after students have observed the life-cycle stages of the darkling beetle in the classroom. This reading is discussed in Step 36 of Guiding the Investigation.

19. **Review videos to introduce anchor phenomenon**

    Preview the videos from the interactive eBook to see how animals interact with their environment. You'll find the videos in the following articles in *FOSS Science Resources*:

    "Two Terrestrial Environments": *Sloth in rain forest, Small frog in rain forest*, and *Bighorn sheep in desert*; "Amazon Rain Forest Journal": *Macaw in rain forest* and *Leaf-cutter ants in rain forest*; and "Isopods": *Isopod in garden*." These videos can also be accessed from Resources by Investigation on FOSSweb.

20. **Plan assessment: notebook entry**

    At the first breakpoint, collect students' notebooks after class to check their answer to the questions on notebook sheet 1, *Mealworm Observations* (see Step 16 of Guiding the Investigation). In Step 25 of Guiding the Investigation, review students' answers to the focus question to see if they understand how structures and behaviors function to help mealworms survive.

Environments Module—FOSS Next Generation

INVESTIGATION 1 – Environmental Factors

FOCUS QUESTION

How do mealworm structures and behaviors help them grow and survive?

Say it · New Word · See it · Hear it · Write it

CROSSCUTTING CONCEPTS

Structure and function

Materials for Step 3
- *Mealworms in containers*
- *Cups*
- *Hand lenses*
- *Plastic spoons*
- *Paper*
- *Microscopes (optional)*

GUIDING *the Investigation*
Part 1: *Observing Mealworms*

1. **Introduce the anchor phenomenon**
   Begin the module by showing students a series of short videos from the *FOSS Science Resources* eBook. Ask them to observe the way each animal interacts with its environment.

   - Sloth in rain forest
   - Small frog in rain forest
   - Bighorn sheep in desert
   - Macaw in rain forest
   - Leaf-cutter ants in rain forest
   - Isopod in garden

   Tell students that they will be investigating firsthand and through media ways that animals and plants interact with their environment. The driving question for the module is how do the structures of an organism allow it to survive in its environment?

2. **Introduce mealworms**
   Tell students that to begin, you have an **organism** for them to study firsthand, **mealworms**.

   *You will observe these organisms to find out about their **structures** and **behaviors**. A structure is any part of an organism. (Some structures of your body are fingers, ears, and legs.) Those structures have a purpose or **function** that help the organism survive.*

   You'll also need to observe what the mealworms do or how they respond—those are their behaviors. You'll also have a chance to closely observe mealworm behaviors and structures over the next 6 weeks.

   Write the new words on the word wall.

   Ask students what they know about mealworms, and make a list of the ideas. Reinforce that mealworms are living organisms and that nothing should be done to harm the animals in any way. Promote an attitude of respect for living organisms. Remind students that mealworms cannot hurt them. They should be handled gently.

3. **Make mealworm observations**
   Ask students to observe the mealworms and to note structures and behaviors. Encourage them to hold the animals carefully—they don't bite or sting. They can put the mealworms on a quarter sheet of paper and observe how they move. Hand lenses will permit close observations.

   Ask the Getters to get a container of mealworms, four hand lenses, four cups, paper, and a spoon. Allow 5–10 minutes for making initial observations. Provide low-power microscopes if available.

Part 1: Observing Mealworms

4. **Set up notebooks**

   Have students get out their notebooks. If this is the first notebook entry in a new notebook, take a few minutes to establish the preliminaries: table of contents, page numbers, and method for attaching notebook sheets to the page.

   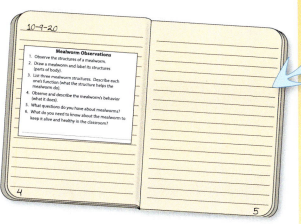

   Ask students to open their science notebooks to a new page and record the date.

5. **Distribute notebook sheet**

   Distribute a copy of notebook sheet 1, *Mealworm Observations*, to each student. Have students glue the sheet on a left-hand page of their notebooks and write their responses on the right-hand page.

6. **Discuss observations**

   After 10–15 minutes, call for attention. Ask students to return their mealworms to the group container. Discuss what students observed about the organisms.

   ▶ *What structures do mealworms have on their body?* [Lines or segments, 6 tiny legs, 2 tiny antennae or feelers, pointed tail.]

   ▶ *What mealworm behaviors did you observe?* [Crawling, feeling the paper, holding onto the paper, recoiling from being touched.]

   ▶ *Did you observe any sensory structures used to get information about the environment?* [Antennae, perhaps two pairs.]

7. **Focus question: How do mealworm structures and behaviors help them grow and survive?**

   Write or project the focus question on the board as you say it aloud.

   ▶ *How do mealworm structures and behaviors help them grow and survive?*

   Have students write the focus question in their notebooks.

   Make a list of ideas students have to answer the focus question. As students suggest needs, write them on the board until the list includes the four most important needs for any organism: oxygen (air), water, food, and space with shelter.

> **TEACHING NOTE**
>
> For additional information, see the Science Notebooks in Grades 3–5 chapter.

> **TEACHING NOTE**
>
> Students should continue to make observations, but now they should record what they find out about mealworm structures and behaviors.

> **EL NOTE**
>
> Suggest students use a table to organize their data for questions 4 and 5 on notebook sheet 1 with headings: structure, function, and behavior.
> During the discussion, draw and label the structures of a mealworm on chart paper for reference.

CROSSCUTTING CONCEPTS

Structure and function

> **TEACHING NOTE**
>
> Students will answer this question at the end of 6 weeks when they have observed all of the life stages of the darkling beetle. (See Step 24 of Guiding the Investigation.)

Environments Module—FOSS Next Generation

INVESTIGATION 1 – Environmental Factors

▶ **SAFETY NOTE**
Students who have wheat allergies should not handle the wheat bran. Be aware that the mealworms shipped to schools often arrive in wheat bran.

Materials for Step 10
- Wheat bran
- Prepared mealworm containers
- Spoons, 5 mL
- Small pieces of fruit or vegetable

EL NOTE
Add these concepts to the mealworm chart. Draw the mealworm environment and label the living and nonliving factors.

8. **Discuss the container setup**
   Reiterate that all living organisms need four things—air, water, food, and space with shelter. Discuss how each will be accommodated in the container.

   a. *The 1/4 L container will provide the space. A self-stick note on the lid will identify the group number.*

   b. *Air will enter the container through holes punched in the lid with a push pin. Students should note that there are 10–15 small holes (made with a push pin) around the label.*

   c. *Mealworms eat grains such as wheat and oat bran. Show students the wheat bran and tell them they can put one 5 mL spoon of bran in the container.*

   d. *Mealworms get water from the moisture in vegetation. Show students the tiny pieces of potato, carrot, or apple. Make sure the piece is small, no bigger than a pea. If there is too much moisture, mold might grow on the bran and harm the mealworms.*

9. **Set up the containers**
   Have Getters get the materials, and have students label and feed the mealworms in the group mealworm containers.

10. **Introduce *environment* and *environmental factors***
    Ask students to describe the parts of the system they created for the mealworms. Make a list on the board. Tell them,

    *All these things taken together describe the mealworm **environment**. Environment is everything both **living** and **nonliving** that surrounds and influences an organism, including other organisms.*

    *Each item on the list, each separate part of an environment, is called an **environmental factor**.*

11. **Introduce air temperature as a factor**
    Tell students,

    We can be more specific in our descriptions of each environmental factor. For example, air is an environmental factor. One way we can describe the air is by measuring its temperature. The air temperature might be 20 degrees Celsius (20°C), or possibly 5°C.

    ➤ What is the temperature of the air in the mealworm environment now? [Room temperature, about 22°C.]

    ➤ What would happen to the mealworms if that environmental factor changed? What if it were colder (5°C)?

102   Full Option Science System

Part 1: Observing Mealworms

12. Design an experiment
Ask,

> How could we find out if the environmental factor of air temperature makes a difference in the structures and behaviors of mealworms?

Ask students to discuss this question in their groups and to write down their ideas in their notebooks. Their ideas should involve setting up additional environments like the ones they have prepared but putting them where the temperature is either warmer or colder than room temperature.

Introduce the idea of testing only one factor at a time. In an experiment, all factors must be the same except for the one factor under investigation—in this case, air temperature.

The air temperature in a refrigerator is about 5°C and so can serve as a colder environment. Set up two additional environments with the same number of mealworms, label them, and put them in a refrigerator (not the freezer). Clarify that air temperature is the cause. The effect will be any change in the mealworms in the cold as compared to those at room temperature. Have students restate in their groups what the cause and the effect are in the experiment.

13. Introduce *Life of a Mealworm* calendar
Tell students that you have a type of calendar they can put in their notebooks to keep track of the mealworm observations they make over the next 6–8 weeks. Distribute two copies of notebook sheet 2, *Life of a Mealworm*, to each student. They should glue one sheet in landscape orientation on a left-hand page and another on the corresponding right-hand page, so that they can have an 8-week calendar. Have students write the dates for each week in the first column (for example, Week 1: March 4–10; Week 2: March 11–17, and so on for 8 weeks). The current date should be written in the small box for the first observation.

Ask students what they should record in today's date on the calendar. One way to record is to write notes about the setup.

- Set up mealworm environment with 10 mealworms and bran at room temperature.

Another way to record is to write a very brief note and the page number where there are more notes about that day. This is what the entry might look like using this method.

- Set up mealworm environment (see page 5)

Encourage students to think about different ways to record the information. Eventually, they will use the information to answer the focus question (at the end of 6–8 weeks).

> **TEACHING NOTE**
>
> If this is students' first experience with controlled experiments, plan to spend some additional time to develop this idea.

SCIENCE AND ENGINEERING PRACTICES
Planning and carrying out investigations

CROSSCUTTING CONCEPTS
Cause and effect

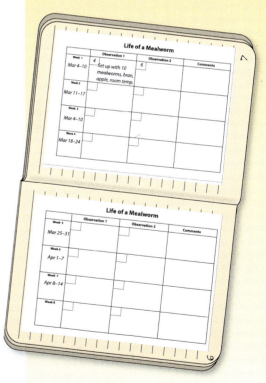

Environments Module—FOSS Next Generation

103

INVESTIGATION 1 – *Environmental Factors*

behavior
environment
environmental factor
function
living
mealworm
nonliving
organism
structure

EL NOTE

Have students record and illustrate new words in their notebooks, using either a glossary or index.

14. Review vocabulary
Review the vocabulary words that have been introduced in the investigation so far. Make sure that these words are on the word wall. Add any other words students may need.

15. Store the mealworm environments
Find a safe place to store the mealworm room-temperature containers, and have Getters put them there. Students might suggest that they should be in a dark location to control the environmental factor of light. The containers in the refrigerator will be in the dark.

16. Assess progress: notebook entry
Have students turn in their notebooks after class, open to the page you will be reviewing. Check students' answers to the questions on notebook sheet 1, *Mealworm Observations*.

What to Look For

- *Structures of a mealworm are drawn and labeled.*
- *Functions of the mealworm structures are described.*
- *Observed mealworm behaviors and how those help the organism survive are discussed.*
- *Students have asked one question of their own about mealworms.*
- *Students describe what they need to know about the mealworms to keep them alive and healthy.*

104   Full Option Science System

Part 1: Observing Mealworms

READING *in Science Resource*

17. Read "Two Terrestrial Environments"

Give students a few minutes to look at and discuss the cover of the *FOSS Science Resources* book. Then have them examine and discuss the table of contents. They should also locate the glossary and the index.

Next, let students preview the text by looking at and discussing the photographs and maps with a reading partner. Ask students what they think they will learn from reading this article. Refer to the questions at the end of the article and discuss how the information in this text is organized [comparison of two environments].

18. Make a concept grid

Suggest that students make a concept grid in their notebooks to help them compare the environmental factors of the two terrestrial environments. Help students get started by making a class concept grid. List only the first two rows now and save the others for extensions. NOTE: See the Interdisciplinary Extensions section.

Environments	Temperature	Rainfall	Soil type	Plants	Animals
Tropical rain forest					
Desert					
Temperate deciduous forest					
Grassland					
Taiga					
Tundra					

First, read the first page aloud or have students read independently. Ask students to take turns summarizing what they have learned from the text just read with a partner. List the six terrestrial environments in the first column of the concept grid. Then ask students what the nonliving environmental factors are and list them in the first row. Next, add the living factors, the plants and animals.

Refer to the photographs of the tropical rain forest and the desert environment and ask students to predict how the environmental factors in these two environments will differ.

Have students read the rest of the article independently and add information to their concept grid. Review with students the strategies they can use to determine or clarify the meaning of unknown words—context clues and word parts, and that the words in bold can be found in the glossary. They should write any words or phrases they are unsure about in their notebooks.

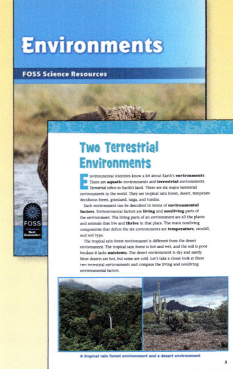

ELA CONNECTION

These suggested reading strategies address the Common Core State Standards for ELA.

RI 2: Determine the main idea of a text and explain how it is supported by key details; summarize the text.

RI 4: Determine the meaning of general academic domain-specific words or phrases.

RI 5: Describe the overall structure of information in a text.

W 8: Gather relevant information from experiences and print, and categorize the information.

L 4: Determine or clarify the meaning of unknown and multiple-meaning words and phrases.

Environments Module—FOSS Next Generation

INVESTIGATION 1 – Environmental Factors

SCIENCE AND ENGINEERING PRACTICES

Obtaining, evaluating, and communicating information

TEACHING NOTE

Go to FOSSweb for *Teacher Resources* and look for the *Science and Engineering Practices—Grade 4* chapter for details on how to engage students with the practice of obtaining, evaluating, and communicating information.

ELA CONNECTION

These suggested reading strategies address the Common Core State Standards for ELA.

RI 1: Refer to details and examples in a text when explaining what the text says explicitly and when drawing inferences from text.

RI 7: Interpret information presented visually, and explain how the information contributes to an understanding of the text.

W 2: Write informative/explanatory text.

TEACHING NOTE

After making the first observation of the mealworms, move ahead and begin Part 2 to design an isopod environment. Continue to make regular observations of the mealworms (Part 1) in the room temperature and cold environments during the rest of the module.

19. Discuss the reading

Fill in the class concept grid together based on the information students found in the article. Review any words or concepts that students have questions about, modeling how to use context clues. This is also an opportunity to point out scientific affixes and root words, such as: *aqu-* water; *terr-* earth or land; *noct-* night; *dorm-*sleep.

Refer students to the maps on pages 10–11 and have them take turns explaining and comparing what they represent with a partner. Review the questions on page 10 and discuss the rainfall measurement in different parts of the United States. Ask students to explain how the information in the map helps them understand the differences between a tropical rain forest and a desert environment.

Refer students to the questions at the end of the article and have them discuss their answers in small groups. Call on a reporter from each group to give an example from the text that supports their answer.

Have students write a summary of the article in their notebooks using the questions as a guide and the information they collected on the concept grid.

➤ *What are the environmental factors that define a tropical rain forest environment?* [It is very wet all year with annual rainfall from 200 to 450 cm. They are hot all year with very little change in temperature from winter to summer. The soil is not very fertile. The rain forest has layers created by plants, with very different amounts of light reaching each layer. Different organisms live in each layer.]

➤ *What are the environmental factors that define a desert environment?* [Annual rainfall is less than 25 cm. Soils are rocky or sandy. Desert summers are very hot. Winters can be cold; organisms in deserts must deal with extremes of temperature and little water.]

➤ *What are some of the structures and behaviors that help organisms survive in the desert.* [See pages 7–9 for specifics such as cacti roots and spines; tortoises store water in bladder, animals are nocturnal, dig burrows, or are dormant when hot and dry.]

20. Make regular observations

Students should make and record mealworm observations on their calendars at least two times each week. When they make observations, they can provide a small fresh piece of apple, carrot, or potato for moisture.

 POSSIBLE BREAKPOINT

Part 1: Observing Mealworms

21. View the video *Deserts*

Prepare the classroom for observing the video, *Deserts* (duration 25 minutes). Ask students to listen for information that was in the reading and for information that is new. Ask them to observe the organisms and how they are interacting with the environment.

Chapter 1: Introduction to Biomes of the World (55 sec)
Biomes are large regions of the world that have similar characteristics, contain distinct communities of plants and animals, and are usually named for the dominant plant life in the area.

Chapter 2: Overview of the Desert Biome (1 min 30 sec)
A desert is a region with low rainfall; most deserts have high temperatures and strong winds; one-fifth of the Earth's land surface is covered with deserts.

Chapter 3: Biomes of the World: Climate Zones
There are three major climate zones: tropical, temperate, and polar; the tropical zone contains tropical rainforests with more species than any other biome on Earth; temperate zones are the largest climate zones and have many deserts; polar zones are home to few plants and animals because of the cold temperatures and lack of food.

Most deserts are formed near a mountain range; the windward side of a mountain is hit by moist air, which drops its moisture before descending down the other side; the leeward side of a mountain is where evaporation exceeds precipitation, near which many deserts are formed.

Chapter 4: Deserts of the World (2 min 24 sec)
Deserts of Africa: the Sahara Desert and Kalahari Desert; the Sahara Desert covers about one-third of Africa.

The Arabian Desert covers most of the Arabian Peninsula and has the largest area of sand dunes in the world.

Deserts of North America: the Mojave Desert, Great Basin, Chihuahuan Desert, and Sonoran Desert; these deserts are home to cacti and sagebrush; Death Valley, in the Mojave Desert, is the hottest place on the North American continent.

The Atacama is a desert along the west coast of South America; it is cooler than most deserts and one of the driest places on Earth.

The Gobi Desert of Northern China and Southern Mongolia is a cool desert because of its high altitude.

Environments Module—FOSS Next Generation 107

INVESTIGATION 1 – Environmental Factors

Chapter 5: Characteristics of the Desert Biome (3 min 18 sec)

Ecosystems are characterized by abiotic and biotic factors—abiotic factors are nonliving components, such as temperature, light, water, and soil; biotic factors are all living things in an ecosystem.

Precipitation: Deserts are the driest biome on Earth; they get less than 25 centimeters (cm) of precipitation each year.

Wind: Deserts are windy because there is little grass and few trees to block the wind; the wind is the desert's most powerful erosional force.

Soil: Desert soil is poor in nutrients because it is so dry.

Sunlight: One reason deserts are so hot is that there are no clouds or fog to block the sunlight; deserts can get very cold at night because there is no insulating layer of moisture to keep the day's heat.

Chapter 6: Animal Adaptations in the Desert Biome (1 min 58 sec)

Animals must adapt to survive in arid desert environments; adaptations are physical features or behaviors that enable a plant or animal to survive in its habitat.

Animal adaptation in deserts

- Many desert animals are nocturnal; they sleep during the day and are active at nighttime.
- Reptiles are ectothermic (cold-blooded); they rely on the temperatures around them to keep their bodies warm or cool; reptiles also have dry, scaly skin that keeps in water.
- Many animals are venomous; they have glands that secrete poisons into prey, usually through a bite or a sting.

Chapter 7: Interview: A Herpetologist Speaks About a Rattlesnake that Lives in the Desert—the Sidewinder (1 min 31 sec)

Craig Ivanyi, herpetologist (an expert on snakes and other reptiles), speaks about a venomous rattlesnake called the sidewinder and its adaptations for the desert.

Chapter 8: Animal Adaptations in the Desert Biome: Mammals (1 min 42 sec)

Mammals that live in the desert include the kit fox, oryx, and camel; few large mammals live in the desert because food and water are hard to find and there are few places big enough for them to hide them from the heat of the Sun.

> **TEACHING NOTE**
>
> If students have completed the **FOSS Soils, Rocks, and Landforms Module**, they should be familiar with the components of desert soils.

Part 1: Observing Mealworms

Chapter 9: Investigation: Plant and Animal Adaptations in the Desert (4 min 22 sec)
Investigation: Students model an experiment that demonstrates the adaptations that plants and animals use to conserve water in the desert.

Chapter 10: Plant Adaptations in the Desert Biome (1 min 33 sec)
Plant adaptations: Desert plants get water by having shallow roots that spread out to absorb rain, having roots that grow deep into underground water sources, or having seeds that lay dormant, waiting for the perfect moisture conditions to begin growth.

Plant adaptations: Many desert plants, such as the Saguaro cactus, store water in their thick, fleshy stems and leaves; many plants also have spines that prevent animals from eating their juicy pulp; these spines also reflect sunlight away from the plants' surfaces.

Chapter 11: The Human Impact on Deserts (1 min 59 sec)
When humans try to build towns and farms near deserts, they must irrigate the land, so there is enough water to grow crops or lawns; this can cause rapidly lowering river and underground water levels.

Desertification is the process by which other ecosystems become deserts due to unwise farming techniques, drought, wind, or overgrazing by animals; for example, the Sonoran Desert was a swamp 97 million years ago.

Chapter 12: Conclusion to Biomes of the World: Deserts (1 min 2 sec)

> **TEACHING NOTE**
> This virtual experience in planning and carrying out an investigation dealing with moisture will help students to design their isopod experiments in Part 2 of this investigation.

22. **Discuss the video**
Have students first talk in their groups about what information was in the reading and for information that is new. Then have a class discussion about some of the structures and behaviors of the desert animals and plants that students observed and how they helped the organisms survive.

BREAKPOINT

Environments Module—FOSS Next Generation

109

INVESTIGATION 1 – *Environmental Factors*

Materials for Step 23
- *Mealworm Larva* poster

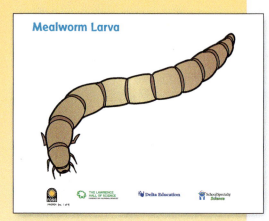

Mealworm Larva poster

SCIENCE AND ENGINEERING PRACTICES

Constructing explanations

23. **Discuss mealworm molting and introduce** *larva*

    One of the first changes students are likely to see is molting, the shedding of the tough, skinlike cuticle so the mealworm can grow. The skin splits down the back, starting at the head, and the mealworm slides out. Students may announce that one of their mealworms turned white, a sure sign that it has molted recently and that the new cuticle has not yet hardened and turned golden. Ask,

    ▶ *Why do you think the mealworm turned white?*

    ▶ *What color do you think it will be tomorrow?*

    ▶ *Why do you think the mealworms shed their skin?*

    Display the *Mealworm Larva* poster, and tell students,

    Mealworms will go through different **stages** *during their* **life cycle**. *The name for this life stage is the* **larva**. *When the mealworm larva needs to grow, it first sheds it outer skin. This is called* **molting**. *Mealworms must molt in order to grow. The old skin, also called a molt, is left behind.*

    Add these new words to the word wall (or the mealworm chart).

    Ask,

    ▶ *Have any of the mealworms in the colder environment molted?*

24. **Answer the focus question**

    Now that students have had a chance to observe the mealworm over time, ask them to answer the focus question.

    ▶ *How do mealworm structures and behaviors help them grow and survive?*

25. **Assess progress: notebook entry**

    Collect the science notebooks after class, and review students' drawings, labels, and responses to the question.

    What to Look For

    - *Students describe mealworm structures and how they help mealworms survive.*

    - *Students describe mealworm behaviors and how they help mealworms survive.*

B R E A K P O I N T

Part 1: Observing Mealworms

26. Observe the first pupae

After 1–3 weeks, the mealworms will undergo a major life transformation—they will turn into pupae. The time to pupation will depend on how mature the mealworm larvae were when they arrived in your classroom. The pupa is the inactive, nonfeeding stage that beetles go through during complete metamorphosis. During this pupal stage the feeding, growing larva transforms into the fertile, reproducing adult.

Students may report that their mealworms got shorter, turned white, or died. When this happens, introduce the poster and the vocabulary associated with the second stage in the life of these insects.

Ask,

➤ *How has your mealworm changed?*

➤ *Why do you think that happened?*

➤ *Why might some people think the mealworm has died?*

Display the *Pupa* poster, and tell students,

This stage in the life cycle of mealworms is called the **pupa**. *We say that the larva has* **pupated**. *Until this time, it has been a larva, but now it is a pupa. It is alive, but it does not eat or drink. Some insects, like mealworms, look very different at different times in their life, but they are still the same insect. We will watch it closely to see what will happen next.*

Put up the *Mealworm Pupa* poster beside the *Mealworm Larva* poster. Write the vocabulary (*pupa* and *pupate*) on the word wall, and incorporate these words into discussions.

Ask,

➤ *Have any of the larvae in the colder environment pupated?*

➤ *Why do you think this is so?*

➤ *How can we display the differences we are observing to make it easy to compare the changes in the two environments?*

Materials for Step 26
- *Mealworm Pupa* poster

▶ **NOTE**
If low-power microscopes are available, this is a good time for students to use them to collect additional information about the organism.

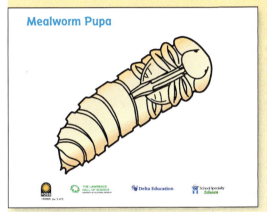

Mealworm Pupa poster

SCIENCE AND ENGINEERING PRACTICES

Analyzing and interpreting data

Constructing explanations

E L N O T E

Provide students with a T-table or other type of graphic organizer to compare and contrast the mealworms in the different temperature environments.

B R E A K P O I N T

Environments Module—FOSS Next Generation

INVESTIGATION 1 – Environmental Factors

TEACHING NOTE

When a number of adult beetles have emerged from the room temperature environments, suggest to students that the mealworms in the cold environment be returned to room temperature to see if the organisms will resume their life cycle.

27. Have a sense-making discussion

The adult beetles will emerge 2–3 weeks after pupation. Some students might be surprised that the pupa gave rise to the beetle. The beetle will be white or tan when it first emerges and then darken over time to a dark brown or black. Explain,

*The beetle is the **adult** stage of the mealworm. It is known as a **darkling beetle**.*

➤ *Have any of the mealworms in the colder environment changed into adults?*

➤ *What do you think will happen if we bring the containers from the cold environment back into the room-temperature environment?*

Tell students they will continue to observe the adult beetles for the next few weeks. Ask,

➤ *What changes do you think will happen?*

28. Observe and draw beetle structures

When enough adult beetles have emerged, have a beetle observation session. Let students put the beetles on paper or in cups and study them closely. Encourage them to determine what structures a beetle has. Students might notice that the beetles don't look exactly alike (different sizes, deformities, color variation). Use this opportunity to discuss variations between individuals.

After students have had time to observe adult beetles, ask them to draw the beetle. Students should look closely at the shape and size of the antennae on the beetle's head.

After students have completed their drawings and added labels, display the *Mealworm Adult (Darkling Beetle)* and the *Mealworm Stages (Darkling Beetle Stages)* poster.

Materials for Step 28
- *Posters, Mealworm Adult (Darkling Beetle) and Mealworm Stages (Darkling Beetle Stages)*

Mealworm Adult (Darkling Beetle) poster

Mealworm Stages (Darkling Beetle Stages) poster

Full Option Science System

Part 1: Observing Mealworms

29. Focus on crosscutting concept

Ask students to think about the functions of each structure they identified on the adult beetle and how the shapes and parts of the substructures relate to that function.

Focus attention on the adult beetle **antennae** as sensory receptors and have students find out what kind of information the antennae collect. There is some information in the reading "Darkling Beetles" (the beetle antennae are primarily used to smell but they may also be used to feel the surface of the physical environment). Students may want to find out more about these structures through research. Students will encounter these sensory structures again on the next organism they investigate, the isopod.

30. Review vocabulary

Review the key vocabulary introduced in this part, and refer to the words on the word wall.

This is a good opportunity to have students make a concept map. Make sets of these vocabulary words on self-stick notes for each group of four students. Each group takes the set of words and decides how to organize them on a big piece of paper. Once the group decides how the words are related, they draw lines connecting the words. On the lines they write words or phrases that describe or explain how the words/concepts are connected. When students have finished their concept maps, post them on the wall so groups can compare their thinking.

31. Establish a multigenerational environment

Toward the end of the module, work with students to design one class environment for all the adult beetles, pupae, and larvae. You can describe the life cycle as an example of complete metamorphosis (egg, larva, pupa, adult, egg).

CROSSCUTTING CONCEPTS

Structure and function

SCIENCE AND ENGINEERING PRACTICES

Obtaining, evaluating, and communicating information

adult
antennae
darkling beetle
larva
life cycle
molting
pupa
pupate
stage

ELA CONNECTION

This suggested strategy addresses the Common Core State Standards for ELA.

L 5: Demonstrate understanding of word relationships.

→ B R E A K P O I N T

Environments Module—FOSS Next Generation   113

INVESTIGATION 1 – *Environmental Factors*

SCIENCE AND ENGINEERING PRACTICES

Constructing explanations

32. Observe the darkling beetle multigenerational environment

Ask students to open their notebooks to the *Life of a Mealworm* calendar. Ask students to go back to the first entry and review their records for the development of the beetle from larva, to pupa, to adult.

Distribute the darkling beetle environment to each group and have them scoop out about a quarter cup (or less) of the bran along with the beetles in their various states, and put the bran and beetles on a piece of white paper. They should observe using hand lenses and see if they can find any young mealworms. Have students discuss in their groups what they observe.

33. Clean up

Have students put the beetles and bran back into the environments and have Getters return the materials to the materials station.

NOTE: The following reading, "Darkling Beetles," is the last reading in the *FOSS Science Resources* book. Have students read it toward the end of the module.

Part 1: Observing Mealworms

READING *in Science Resources*

34. Read "Darkling Beetles"

Read the article "Darkling Beetles" to conclude Part 1 of this investigation toward the end of the module, using the strategy that is most effective for your class. Start by having students preview the text by looking at and discussing the photographs and the bar graph.

Review what students have learned from observing beetles go through their life cycle over 6–8 weeks. Discuss what students know so far about the difference between inherited characteristics and those caused by the environment. Explain that this article will provide more information about darkling beetles and their characteristics.

35. Use a 3-2-1 strategy

To support reading comprehension, have students do a 3-2-1 strategy. In their notebooks, they should write down three important details, two interesting facts, and one question they have about the reading.

3	things that are important
2	things I learned
1	question I have

For reading strategies to support English learners and below-grade-level readers, see the Science-Centered Language Development chapter.

36. Discuss the reading

Ask students to study the bar graph at the end of the article.

▶ *What does this bar graph tell you about the numbers of different kinds of organisms on Earth?* [There are far more insects than all of the other organisms.]

▶ **NOTE**
Students will wait to read the article "Darkling Beetles" until the beetles have gone through the life cycle.

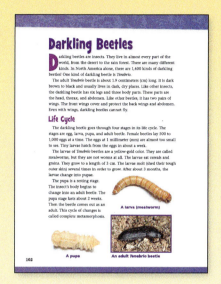

SCIENCE AND ENGINEERING PRACTICES

Obtaining, evaluating, and communicating information

ELA CONNECTION

These suggested reading strategies address the Common Core State Standards for ELA.

RI 2: Determine the main idea of a text and explain how it is supported by key details; summarize the text.

RI 7: Interpret information presented visually, and explain how the information contributes to an understanding of the text.

SL 1: Engage in collaborative discussions.

Environments Module—FOSS Next Generation

INVESTIGATION 1 — *Environmental Factors*

➤ *Why do you think the author used a bar graph to show the number of different kinds of organisms?* [Bar graphs make the comparison of numbers easier to see and understand.]

Give students a few minutes to share their 3-2-1 notes in their small groups. If students need practice with determining the main idea and key details, use the section, "Other Beetles" on page 104 to model locating the topic sentence and key details for each paragraph.

Ask for a few volunteers to share their questions and record them on the board or chart paper. Choose an interesting question to have students discuss as a class.

WRAP-UP/WARM-UP

37. Share notebook entries

Conclude this study of beetles by having students share notebook entries. Ask students to open their science notebooks to the original focus question dealing with mealworms and their environment. Read the focus question together.

➤ *How do mealworm structures and behaviors help them grow and survive?*

Ask students to pair up with a partner to

- discuss their current thoughts about the answer to the focus question;
- discuss how the beetles and their environment changed over time.

Have students give each other constructive feedback on how well information is organized and presented in their entries. For example, Student A can tell Student B one thing that is good about the entry and one thing that could make it better, such as more details in the drawings or being more precise in their use of words. Students then switch roles. Allow time for students to revise their entries based on feedback from their peers.

ELA CONNECTION

This suggested strategy addresses the Common Core State Standards for ELA.

W 5: Strengthen writing by revising.

Part 2: Designing an Isopod Environment

MATERIALS for
Part 2: Designing an Isopod Environment

For each student
- ☐ 2 Notebook sheet 3, *Isopod Investigation*
- ☐ 1 Notebook sheet 4, *Isopod Environment Map*
-   1 *FOSS Science Resources: Environments*
    - "Setting Up a Terrarium"
    - "Isopods"

For each group
- 10 Isopods ★
- 1 Basin, clear plastic, with lid, 6 L
- 1 Plastic spoon
- 1 Spoon, 5 mL
- 4 Plastic cups with lids
- 1 Craft stick
- 4 Hand lenses
- 1 Sheet of recycled paper ★
- 1 Beaker, 100 mL
- 1 Quarter piece of black paper (See Step 6 of Getting Ready.)

For the class
- Dry potting soil (See Step 5 of Getting Ready.) ★
- 1 Zip bag, 4 L
- Moist soil, 2 L (See Step 4 of Getting Ready.) ★
- 2 Basins, 6 L
- 1 Pitcher
- 64 Radish seeds
- Clover seeds
- 8 Minispoons
- Water ★
- Paper towels ★
- Leaves ★
- Low-power microscopes (optional) ★

For embedded assessment
- ☐ Notebook sheet 5, *Response Sheet—Investigation 1*
- ☐ Performance Assessment Checklist
- ☐ Embedded Assessment Notes

★ Supplied by the teacher. ☐ Use the duplication master to make copies.

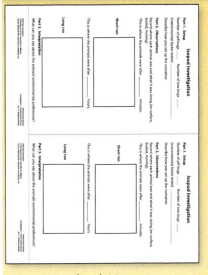

No. 3—Notebook Master

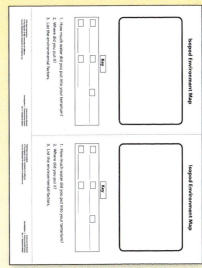

No. 4—Notebook Master

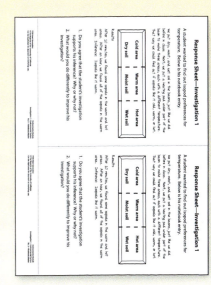

No. 5—Notebook Master

INVESTIGATION 1 – Environmental Factors

> **NOTE**
> Start Part 2 after Step 20 of Investigation 1, Part 1.

> **NOTE**
> In most regions of the country, isopods are common and easily collected by students outdoors. Isopods are also easy to maintain in classroom habitats. A layer of moist soil covered with some dead leaves, twigs, and bark is great, but isopods will do fine with moist paper towels or newspaper.

GETTING READY for
Part 2: Designing an Isopod Environment

1. **Schedule the investigation**
   This part will take four sessions of active investigation. The two readings will be done during the active sessions while students are investigating isopod preferences.

2. **Preview Part 2**
   The class conducts two different investigations to find out how isopods respond to the environmental factors of water and light. Based on their findings, students design an isopod environment in a terrarium. The focus questions are **What moisture conditions do isopods prefer?** and **What light conditions do isopods prefer?**

3. **Acquire isopods**
   Each group of students will need at least 10 isopods, so get about 150. If possible, get both pill bugs (they roll up) and sow bugs (they are flatter and cannot roll up). Store the isopods in a container with moist paper towels and dead leaves. Poke air holes in the lid.

4. **Prepare potting soil for moisture factor**
   This investigation uses about 3 L of potting soil. Each group needs 200 mL of moist potting soil (straight from the bag) and 100 mL of dry soil (see Step 5). For easy distribution, put the soil into basins just before the investigation.

5. **Prepare dry soil**
   Spread 1 L of the potting soil on newspaper. Let it sit exposed to air for 2–3 days. When it is all dry, store it in a plastic bag.

6. **Prepare quarter pieces of black paper**
   Cut each sheet of black construction paper into four pieces.

118   Full Option Science System

Part 2: Designing an Isopod Environment

7. **Put isopods in cups**
   For each group, put 10 isopods in a plastic cup with a small piece of moist paper towel. Include both pill bugs and sow bugs.

8. **Plan to read *Science Resources*: "Setting Up a Terrarium" and "Isopods"**
   These two short articles should be read during the active investigation while students are waiting to gather data from their isopod investigations.

9. **Plan assessment: performance assessment**
   Starting in Step 20 of Guiding the Investigation, students set up a controlled experiment to investigate isopod preferences for light. This will provide a good opportunity for you to observe students' science and engineering practices. Carry the *Performance Assessment Checklist* with you as you visit the groups while they conduct their investigations. For more information about what to look for during observations, see the Assessment chapter.

10. **Plan assessment: response sheet**
    Use notebook sheet 5, *Response Sheet—Investigation 1*, for a closer look at students' understanding of fair tests when investigating animal preferences. (See Step 36 in Guiding the Investigation.) Have students reflect on their responses after you have reviewed them. For more information about next-step strategies, see the Assessment chapter.

Environments Module—FOSS Next Generation

INVESTIGATION 1 – Environmental Factors

FOCUS QUESTION

What moisture conditions do isopods prefer?
What light conditions do isopods prefer? (or student-developed question)

Materials for Step 1–2
- Isopods in plastic cups
- Plastic spoons
- Recycled paper
- Hand lenses
- Microscopes (optional)
- Extra plastic cups

EL NOTE

Start a new chart to list observations of structures.

TEACHING NOTE

If you have both species, introduce the common names for the types of isopods. **Pill bugs** *have a highly domed shape, short legs, and inconspicuous antennae. Pill bugs move slowly and have a difficult time righting themselves if they roll onto their backs on a smooth surface.* **Sow bugs** *are relatively flat with legs that extend a little bit beyond the edge of the shell, and they have powerful antennae to sense their environment. Sow bugs move rather quickly and will use their long antennae and little spikelike tail projections to right themselves if they happen to roll onto their backs.*

SCIENCE AND ENGINEERING PRACTICES

Asking questions

GUIDING *the Investigation*
Part 2: Designing an Isopod Environment

1. **Introduce isopods**
   Tell students you have another organism for them to study—**isopods**. Bring out the isopods in the cups for each group. Reinforce that they are living organisms and that nothing should be done to harm the animals. The isopods will not harm students in any way. Promote an attitude of respect for living organisms.

2. **Make isopod observations**
   Ask students to observe the isopods and to note structures and behaviors. They can compare the isopods to the mealworms. Encourage them to hold the animals carefully—they don't bite or sting. They can put the isopods on a quarter sheet of paper and observe how they move. Hand lenses will permit close observations. Have students make drawings of the isopods and label the structures they can identify.

   Ask Getters to get a cup of isopods, four hand lenses, three additional cups, a sheet of recycled paper, and a spoon. Allow 5–10 minutes for making observations.

   If low-power microscopes are available, this is a good time for students to use them to collect additional information about the organism.

3. **Discuss isopod observations**
   Ask students to put the isopods back in the cups. Ask,

   ➤ *What did you know about isopods before you observed them?*

   ➤ *What did you find out about isopods by observing them?*

   ➤ *What differences, if any, did you observe in the structures or behaviors of the isopods?* [They may vary in size or color. If students observed both pill bugs and sow bugs, they might notice the differences in the shape and the ability to roll into a ball.]

   ➤ *What questions do you have about isopod structures and their functions or about isopod behavior?*

   Explain,

   To determine if an isopod is a **sow bug** *or a* **pill bug**, *place it in your hand and shake it gently. If it rolls into a tight ball or is unable to flip over from an upside-down position, you have a pill bug. If the isopod has two tail-like appendages and quickly flips from an upside-down position onto its feet, it is a sow bug.*

Part 2: Designing an Isopod Environment

Tell students,

We can keep these isopods in our classroom if we can provide a suitable environment. Let's find out what kind of environment the isopods prefer.

4. Discuss environmental factors
Tell students that each group will design an environment for their isopods in a 6 L basin. Hold up the basin for all to see. But first they need to think about what isopods might need from their environment. Ask,

➤ *What living and nonliving environmental factors should you consider as you design the environment for isopods?* [Food, soil, plants, light, air, air temperature, moisture and space.]

Suggest they first conduct an investigation to find out what amount of moisture the isopods prefer.

5. Focus question: What moisture conditions do isopods prefer?
Write or project the focus question as you say it aloud.

➤ *What moisture **conditions** do isopods prefer?*

Have students write the focus question in their notebooks. Discuss and clarify the term *conditions*.

6. Test the factor of moisture
Reinforce that all other factors must be the same in order to test the factor of water. Ask students how they might design such an investigation.

Build on their ideas as you describe the investigation.

a. Provide different conditions of moisture for the isopods, and see where they go.

b. One option would be to use three equal piles of soil separated from one another by a bit of open space. One measure of the soil should be dry, another moist, and a third wet. Students may want to use paper towels with different amounts of moisture (dry, moist, wet) instead of soil and that would be another way to provide different conditions.

c. The isopods need plenty of time to explore the environment in order to show preference.

7. Identify the sources of soil
Show students the dry and moist potting soil. Explain that the two kinds of soil are exactly the same except that one portion has been allowed to dry out.

TEACHING NOTE

Go to FOSSweb for Teacher Resources and look for the Science and Engineering Practices—Grade 4 chapter for details on how to engage students with the practice of planning and carrying out investigations.

SCIENCE AND ENGINEERING PRACTICES
Planning and carrying out investigations

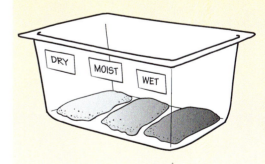

▶ **NOTE**
The soil should be in basins for easy distribution.

Environments Module—FOSS Next Generation

121

INVESTIGATION 1 – *Environmental Factors*

SCIENCE AND ENGINEERING PRACTICES

Planning and carrying out investigations

Materials for Step 10
- *Basins*
- *Beakers*
- *Spoons, 5 mL*
- *Craft sticks*
- *Soil*
- *Water*
- *Newspaper*

CROSSCUTTING CONCEPTS

Cause and effect

Materials for Step 12
- *Basin lids*

8. Outline a procedure
Write the investigation setup on the board to guide students as they set up their group basins.

- *Soil samples will all be 100 mL. Measure soil with the beaker.*
- *Wet soil is made by adding four 5 mL spoons of water to 100 mL of moist soil and mixing thoroughly.*
- *The samples of soil with different moisture content should not touch each other when they are placed in the basin.*
- *Using a self-stick note, label the type of soil in each pile and place it on the outside of the basin.*

9. Introduce the notebook sheet
Have Getters get four copies of notebook sheet 3, *Isopod Investigation*. Have students note the number of isopods they will use. Point out where to record the environmental factor investigated (in this case water) and where to describe the setup for their investigation. Tell students that after about 10 minutes, they will record what took place in the basin by drawing the location of the animals and indicating what each animal was doing (sitting on the surface, buried in the soil, moving on the soil, etc.).

10. Start the moisture investigations
Have the Getters get the materials their groups will need to conduct the investigations.

When the basins are ready, have everyone fill in part 1 of the *Isopod Investigation* sheet. At your signal, have the teams put the isopods in the basins and note the time. Each team can determine the best location(s) to place the isopods in the basin.

You could start reading "Setting Up a Terrarium" at this time.

11. Observe short-run results
After 10 to 15 minutes, call "Stop," and have the teams draw the locations of all the isopods in the basin on the *Isopod Investigation* sheet. Students might need to use the craft sticks to look for the isopods in the soil. These short-run results might not be conclusive.

12. Go for long-run results
Ask if students think that the isopods might need more time to explore the environment. Suggest leaving the basins for an hour or two or overnight, and then to observe where the isopods are to determine the effect of moisture. Provide a lid for each basin.

13. Clean up
Have Getters clean and dry the materials that are no longer needed. Store unused soil. Save dry soil for Investigation 4.

Part 2: Designing an Isopod Environment

READING *in Science Resources*

14. Read "Setting Up a Terrarium"

Read the article "Setting Up a Terrarium," using the strategy that is most effective for your class. Start by letting students preview the text by looking at the photographs and the diagram, and reading the captions. Have students share what they know about **terrariums** and then make predictions about what they will learn from reading the article. Confirm that the article explains how to set up a terrarium, something that they will also be doing in class.

15. Use a "questions-and-responses" strategy

Have students divide a page in their notebooks into two columns—one labeled "questions" and the other "responses." Give them a few minutes to brainstorm questions they have about how to set up a terrarium with their small group and record them in their notebook. Then let students read the article independently to find the answers. For reading strategies to support English learners and below-grade-level readers, see the Science-Centered Language Development chapter.

16. Discuss the reading

Call on a few students to share what questions they found answers to in the text. Review the meanings of the words in bold.

Point out the sentence "A terrarium can also be a habitat for small animals found in the garden." Ask students to discuss why a terrarium can be called a habitat.

Point out the word "crustacean" on page 14 and model with students how you might figure out the meaning of this word. For example, "I know it's an animal, and it's not a reptile, amphibian, insect, worm, or spider. What other type of animal is small enough to live in a terrarium? I think I'll look it up in the glossary."

Refer students to page 15 and ask how the text is structured differently from the rest of the article. Students should point out the text features of procedural writing, (e.g., list of materials and steps).

Ask students to review the question at the end of the article.

> ➤ *The word* terra *means "earth" or "land." The suffix* -arium *means "a place." What do you think* aquarium *means?* [A watery place or a small container with water for keeping aquatic plants and animals.]

Questions | Responses

ELA CONNECTION

These suggested reading strategies address the Common Core State Standards for ELA.

RI 3: Explain procedures, ideas, or concepts in a scientific text.

RI 4: Determine the meaning of general academic domain-specific words or phrases.

L 4: Determine or clarify the meaning of unknown and multiple-meaning words and phrases.

SCIENCE AND ENGINEERING PRACTICES

Obtaining, evaluating, and communicating information

POSSIBLE BREAKPOINT

Environments Module—FOSS Next Generation

INVESTIGATION 1 – Environmental Factors

SCIENCE AND ENGINEERING PRACTICES
Planning and carrying out investigations

SCIENCE AND ENGINEERING PRACTICES
Analyzing and interpreting data

EL NOTE
Provide a sentence frame for students who need scaffolding such as, I think _____. My evidence is _____.

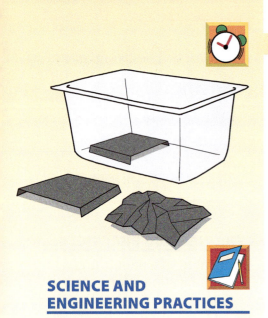

SCIENCE AND ENGINEERING PRACTICES
Asking questions

17. Record long-run results
After 1–2 hours (or overnight), have the groups record the locations of the isopods in the basin on the *Isopod Investigation* sheet (long run) and note in their notebooks any changes in the moisture in the soil.

18. Introduce *preferred environment*
Ask students to review the results of the isopod soil-moisture investigations. Then say,

*Animals move to the conditions of the environmental factor that are most suitable for their needs. We were studying the factor of water. We had dry conditions, moist conditions, and wet conditions. The conditions that an animal moves to and remains in are the animal's **preferred environment** for that factor.*

Add the term to the word wall.

Have students record their conclusions about the isopods' environmental preference for water conditions in their notebooks using the prompt in Part 3 of the notebook sheet.

19. Answer the focus question
Have students glue notebook sheet 3, *Isopod Investigation*, in their notebooks and answer the focus question.

▶ *What moisture conditions do isopods prefer?*

Students should indicate the preferred environment for the condition they tested and provide evidence to support their answers.

BREAKPOINT

20. Investigate a different environmental factor
Propose that students investigate the environmental factor of light. Ask students for input on how to design the experiment. They will most likely suggest covering one side of the basin. Explain that a quarter piece of black construction paper with the edges folded down will provide a canopy to place in the basin. Any other way they can make a dark area with the quarter piece of black paper would be fine.

21. Write a focus question
Have each group come up with the wording for their own focus question for this investigation and write it in their notebook. (For example, What light conditions do isopods prefer?) Once they have the focus question, they can get their equipment for the investigation.

Part 2: Designing an Isopod Environment

22. Assess progress: performance assessment
Circulate from group to group as students begin to work on the light factor experiment. Carry the *Performance Assessment Checklist* with you as you observe them conducting the investigation.

What to Look For
- *Students write a focus question to guide their investigation. (Asking questions.)*
- *Students control all variables except light in their investigation. (Planning and carrying out investigations.)*
- *Students analyze and interpret their observational data to determine light preference. (Analyzing and interpreting data; LS1.D: Information processing.)*
- *Students base their explanations and arguments for light preference on observational evidence. (Constructing explanations; engaging in argument from evidence; cause and effect.)*

23. Set up basins
Have students collect the isopods into a cup while setting up new conditions. Remind students that the basins need to be set up so that all the factors are the same except for the environmental factor of light. They should discuss whether they will use soil or not, and if so, what kind of soil (dry, moist, wet).

Have Getters get a small piece of black construction paper (quarter sheet) and new *Isopod Investigation* sheets from the materials station.

24. Start the light investigations
When the basins are ready, have students fill in Part 1 of the second *Isopod Investigation* sheet. At your signal, have the teams put the isopods in the basins. Note the time. You should start reading "Isopods" now.

25. Observe short-run results
After 10 to 15 minutes, call "Stop," and have the teams draw the locations of all the isopods in their basins. These short-run results might not be conclusive.

26. Go for long-run results
Ask if students think that the animals might need more time to explore the environment. Have students decide how long they would like to leave the isopods in the basins. They can let it go for an hour or so, but they shouldn't let it go overnight if there is no moisture for the isopods in the basin.

SCIENCE AND ENGINEERING PRACTICES

Asking questions

Planning and carrying out investigations

Analyzing and interpreting data

Constructing explanations

Engaging in argument from evidence

DISCIPLINARY CORE IDEAS

LS1.D: Information processing

CROSSCUTTING CONCEPTS

Cause and effect

Materials for Step 23
- *Quarter pieces of black paper*

SCIENCE AND ENGINEERING PRACTICES

Analyzing and interpreting data

Environments Module—FOSS Next Generation

INVESTIGATION 1 – Environmental Factors

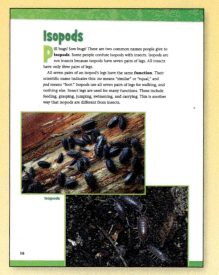

ELA CONNECTION

These suggested reading strategies address the Common Core State Standards for ELA.

RI 6: Compare and contrast a firsthand and secondhand account of the same topic.

W 8: Gather relevant information from experiences and print, and categorize the information.

SCIENCE AND ENGINEERING PRACTICES

Obtaining, evaluating, and communicating information

CROSSCUTTING CONCEPTS

Structure and function

READING in Science Resources

27. Read "Isopods"

The article gives background information on the organisms in the investigation. It provides detailed information that cannot be learned firsthand in the classroom.

Before reading, discuss what students know about isopods so far. Have them look back in their notebooks at their drawing of the isopods and compare them to the photographs in the article. Ask students if they can identify other structures or substructures they couldn't see before and to make inferences about their functions.

28. Use a summary chart to record notes

Ask if they know what type of animal isopods are. Tell them that many people think they are insects, but they are not. Invite students to read the text to find out more about these organisms. To help students keep track of the information, suggest using a summary chart to record their notes about isopods.

Have students read the article on their own or with a partner.

Isopods	
Structures	Food
Habitat	Interesting facts

29. Discuss the reading

Use these questions to check for understanding.

▶ *Isopods are crustaceans. What other organisms do you know that are crustaceans?* [Crayfish, crabs, lobsters, shrimp.]

▶ *What structures do isopods use to survive in their environment?* [Legs to move, gill-like breathing structures to get oxygen, shell for protection, two pairs of antennae to sense the environment.]

▶ *What kind of environment do isopods need to survive well? Why?* [An isopod requires a moist environment to keep its breathing structures moist, decaying vegetation for food.]

Ask students to compare their experiences with the isopods in class (and outside) to those described in the article. What have they observed firsthand? What was new information they gathered from the text? Ask about how the isopods use their antennae (structure and function relationship).

Part 2: Designing an Isopod Environment

30. Record long-run results

After an hour, have the teams record the locations of the isopods in the basin.

31. Have a sense-making discussion

Have students discuss the results of the light-preference investigation with their group. Ask,

➤ *What evidence have you gathered that indicates your animals' environmental preference for light?* [If most of the animals were under the black canopy after a long time passed, then they probably prefer a darker environment.]

➤ *Describe the preferred environment of isopods from the evidence we have so far.* [Isopods usually prefer dark, moist environments.]

➤ *How do our terrariums serve as a model of the natural world?* [The terrariums have different light and moisture conditions just like the environment, but they don't have all the living and nonliving environmental factors.]

SCIENCE AND ENGINEERING PRACTICES

Developing and using models

32. Discuss *observations* and *inferences*

This is a good time to have students reflect on the difference between **observations** (information you obtain through your senses) and **inferences** (what meaning you make of those observations). Tell students they will record both their observations and their inferences from those observations.

E L   N O T E

Write **observations** *and* **inferences** *on the board. Have students come up with examples of each related to the isopod and beetle investigations.*

33. Answer their own focus question

Have students answer the focus question they wrote for this investigation. It might be something like

➤ *What light conditions do isopods prefer?*

Students should indicate the preferred environment for the condition they tested and provide evidence to support their answers.

POSSIBLE BREAKPOINT

Environments Module—FOSS Next Generation

127

INVESTIGATION 1 – Environmental Factors

Materials for Step 34
- *Basins*
- *Soil*
- *Water*
- *Radish seeds*
- *Clover seeds*
- *Minispoons*
- *Plastic spoons*
- *Beakers, 100 mL*
- *Leaves*

CROSSCUTTING CONCEPTS

Systems and system models

condition
inference
isopod
observation
pill bug
preferred environment
sow bug
terrarium

34. Design an isopod environment

Tell students that they can use the data they have gathered from their investigations to design a suitable environment for the isopods to live in the classroom. Ask,

▶ *How will the environment (the system) that you design meet the basic needs of the organism?*

Have students design a terrarium environment for their isopods in the basin. Students might feel that one set of conditions is not good for both types of isopod, so different terrariums can be maintained for each kind of isopod. Let students set up terrariums as they see fit. They should monitor the terrariums with the isopods over time.

Make radish and clover seeds available for students to plant in their terrariums if they choose to do so. Provide eight radish seeds and a minispoon of clover seeds for each group. Distribute notebook sheet 4, *Isopod Environment Map*, for students to draw and describe how they designed the environment, and glue the sheet in their notebooks.

35. Review vocabulary

Review key vocabulary introduced in this part, and refer to the words on the word wall.

- *Condition refers to the existing state of something. We can refer to dry conditions or wet conditions when describing the environmental factor of water.*

- *A preferred environment is the set of environmental conditions that an organism appears to choose over other conditions.*

- *An observation is information obtained through your senses (sight, hearing, smell, touch, and taste).*

- *An inference is the meaning that you make from your observations. For example, you observe that all the isopods go to one corner of the basin. You infer that isopods prefer the conditions in that corner.*

Part 2: Designing an Isopod Environment

36. Assess progress: response sheet

Distribute a copy of notebook sheet 5, *Response Sheet—Investigation 1*, to each student. Students should glue the sheet on a left-hand page and write their responses on a blank right-hand page. After class, review students' responses to determine if they understand that only one factor at a time should be changed in a controlled experiment. Discuss the sheets with students after you review them. See the Assessment chapter for student self-assessment and next-step strategies.

TEACHING NOTE

This response sheet will let you know if students need more instruction on how to set up a controlled experiment. Provide class time to respond to student needs.

What to Look For

- *This student did not conduct a controlled experiment.*
- *The recorded observations support the conclusions, but you can't be sure because more than one factor was changed.*
- *The investigation would be improved if the soil moisture was kept the same in all three areas (moist based on earlier classroom investigations) and only the temperature varied.*

SCIENCE AND ENGINEERING PRACTICES

Engaging in argument from evidence

WRAP-UP/WARM-UP

37. Share notebook entries

Conclude Part 2 or start Part 3 by having students share notebook entries. Ask students to open their science notebooks to the last two focus questions.

➤ *What moisture conditions do isopods prefer?*

➤ *What light conditions do isopods prefer?*

Ask students to pair up with a partner from another group to

- discuss their current thoughts about the answer to the focus question;
- discuss the environmental factors they tested with isopods and what they observed.

Have students extend their understanding to other organisms they have experience with and discuss the environmental preferences of those organisms.

Environments Module—FOSS Next Generation

129

INVESTIGATION 1 – *Environmental Factors*

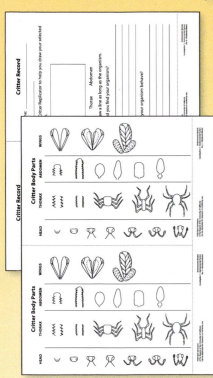

Nos. 6–7—Notebook Masters

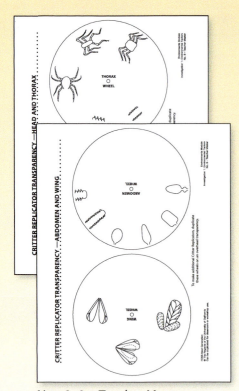

Nos. 8–9—Teacher Masters

MATERIALS for
Part 3: *Leaf-Litter Critters*

For each student
- 1 Plastic cup with lid
- 1 Hand lens
- ❑ 1 Notebook sheet 6, *Critter Record*
- ❑ 1 Notebook sheet 7, *Critter Body Parts* (optional)
- 1 *FOSS Science Resources: Environments*
  - "Amazon Rain Forest Journal"

For each group
- 1 Basin, not clear, 8 L
- 2 Spoons or index cards
- 1 Critter Replicator
- 1 Container, 1/2 L
- 1 Screen, large mesh
- 1 Sheet of recycled white paper ★
- 4 Zip bags, 1 L (optional)

For the class
- Moist leaf and ground litter (lots of it) ★
- 1 Camera (optional) ★
- 1 Chart paper with 4 × 6 grid (See Step 11 of Getting Ready.) ★
- *FOSS Outdoor Safety* poster
- Self-stick notes and scissors (for review) ★
- Clipboards ★
- Carrying bag or box ★
- ❑ 1 Teacher master 6, *Critter Replicator Instructions* (optional)
- ❑ 1 Teacher master 7, *Critter Replicator* (optional)
- ❑ 1 Teacher master 8, *Critter Replicator Transparency—Head and Thorax* (optional)
- ❑ 1 Teacher master 9, *Critter Replicator Transparency—Abdomen and Wing* (optional)

For benchmark assessment
- ❑ *Investigation 1 I-Check*
- ❑ *Assessment Record*

★ Supplied by the teacher.   ❑ Use the duplication master to make copies.

Full Option Science System

Part 3: Leaf-Litter Critters

GETTING READY for
Part 3: Leaf-Litter Critters

1. **Schedule the investigation**
   This part will take one outdoor session, two video/discussion sessions, one reading session, and two assessment sessions (one for review and one for the I-Check).

2. **Preview Part 3**
   Students go to the schoolyard to collect, observe, and sort small animals living in natural ground litter. They use a Critter Replicator to become familiar with the anatomical parts of animals they find in the leaf litter. Students consider adding found organisms to their group's isopod environment. The focus question is: **What are the characteristics of animals living in the leaf-litter environment?**

3. **Plan for seasonal weather**
   This lesson is best conducted in the fall or spring. In the winter, animals will hide deeper in the leaf litter and will move more slowly. Students will need to be more patient and look more carefully for the animals.

4. **Select your outdoor site**
   Select a site with a thick layer of natural ground litter (decaying leaves, sticks, and bark). The leaf litter is likely to be dry and barren on top, but students might discover a variety of small animals living in the moist underlayers. Make sure the leaf litter harbors a diversity of animals such as insects, slugs, spiders, and salamanders.

   If leaf litter has been cleared from your schoolyard, students may search under rocks, large pieces of wood, flowerpots, and so forth. Or place a white sheet (or chart paper) on the ground under a large bush and shake. Observe what falls out.

   If your schoolyard is devoid of natural ground litter, speak with your custodians and administration about leaving an out-of-the-way corner for some decomposition activity. Create a compost pile of leaves, twigs, bark, and other plant parts, and moisten the material. Small critters will soon call this environment home and join the decomposition fun.

   > **TEACHING NOTE**
   >
   > Consider having students design and build a compost pile before or after doing this investigation. See the Interdisciplinary Extensions section.

   Also important is to make sure the site will have some shady spots during the activity. Organisms isolated in closed cups can overheat quickly in the sunshine.

Environments Module—FOSS Next Generation

131

INVESTIGATION 1 – *Environmental Factors*

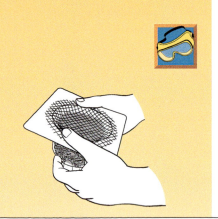

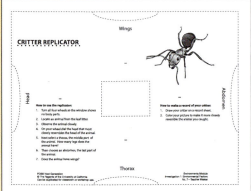

No. 7—Teacher Master

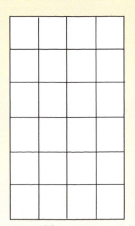

A 4 × 6 grid for sorting activities

5. Check the site
Tour the outdoor site on the morning of your outdoor activity. Do a quick search for potentially distracting or unsightly items.

6. Plan to review outdoor safety rules
Have the *FOSS Outdoor Safety* poster on hand for a review of rules and expectations for outdoor investigations.

7. Practice using a litter shaker
A litter shaker has two parts, a 1/2 L container and a large-mesh screen. To use the shaker, fill the 1/2 L container with leaf litter and place the screen on top of the container. Hold the screen securely over the top of the container, and then invert the system and shake it gently over a basin or sheet of white paper. Transfer the critters that fall through the screen to a cup. Check the screen to make sure critters are not stuck in the mesh and are not too big to pass through. See Step 5 of Guiding the Investigation.

8. Explore the Critter Replicator
Eight Critter Replicators are in the kit, assembled and ready to use. Practice using a Critter Replicator. Try to create an ant or an organism you know well. Students will use these after they have collected organisms to help them observe more carefully.

Students can make additional Critter Replicators by following the directions on teacher master 6 and using copies of teacher masters 7–9. See the Interdisciplinary Extensions for details.

9. Watch for overheating
When closed clear plastic cups are placed in the sun, the air inside heats up rapidly and could harm the organisms. Keep containers with organisms in the shade.

10. Consider identifying organisms
Although it is not necessary to identify the organisms, you might feel more comfortable knowing what students might find. Check the Resources by Investigation section on FOSSweb for links to appropriate guides to common invertebrates.

11. Prepare grid on chart paper
In Step 9 of Guiding the Investigation, students sort the animals they found in the ground litter. Draw a 4 × 6 grid on a piece of chart paper for this sorting activity. Each cell should be big enough for a plastic cup to fit in it.

12. Take a camera (optional)
If possible, have on hand a camera to capture students collecting organisms. Have students photograph their discoveries.

Part 3: Leaf-Litter Critters

13. Enlist additional adults
If possible, seek out an additional adult to join you outdoors. Remind the adult that students will need time to struggle with the challenges and that his or her job is to lend support but not to solve the problems for students.

14. Preview the videos
Preview the first video, *Animals of the Rain Forest* (duration 24 minutes). The video introduces the diversity of animals that live in the environment and the structures and behaviors that allow them to survive. Look for the detailed video summary in Step 12 of Guiding the Investigation.

Preview the second video, *Animal Needs* (duration 23 minutes). This video presents the basic needs of all animals with a focus on oxygen, food, and water. Throughout the module, students will compare how a diversity of animals use their structures to meet these needs. Look for the detailed video summary in Step 18 of Guiding the Investigation.

The links to both videos are in Resources by Investigation on FOSSweb.

15. Plan to read *Science Resources*: "Amazon Rain Forest Journal"

Plan to read the article "Amazon Rain Forest Journal" during reading periods before or after the active investigation.

16. Plan benchmark assessment: I-Check

When students have completed all the activities and readings for this investigation, they will be ready for a review and *Investigation 1 I-Check*. These activities begin with the review session suggested in Step 20 of Guiding the Investigation.

Print or copy the assessment masters or schedule students on FOSSmap. Have students take the I-Check a day or two after the review session. You can read the questions aloud, if students need this scaffolding; you want to measure their science understanding, not their reading ability. Students should work individually to mark or write their responses to each item.

The Assessment Coding Guides are on FOSSweb. When you review and code the completed I-Checks, note any concepts students might be struggling with. Plan self-assessment activities to prompt further reflection on those concepts. See the Assessment chapter for guidance and information about online tools.

Environments Module—FOSS Next Generation

INVESTIGATION 1 – Environmental Factors

FOCUS QUESTION

What are the characteristics of animals living in the leaf-litter environment?

Materials for Step 1
- *Critter Replicators*

Materials for Step 4
- *Basins*
- *Cups with lids*
- *Plastic spoons or index cards*
- *Hand lenses*
- *Containers, 1/2 L*
- *Screens*
- *Sheets of white paper*
- *Zip bags (optional)*
- *Chart paper*
- *Clipboards*
- *Carrying bag or box*

GUIDING *the Investigation*
Part 3: *Leaf-Litter Critters*

1. **Explain the outdoor exploration**

   Explain to students that they are going to search for small animals that live in the ground-litter or leaf-litter environment. Explain that ground litter is the natural accumulation of dead leaves and twigs on the ground near and under plants. Ask students to brainstorm suggestions for leaf-litter locations on the school grounds. Make a list on the board.

   Explain that students will have about 15 minutes to search the outdoor location. Each student will select one critter and record information about it on notebook sheet 6, *Critter Record*.

   Distribute a Critter Replicator to each group, and give them a minute to determine how to rotate the wheels to bring various heads, thoraxes, abdomens, and wings together in the view window. Have students try to create a replica of an ant. Explain that they will use the replicators during their search.

2. **Focus question: What are the characteristics of animals living in the leaf-litter environment?**

   Write or project the focus question on the board as you say it aloud.

   ▶ *What are the characteristics of animals living in the leaf-litter environment?*

   Have students write the focus question in their notebooks. Make sure students understand the word *characteristics*.

3. **Review outdoor safety**

   Remind students of the rules and expectations for the outdoor work and review the *FOSS Outdoor Safety* poster. Establish the boundaries for their work outdoors.

   Students may not know if critters bite or sting. Tell them,

   If you don't know it, don't touch it.

   Students might scoop the critters into their cups using a spoon or index cards and quickly cover the cup with a lid. Students should protect animals from overheating by putting closed containers in the shade.

4. **Go outdoors**

   Distribute notebook sheet 6, *Critter Record,* for students to place on clipboards. Have students place their clipboards and pencils in a box or large bag. The box of clipboards will go outdoors and the

134 Full Option Science System

Part 3: Leaf-Litter Critters

clipboards will be distributed after students have had a chance to search for critters. Have Getters from each group gather materials into the basins. Make sure students are dressed appropriately, and head outdoors to your home base.

5. Model collection techniques

There are two methods for collecting ground-litter animals. Gather students in a circle around an area with leaf litter, and demonstrate these methods.

- Show students how to use a litter shaker. They will fill a 1/2 L container with leaf litter and place the large-mesh screen on top of the container. Model how to hold the screen securely over the top of the container, and then invert the system and shake it gently over a basin or sheet of white paper. Transfer critters to a cup. Check the screen to make sure critters are not stuck in the mesh and are not too big to pass through. After observing the scurrying critters carefully, students may repeat this process.

- Show students how to scoop up a small amount of ground litter into the basin and search for critters by pushing the leaf litter around with a spoon. Transfer critters to a cup for close observation.

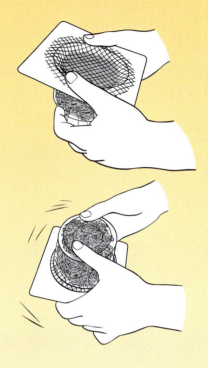

6. Search for critters

Encourage students to learn as much as they can about the animals. They should note structures, color, behavior, and other characteristics. Students can take turns using the Critter Replicators to create animals similar to those they find in the ground litter. Give students about 15 minutes to search.

You might want to encourage students to collect a few ground-litter items to provide a more natural setting for the litter critters if the animals will be added to the isopod environment (short sticks, leaves, small rocks, and bark are all great additions to the terrarium).

7. Collect and record

Call for attention, distribute the clipboards, and ask students to fill out a notebook sheet for one organism. Remind students to keep the cups in the shade to protect the critters.

8. Return to class

Have students gather all the materials they used and return to the classroom. Each student might want to bring one organism to the classroom for a short visit (make sure lids are on tight). Some students might decide to add the organism to their isopod environments.

SCIENCE AND ENGINEERING PRACTICES

Planning and carrying out investigations

CROSSCUTTING CONCEPTS

Structure and function

▶ **NOTE**
All other organisms and leaf litter collected should be returned to where they were found.

Environments Module—FOSS Next Generation

INVESTIGATION 1 – *Environmental Factors*

Materials for Step 9
- *Chart paper with grid*

Chart paper grid for sorting based on the location where each organism was found.

SCIENCE AND ENGINEERING PRACTICES

Asking questions

Analyzing and interpreting data

9. Have a sense-making discussion

Gather students in a half circle. In the center of the circle, place a piece of chart paper with a 4 × 6 grid drawn on it. Conduct a few sorting activities based on different characteristics of the organisms in the cups.

Write the sorting characteristic on a self-stick note and place it at the top of the chart paper grid. For example, location where the organism was found might be the sorting characteristic. Then ask students to offer ways to sort on that characteristic (location, in this example). Write the information the students provide on self-stick notes and place them as column heads on the grid. Then ask students (maybe from two or three groups) to place their cups containing organisms on the chart under the appropriate heading for location where the organism was found.

Here are some ideas for sorting characteristics.

Ask questions such as,

▶ *Could we sort our critters based on where we found them?* [For example, students found critters in the leaf litter by the compost pile, in the leaves under the maple tree, and so forth.]

▶ *Could we sort them by kind? (Are all the ants the same kind? Are there variations among them?)*

▶ *Could we sort them by how they behave?* [Slow movers, fast movers, climbers, and so forth.]

▶ *Could we sort them by color? What are the most common colors? Why do you think that is the most common color for leaf-litter critters?*

▶ *Could we sort them into broader groups such as insects, isopods, spiders, and so forth?*

Throughout the sorting activity, ask questions such as,

▶ *What does this sorting activity tell us about the litter critters?*

▶ *Do litter critters share certain characteristics?* [Smell, dark color.]

▶ *What might the litter critters eat?* [Dead plants or small animals in the litter.]

▶ *What do litter critters need from their environment to stay alive?* [Moisture, air, food, protection.]

▶ *What larger animals might eat the litter critters?* [Lizards, toads, salamanders, spiders, birds.]

▶ *What questions do you have about the organisms collected?*

▶ *What structures do leaf litter critters have that help them survive?*

Part 3: Leaf-Litter Critters

10. Answer the focus question
Have students answer the focus question in their notebooks.

➤ *What are the characteristics of animals living in the leaf-litter environment?*

Have students glue notebook sheet 6, *Critter Record*, into their notebooks and answer the focus question.

11. Decide if animals could be added to terrariums
Ask students to talk in their groups and decide if it would be appropriate to add their leaf-litter animals to the terrariums. Students will need to consider if the animal will get everything it needs from the terrarium.

After the discussion, students may add the animals and leaf-litter items collected in the schoolyard to the terrariums. Organisms not added to terrariums should be returned to their natural environment during recess or at the next available opportunity.

12. View the video
Prepare the classroom for observing the video, *Animals of the Rain Forest* (duration 23 minutes). Ask students to observe the organisms and their structures and how those structures function to help the animals survive. Here is a detailed summary of the video.

Chapter 1: Introduction to Animals of the Rainforest (2 min 51 sec)
Tropical rainforests are located near Earth's equator. They are a type of biome. Rainforests are home to over half of all the animals found around the world.

Chapter 2: The Geography, Climate and Biodiversity of the Rainforest (2 min 27 sec)
The tropical climate in the rainforest biome encourages the growth of many different plants and animals (biodiversity).

The rainforest in Costa Rica (Central America) is home to many different animal species; many species found in the rainforest are endemic (found nowhere else in the world).

The rainforest is an ecosystem—an environment where many plants and animals live together, feeding and breeding new animals; the rainforest does not have a hibernation season, which means that animals must always be on the watch for predators.

EL NOTE

For students who need scaffolding, provide a sentence frame such as, I observed _____ .

TEACHING NOTE

Make sure students understand when it is appropriate to release organisms into the local environment (they were just collected from that environment and will not be harmed by releasing them back into the local environment). Students should understand that organisms from other environments should not be released locally.

Environments Module—FOSS Next Generation

INVESTIGATION 1 — *Environmental Factors*

Chapter 3: The Five Layers of the Rainforest: Emergent, Canopy, Understory, Shrub and Forest Floor (2 min 26 sec)
The emergent layer of the rainforest is the highest level—made up of the tops of growing trees and is home to many birds of prey.

The canopy layer is where the leaves of trees let out their long leafy branches—home to parrots, macaws, and anteaters.

The understory layer is located just below the tree tops—home to many tree frogs, insects, and monkeys.

The shrub layer is dark because plants above block out the light; this layer is home to many different species of spiders.

The forest floor is the ground level of the rainforest; it is dark, damp, and home to many different lizards, insects, and mammals.

Chapter 4: Animals of the Rainforest Develop a Niche to Survive in Their Habitat (2 min 23 sec)
Animals in the rainforest have structures and behaviors to survive; for example, the Jesus lizard can run on top of water, and the white-faced monkey can jump from tree to tree.

The three-toed sloth is adapted to the rainforest in many ways: it is able to digest tree leaves, it moves slowly in order to concentrate body energy on digestion, and it is able to camouflage itself with the green mold that grows on its shaggy fur.

Chapter 5: Animals of the Rainforest: Behavior and Camouflage (2 min 36 sec)
Rainforest animals: Snakes use their muscles to move; most venomous snakes are camouflaged; have triangular shaped heads where they store their venom; use their tongues to smell and taste.

Rainforest animals: Many insects have marks on their bodies that camouflage them with their surroundings (cryptic coloration); for example, caterpillars and katydids.

The iguana moves very slowly but uses sudden bursts of speed to capture its prey; it is also able to change color in order to camouflage with its surroundings.

Many butterflies have spots on their wings that resemble eyes; for example, owl butterflies.

Chapter 6: Animals of the Rainforest: Behavior, Communication and Defense (2 min 53 sec)
Many poisonous frogs are brilliantly colored to communicate to other animals that they are dangerous; for example, arrow poison frogs.

Part 3: Leaf-Litter Critters

Some animals communicate using sound; for example, howler monkeys and spider monkeys.

Many birds raise their young high up in the emergent layer of the rainforest; their bright colors attract mates and camouflage them with orchids and liana blossoms; for example, macaws and aricaris.

Chapter 7: Interdependent Relationships Between Animals, Plants and People of the Rainforest (3 min 4 sec)
A symbiotic relationship occurs when animals live close together and benefit from their association; for example, egrets eat the bugs that cattle attract.

Birds and bats help to spread seeds and pollinate plants; for example, hummingbirds.

Insects, such as leaf cutting ants, break down existing trees and plants to make room for new growth.

Chapter 8: People and the Destruction of the Rainforest (1 min 36 sec)
The destruction of the rainforest is attributed to people who are clearing the land for farming, cutting the trees to sell, and digging for minerals and oil; this has already resulted in the loss of millions of animals.

About one quarter of all medicine for humans come from plants that are only found in the rainforest; some are used to fight diseases such as high blood pressure, malaria, and cancer.

The destruction of the rainforest is considered one of the world's worst ecological disasters; further destruction could result in catastrophic changes in the atmosphere.

Chapter 9: Conclusion to Animals of the Rainforest (1 min 32 sec)

13. **Discuss the video**

    Have students first talk in their groups about some of the structures and behaviors of the rain forest animals that students observed and how they helped the organisms survive. Then make a class list of some of their observations.

    The video will prepare students for the next reading, "Amazon Rain Forest Journal."

Environments Module—FOSS Next Generation

INVESTIGATION 1 – Environmental Factors

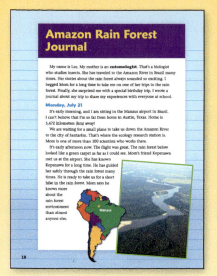

READING in Science Resources

14. Read "Amazon Rain Forest Journal"

Read the article "Amazon Rain Forest Journal," using the strategy that is most effective for your class. The article describes the observations of a student who takes a 6-day trip to a tropical rain forest.

Students can read the article on their own at home or in class. Or you could read the article aloud. The article is divided into six daily journal observations and could be read over several days.

Before starting, ask students to flip through the pages and describe the structure of this article. Students should notice that the information is organized chronologically. Explain that it is, in fact, a journal, and that daily journal writing is much like keeping a diary of their experiences. Ask students to think about how they would use a diary or journal if they traveled to a new place.

15. Use a pair-reading strategy

If students will be reading in class, one strategy is to read in pairs. Each one takes a turn to read and pretends to be Lee. Student A reads one journal entry while Student B listens. Student B then summarizes what she heard and thinks of a question she would like to ask Lee. Students can then discuss what Lee might answer, or move on to the next entry, switching roles.

For other reading strategies to support English learners and below-grade-level readers, see the Science-Centered Language Development chapter.

16. Discuss the reading

Give students time to discuss and answer the questions at the end of the article in their small groups. They should be prepared to refer to details and give examples from the text in support of their ideas. Students can use self-stick notes to mark places in the text they want to refer to as evidence when they share with the group.

Ask students to review the questions at the end of the article. Remind students of the article they read previously about the tropical rain forest and discuss how reading a journal—a firsthand account—compares to reading about the rain forest from a secondhand account. What is similar and what is different about how the information is provided?

ELA CONNECTION

These suggested reading strategies address the Common Core State Standards for ELA.

RI 1: Refer to details and examples in a text when explaining what the text says explicitly and when drawing inferences from text.

RI 2: Determine the main idea of a text and explain how it is supported by key details; summarize the text.

RI 5: Describe the overall structure of information in a text.

RI 6: Compare and contrast a firsthand and secondhand account of the same topic.

Part 3: Leaf-Litter Critters

Call on a reporter from each group to share the group's answer to a question along with details and examples from the text. Allow other group reporters to respond by either agreeing, disagreeing and offering alternative explanations or evidence, or adding on more information.

After the discussion, ask students to refine or add to their original answers and record them in their notebooks.

▶ *What did Lee learn about ants on the rain forest adventure?* [There are different kinds of ants. Ants live in social groups called colonies. They communicate with one another using chemicals they leave on their trails—pheromones. Ants reproduce with complete metamorphosis, like beetles (egg, larva, pupa, adult).

Leaf-cutter ants carry leaves to their underground nests, where they chew the leaves and make them into a pulp to grow fungus, which they eat. Red army ants eat other insects, such as wasps. Army ants make temporary nests because they need to keep moving to new places to get food. One kind of ant protects the acacia tree from predators.]

▶ *How do ants communicate with each other about navigating through the forest?* [Ants put down drops of a chemical, called a pheromone. The pheromones mark the trails for other ants to follow.]

Reinforce that the environment can be thought of as a system of interacting parts. The interdependence of the plants and the animals is one way that the system can be described.

▶ *How does the rain-forest system help different kinds of ants to survive?* [The rain-forest system provides for many different kinds of food and moisture for the different kinds of ants.]

▶ *In what ways do animals depend on plants in the rain forest environment? How do the plants depend on the animals in the rain forest environment?* [Animals depend on plants for food and shelter. Plants depend on animals for pollination, seed dispersal, and sometimes protection.]

▶ *What environmental factor changes as you go from the rain forest canopy to the rain forest floor?* [The amount of sunlight.]

17. **Read more about the rain forest**
This is a good opportunity for students to learn more about the rain forest environment by reading another text. See FOSSweb for a list of recommended books. One suggestion is *Explore the Tropical Rain Forest* by Linda Tagliaferro.

SCIENCE AND ENGINEERING PRACTICES

Obtaining, evaluating, and communicating information

TEACHING NOTE

This is a good time to focus on sensory structures and the ways animals receive information about their environment to survive.

CROSSCUTTING CONCEPTS

Systems and system models

ELA CONNECTION

These suggested reading strategies address the Common Core State Standards for ELA.

RI 9: Integrate information from two texts on the same topic.

SL 1: Engage in collaborative discussions.

Environments Module—FOSS Next Generation

INVESTIGATION 1 – *Environmental Factors*

POSSIBLE BREAKPOINT

18. View the video
Prepare the classroom for observing the video, *Animal Needs* (duration 22 minutes). Explain that this video isn't about a specific environment, but is a look at what most animals need to survive. Again, ask students to observe the organisms and their structures and how those structures function to help the animals survive. Here is a detailed summary of the video.

Chapter 1: Introduction to Animal Needs (45 sec)

Chapter 2: Basic Animal Needs: Oxygen, Food, Water and Shelter (1 min 12 sec)

Animals need food, water, and oxygen to survive.

Animals need a safe shelter from the elements and other animals; animals have structures and behaviors that help them survive.

Chapter 3: Animal Needs: Oxygen (54 sec)
Animals need oxygen in order to survive; humans and other animals have lungs that take the oxygen from the air.

Many aquatic animals use gills to derive oxygen from the water that surrounds them.

Chapter 4: Animal Needs: Food (2 min 26 sec)
Animals need to eat in order to get the energy that their bodies need; energy is transferred from one living thing to another.

Animals derive energy from the food that they eat; animals get energy by consuming plants, animals, or both.

Animals transform their food into energy, and the energy into activities that are necessary for survival. They constantly need oxygen, food, and water, which enable them to use energy efficiently and survive.

Chapter 5: Animal Needs: Water (1 min 43 sec)
Animals need water to survive; water helps digestion, cools the body, and is the main component in blood.

Animals have adaptations that allow them to get and purify water.

Camels store fat in their humps and are very efficient at retaining and using water.

Chapter 6: Animal Needs: Habitat (1 min 58 sec)
An animal's habitat has everything that it needs to survive.

Animals compete within their habitats for food, water, and space.

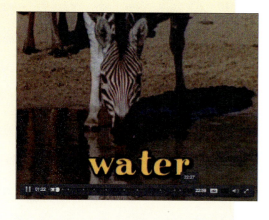

Part 3: Leaf-Litter Critters

Many animals' natural habitats are being destroyed by humans; human destruction of rain forests has caused many species to become endangered.

Chapter 7: Animal Husbandry: Taking Care of Animals' Needs (2 min 53 sec)
People who practice animal husbandry tend to the needs of animals kept in captivity. They must understand and meet the needs of each individual animal that resides in captivity and replicate the natural habitats of the animals in their care.

Chapter 8: Animal Needs: Shelter and a Stable Body Temperature (2 min 10 sec)
Animals need shelter that provides them with a safe place to raise their young and protects them from predators and extreme temperatures. Some animals create their own shelters.

Animals need to keep their bodies from becoming too hot or too cold; humans are warm-blooded—able to regulate body temperature.

Warm-blooded animals regulate their body temperature without help from their surroundings; cool-blooded animals adjust to the temperature around them.

Chapter 9: Investigation: Cold-Blooded Animals Respond to Changes in Their Environment (3 min 46 sec)
All animals depend on their environment for survival, including the cold-blooded fruit fly.

Investigation: Students test to see which type of environment is most beneficial to fruit flies.

Chapter 10: Animal Needs: Hibernation, Estivation and Migration (2 min 15 sec)
Animals hibernate, estivate, and migrate to avoid harsh environmental conditions.

Some animals meet their needs by migrating, or traveling long distances to avoid extremes of seasonal weather changes. Some animals estivate, or enter a deep sleep, to avoid extreme heat.

Chapter 11: Animal Needs: Interdependency (49 sec)
Animals are interdependent; animals depend on plants and other animals to help them satisfy their needs. If a species' needs are consistently met, those animals will flourish, and future generations will survive.

Chapter 12: Conclusion to Animal Needs (38 sec)
All animals face the same struggles to ensure their needs are met.

> **TEACHING NOTE**
> Ask students to compare the fruit fly investigation to the firsthand investigations they conducted with isopods and mealworms.

Environments Module—FOSS Next Generation

INVESTIGATION 1 – Environmental Factors

SCIENCE AND ENGINEERING PRACTICES

Obtaining, evaluating, and communicating information

CROSSCUTTING CONCEPTS

Structure and function

▶ **NOTE**
Go to FOSSweb for *Teacher Resources* and look for the Crosscutting Concepts—Grade 4 chapter for details on how to engage students with the concept of structure and function.

TEACHING NOTE

See the **Home/School Connection** for Investigation 1 at the end of the Interdisciplinary Extensions section. This is a good time to send it home with students.

19. Use a concept grid

Review the animal needs that were described in the video. Students might talk in their groups for a few minutes before having a class discussion. Ask,

▶ *What were the three main needs of most animals?* [Oxygen, food, and water.]

▶ *How do animals use oxygen, food, and water?* [Animals transform their food into energy, and the energy into activities that are necessary for survival. Water and oxygen are needed in those processes.]

▶ *What other things do most animals need?* [Keep their body temperature stable or within good range for the organism; space to get resources; habitat or home, and protection.]

Suggest that students use a concept grid to keep track of different animal structures and how those structures help the animal survive in their environments. The structures can be external or internal.

Have students draw a table in their notebooks to summarize what they have learned in this first investigation from firsthand experiences, from the readings, and from the videos. They should use their notebooks as reference. See the sample concept grid below. You might get the students started by using this grid with the land isopod as an example.

Animal	Structure (external or internal)	Function of structure	How structure helps in survival
land isopod	legs, seven pairs	only walking	move to moist environments
land isopod	two pairs of antennae	sense environment	find food
land isopod	gill-like structures	breathe when wet	get oxygen into body

Students can make a second grid to summarize information they have gathered about plants and how they meet their needs. Ask students to discuss how plant needs are the same and different from animal needs.

144   Full Option Science System

Part 3: Leaf-Litter Critters

WRAP-UP

20. Review Investigation 1

Distribute one or two self-stick notes to each student. Ask students to cut each note into three pieces, making sure that each piece has a sticky end.

Ask students to take a few minutes to look back through their notebook entries to find the most important things they learned in Investigation 1. Students should include at least one science and engineering practice, one disciplinary core idea, and one crosscutting concept. They should tag those pages with self-stick notes. They might use a highlighter or colored pencils to call out the key points.

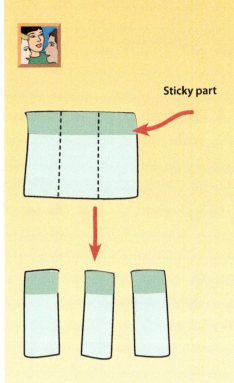

Sticky part

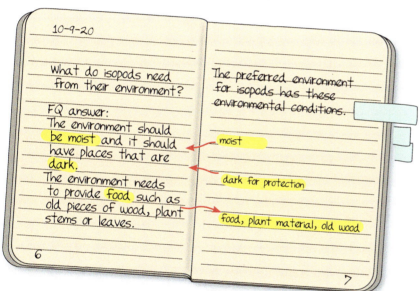

Lead a short class discussion to create a list of three-dimensional statements that summarize what students have learned in this investigation. Here are examples of the big ideas that should come forward in this discussion.

- We observed small animals to study their structures and behaviors that function to support survival, growth, and reproduction. This includes sensory structures and systems. (Planning and carrying out investigations; structure and function.)
- Through our use of readings and video, we know an environment is everything living and nonliving that surrounds and influences an organism. An environmental factor is one part of an environment. Food, oxygen, moisture, and temperature are important factors. (Obtaining, evaluating, and communicating information.)

DISCIPLINARY CORE IDEAS

LS1.A: Structure and function

LS1.D: Information processing

LS2.C: Ecosystem dynamics, functioning, and resilience

LS4.D: Biodiversity and humans

Environments Module—FOSS Next Generation

INVESTIGATION 1 – Environmental Factors

- We argued from evidence that organisms have certain conditions they prefer within an environment. (Engaging in argument from evidence; cause and effect.)

21. Discuss investigation guiding question

Students should discuss the investigation guiding question with a partner before responding to it in their notebooks.

➤ *How do the structures of terrestrial organisms function to support the survival of the organisms in that environment?*

BREAKPOINT

22. Assess progress: I-Check

Give the I-Check assessment at least one day after the wrap-up review. Distribute a copy of *Investigation 1 I-Check* to each student. You can read the items aloud, but students should respond to the items independently in writing. Alternatively, you can schedule the I-Check on FOSSmap for students to take the test and then you can access data through helpful reports to look carefully at students' progress.

If students take the assessment on paper, collect the I-Checks. Code the items using the coding guides on FOSSweb, and plan for a self-assessment session, identifying disciplinary core ideas, science and engineering practices, and crosscutting concepts students may need additional help with. (If you are using FOSSmap, you will only need to code open-response items and confirm codes for short-answer items as other items are automatically coded.) Refer to the Assessment chapter for more information.

TEACHING NOTE

During or after these next steps with the I-Check, you might ask students to make choices for possible derivative products based on their notebooks for inclusion in a summative portfolio. See the Assessment chapter for more information about creating and evaluating portfolios.

Interdisciplinary Extensions

INTERDISCIPLINARY EXTENSIONS

Language Extension

- **Write organisms booklets**
  Have students research some of the organisms that live in the rain forest. Then assign groups to write a booklet about one of the organisms. Each group member might write a topical chapter (page) for the booklet. Topics might include what the animals eat, how they reproduce, where they live, what environment they prefer, physical structure, and so forth.

Math Extension

- **Problem of the week**

  1. A girl made a ladybug cage. Her cage has five sides. Each side has a length of 28 centimeters (cm). The cage is 20 cm high. What is the shape of her cage's base called? What is the perimeter of the base?

  2. A boy wants to build rectangular beetle cages to study beetle behavior. He wants all the sides on the base of his cages to be at least 10 cm long. He also wants the base area to equal 576 cm².

  The boy needs help to find all the possible lengths and widths for the base of the cages. He used only whole centimeters to measure, with no fractional parts.

  What are all the possible dimensions for the rectangular base?

  The boy has decided to make the height of each cage 10 cm. If he fills the container to capacity, how much soil will he need for each cage?

  Notes on the problem. 1) Students need to recall that a polygon with five sides is a pentagon and that a perimeter is the distance around all the sides of a polygon. Perimeter is calculated by adding the lengths of all the sides.

  Perimeter (pentagon) = $s_1 + s_2 + s_3 + s_4 + s_5$

  In this case all sides are equal, so students can also use
  Perimeter = $5 \times s$

  The solution is 28 cm + 28 cm + 28 cm + 28 cm + 28 cm = 5×28 cm = 140 cm.

> **TEACHING NOTE**
>
> Refer to the teacher resources on FOSSweb for a list of appropriate trade books that relate to this module.

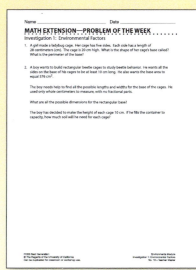

No. 10—Teacher Master

> **TEACHING NOTE**
>
> When discussing the problem, review the names of the other polygons with six or more sides: hexagon (6 sides), septagon (7 sides), octagon (8 sides), nonegon (9 sides), decagon (10 sides), undecagon (11 sides), dodecagon (12 sides), and n-gon for a polygon with n sides.

Environments Module—FOSS Next Generation

INVESTIGATION 1 — *Environmental Factors*

2) To determine different pairs of dimensions that will produce the same area for beetle cages, students are likely to use a guess-and-check strategy. Students who are facile with numbers might be able to estimate a starting point for pairs of numbers quickly. For students who have difficulty approaching the problem, guide them to begin with simple numbers for length and width such as 1 and 576. Continue with 2 and 288. Once they understand the concept, they can continue to find whole numbers whose product is 576.

As they calculate the dimensions, students will discover that in one case, the length is equal to the width or that the base can be a square. This is an opportunity to refresh definitions of these two quadrilaterals.

Squares are polygons with four equal sides and four equal angles of 90° (right angles). More formally: A square is a quadrilateral with equilateral sides and four right angles.

Rectangles are polygons with four sides, opposite sides equal in length, and four equal angles of 90° (right angles). More formally: A rectangle is a quadrilateral that has four right angles and whose opposite sides are equal in length.

From these definitions, it follows that

- All squares are rectangles.
- Not all rectangles are squares. (Some rectangles are squares.)

Encourage discussion of how students arrived at their solutions and how they knew when they had found all the possible solutions.

Here is the solution to the problem:

$A = l \times w$

576 cm^2 = 12 cm × 48 cm

576 cm^2 = 16 cm × 36 cm

576 cm^2 = 18 cm × 32 cm

576 cm^2 = 24 cm × 24 cm

Interdisciplinary Extensions

In the second part of the problem, students determine the capacity or volume for one habitat. If students are unfamiliar with the formula for volume, they might use cubes to help construct the formula.

A set of 20 uniform cubes (nonlinking), such as base ten units, 1" color cubes, or wooden cubes, can serve as a model.

Review the dimensions of a cube—length, width, and height. In the case of a cube, all the dimensions are equal. It is a cubic unit. Review that linear units measure length/distance, and square units measure area.

Put two cubes side by side so that a pair of square faces is touching.

If you had a bird's-eye view of the shape, what would it be? [A rectangle.] The area of the top of the shape is two square units. If you looked at the shape as a whole, it would be two cubes or have a volume of 2 cubic units.

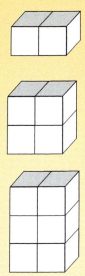

Add another layer of two cubes on top. Again, with a bird's-eye view, the area of the top of the shape is still two square units. However, now that the structure is two units high, the number of cubes is four. If it is three levels of two cubes, a total of six cubes is needed.

Continue to build layers. The area is being duplicated by a multiple of the height of the building. This lends itself to looking at the formula for volume as $V = A \times h$.

Students can then connect the case of the soil needed to fill the habitat as follows:

$V = A \times h$
$V = 576 \text{ cm}^2 \times 10 \text{ cm} = 5{,}760 \text{ cm}^3$

Environments Module—FOSS Next Generation

INVESTIGATION 1 – Environmental Factors

TEACHING NOTE

Review the online activities for students on FOSSweb for module-specific science extensions.

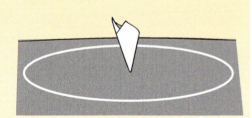

▶ **NOTE**
Go to FOSSweb for insect (invertebrate) identification guides.

Science and Engineering Extensions

- ### Investigate local isopods
  After investigating and reading about isopods in the classroom, take students to the schoolyard to hunt for and collect local isopods. Ask students to apply what they learned and read about isopods to their search. Look under boards, logs, rocks, in the litter beneath trees, or in compost piles.

  Remind students that to determine if an isopod is a sow bug or a pill bug, place it in your hand and shake it gently. If it rolls into a tight ball or is unable to flip over from an upside-down position, you have a pill bug. If the isopod has two tail-like appendages and flips from an upside-down position onto its feet, it is a sow bug.

- ### Conduct isopod races
  Have an isopod race. Each group will need to find five to ten of each kind of isopod and place them in a container. Each group will need a funnel made from a rolled piece of paper (8.5" × 11") and tape. Cut the tip of the funnel to leave a 2 cm opening at the top.

  The group draws a circle with chalk on the asphalt about 60 cm in diameter. To start the race, students place one of each kind of isopod through the funnel in the center of the circle. The first isopod to reach the outside of the circle is the winner. (It is not fair to right an upside-down isopod.) Students should consider how the structures of the isopods affect the results of the races and how these structures might help the isopods survive in their natural environment. Encourage students to have challenge races between winning isopods.

- ### Sample terrestrial environments
  Have students use collecting nets (sturdy sweep nets) to sample the organisms in different terrestrial environments (grassy area, weedy area, brush area). A collecting or sweep net is used like a broom to sweep back and forth just at the tops of plants. Record on a chart the kinds and numbers of organisms found in each environment. Provide insect field guides for identification of the organisms.

- ### Make a terrarium of local organisms
  Ask students to bring in earthworms, snails, salamanders, and other organisms found around the school or in students' neighborhoods. Plan what kinds of materials and environmental conditions will be needed to keep the animals healthy.

  To be sure the animals stay healthy, keep them in the classroom for only 2 weeks, and then return them to their natural environments.

Interdisciplinary Extensions

- **Investigate beetle complete metamorphosis**
  Have students investigate other conditions that are favorable to mealworm metamorphosis. Environmental factors to consider are moisture and light. (Temperature is the most significant and was investigated in class.)

- **Make Critter Replicators**
  Have students make their own Critter Replicator following the instructions on teacher master 6. For each replicator you will need a paper copy of teacher master 7 (cover), a copy of teacher masters 8 and 9 on transparencies, and four paper fasteners, 1/2". For tools, you will need a pair of scissors.

- **Research pheromones**
  Animals have different sense receptors that are specialized for particular kinds of information. Have students use print and/or media to research animals that use pheromones and what kind of information they convey.

- **Research other terrestrial environments**
  Have students complete the concept grid they started with the reading of "Two Terrestrial Environments" by researching the factors in other environments (temperate deciduous forest, grassland, taiga, and tundra).

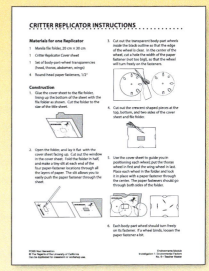

No. 6—Teacher Master

Environmental Literacy Extensions

- **Build a compost pile**
  A compost pile made of leaves, twigs, bark, and other plant parts can become a great environment for small animals (leaf-litter critters). Have a group of students design and build a compost pile at the school. Students should research different methods for making a compost pile, and work with the school custodian or grounds-keeper to complete the project.

- **Go on a local field trip**
  Seek out an environmental educator in your area (see the Regional Resources section on FOSSweb) offering field trips to natural spaces featuring a local environment. Encourage the environmental educator to focus on the environmental factors that influence the living things in the area. Ask students to consider how the natural environment benefits humans as well as other living things. Also, give time for students to appreciate the beauty and peacefulness of the area, possibly with some independent nature journal time.

> **TEACHING NOTE**
>
> *Encourage students to use the Science and Engineering Careers Database on FOSSweb.*

Environments Module—FOSS Next Generation

151

INVESTIGATION 1 – *Environmental Factors*

- ### Research local environments
  Have students conduct research about local environments or features that affect or define these environments such as flood plains, wetlands, bogs, marshes, beaches, rivers, damns, prairies, woodlands, sand dunes, etc. Ask students to consider the benefits and drawbacks of these areas to human populations. Students can also consider the ecosystem services these areas provide for humans.

Home/School Connection

Students go on a safari in and around their homes or into the local neighborhood to look for insects. Students organize the results of their safaris and bring them to class.

Print or make copies of teacher master 11, *Home/School Connection for Investigation 1*, and send it home with students after Part 3.

▶ **SAFETY NOTE**
While most insects, spiders, and other small animals are harmless, some can sting (ants, wasps, bees), and some can bite (spiders, centipedes). Caution students to observe the animals without touching them.

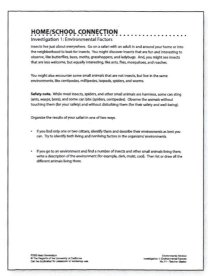

No. 11—Teacher Master

152   **Full Option Science System**

Investigation 2: Ecosystems

INVESTIGATION 2 – *Ecosystems*

Part 1
Designing an Aquarium 164

Part 2
Food Chains and
Food Webs 175

Part 3
Population Simulation 190

Part 4
Sound Off 202

Guiding questions for phenomenon:
How are the structures of aquatic organisms similar and different from land animals? How do organisms sense and interact with their environment?

Science and Engineering Practices
- Developing and using models
- Planning and carrying out investigations
- Analyzing and interpreting data
- Constructing explanations
- Engaging in argument from evidence
- Obtaining, evaluating, and communicating information

Disciplinary Core Ideas
LS1: How do organisms live, grow, respond to their environment, and reproduce?
LS1.A: Structure and function
LS1.D: Information processing
LS2: How and why do organisms interact with their environment and what are the effects of these interactions?
LS2.C: Ecosystem dynamics, functioning, and resilience
LS4: How can there be so many similarities among organisms yet so many different kinds of plants, animals, and microorganisms?
LS4.D: Biodiversity and humans

Crosscutting Concepts
- Systems and system models
- Energy and matter
- Structure and function
- Stability and change

PURPOSE

Students investigate the phenomenon of life in water and how organisms' needs are the same and different from life on land, and obtain information about structures that allow animals to use sound to communicate and sense their environment.

Content
- The interaction of organisms with one another and with the nonliving environment is an ecosystem. Organisms have structures that allow them to interact in feeding relationships and obtain oxygen; they may compete for resources.
- Producers (plants, algae, phytoplankton) make their own food, which is also used by animals (consumers). Decomposers eat dead plant and animal materials and recycle the nutrients in the system.
- Organisms on land and in water have sensory systems to gather information about their environment and act on it.

Practices
- Use modeling to construct representations of the natural world and make predictions.

Full Option Science System

INVESTIGATION 2 – Ecosystems

Investigation Summary	Time	Focus Question for Phenomenon, Practices
PART 1 — Designing an Aquarium Students review the environmental factors in a terrestrial environment and compare them to environmental factors in aquatic environments. They observe guppies and goldfish and add them separately to two class aquaria. They add other organisms to both aquaria and monitor the living and nonliving factors in each environment over time.	**Active Inv.** 2 Sessions* **Reading** 1 Session	**What are the environmental factors in an aquatic system?** **Practices** Planning and carrying out investigations Constructing explanations Obtaining, evaluating, and communicating information
PART 2 — Food Chains and Food Webs Students work with organism cards to create food chains and food webs in a woodland ecosystem that includes terrestrial and aquatic environments. Students compare the structures of land and water organisms and the ways the structures function to meet the organisms' needs. Students gather and compare information on how different animals obtain one basic need—oxygen.	**Active Inv.** 2–3 Sessions **Reading** 2 Sessions	**What are the roles of organisms in a food chain?** **Practices** Analyzing and interpreting data Engaging in argument from evidence Obtaining, evaluating, and communicating information
PART 3 — Population Simulation Students go to the schoolyard to simulate a population of deer foraging for food in its home range. Students are introduced to the concept of carrying capacity, the greatest number of organisms that can be supported (carried) by an area without damaging it. Students gather and compare information on how different animals obtain one basic need—oxygen.	**Active Inv.** 1 Session **Reading** 1 Session	**How does food affect a population in its home range?** **Practices** Developing and using models Analyzing and interpreting data Constructing explanations Obtaining, evaluating, and communicating information
PART 4 — Sound Off Students go to the schoolyard and pretend to be animals who have poor vision or are active at night. The animals communicate with one unique sound and try to find others of their kind before being "captured" by a predator. After three rounds of this activity, students sit silently to listen to animals in the schoolyard. This experience is an introduction to the use of sound by animals. Students gather information about the structure and function of animals' systems that involve sound.	**Active Inv.** 1–2 Sessions **Reading** 1–2 Sessions **Assessment** 2 Sessions	**How do animals use their sense of hearing?** **Practices** Planning and carrying out investigations Constructing explanations Obtaining, evaluating, and communicating information

* A class session is 45–50 minutes. **Full Option Science System**

At a Glance

Content Related to DCIs	Writing/Reading	Assessment
• Aquatic environments include living and nonliving factors (water and temperature). • Organisms that live in water have structures that function to meet their needs. • The interaction of organisms with one another and with the nonliving environment is an ecosystem.	**Science Notebook Entry** *Living and Nonliving Factors* *Aquarium Observation Log* **Science Resources Book** "Freshwater Environments"	**Embedded Assessment** Science notebook entry
• Organisms have structures that allow them to interact in feeding relationships; they may compete for resources. • Producers make their own food, which is also used by animals (consumers). • Decomposers eat and recycle the nutrients in the system. • Animals have different systems for obtaining oxygen.	**Science Notebook Entry** *Practice with Food Chains* **Science Resources Book** "What Is an Ecosystem?" "Food Chains and Food Webs"	**Embedded Assessment** Science notebook entry
• Organisms interact in feeding relationships in ecosystems. • When the environment changes, some plants and animals survive and reproduce; others move to new locations, and some die.	**Science Notebook Entry** *Population Simulation Results* **Science Resources Book** "Human Activities and Aquatic Ecosystems" "Comparing Aquatic and Terrestrial Ecosystems" **Online Activities** Virtual Terrarium and Aquarium	**Embedded Assessment** Science notebook entry
• Animals communicate to warn others of danger, scare predators away, or locate others of their kind, including family members. • Organisms have sensory systems to gather information about their environment and act on it. • Animals detect sounds, interpret, and act on them.	**Science Notebook Entry** Answer the focus question **Videos** *Animal Language and Communication* *All about the Senses* **Science Resources Book** "Animals Sensory Systems" "Saving Murrelets through Mimicry"	**Embedded Assessment** Response sheet **Benchmark Assessment** Investigation 2 I-Check **NGSS Performance Expectations addressed in this investigation** 4-LS1-1 4-LS1-2 3-LS4-4

Environments Module—FOSS Next Generation

INVESTIGATION 2 – *Ecosystems*

BACKGROUND *for the Teacher*

What Are the Environmental Factors in an Aquatic System?

Important factors in aquatic systems include temperature, transparency or light penetration (turbidity), nutrients, currents, concentrations of dissolved gases (carbon dioxide and oxygen), acidity, concentration of salts, and organisms. Some of the organisms you might expect to find in natural **freshwater environments** and freshwater **aquaria** include **algae**, bacteria, vascular plants (such as **elodea**), crustaceans (such as crayfish, amphipods, or scuds), insects, mollusks (snails and clams), fish, frogs, salamanders, and turtles. Microscopic **phytoplankton** and **zooplankton** are also found in freshwater environments.

In both terrestrial and **aquatic environments**, oxygen and carbon dioxide cycle among plants, animals, and the physical environment. In terrestrial systems, the proportion of these gases in the air is fairly constant, with oxygen representing about 21 percent and carbon dioxide about 0.03 percent. In freshwater systems, however, there are great variations in the proportion of these gases, including daily and seasonal fluctuations.

The concentration of oxygen dissolved in water is much lower than in the terrestrial atmosphere. Oxygen is very soluble in water but is absorbed from the air quite slowly. Agitation of water caused by wind and currents increases oxygen absorption. This is why bubblers are often used in aquaria. Also, the colder the water, the more dissolved oxygen the water will hold.

Carbon dioxide (CO_2) is more soluble in water than is oxygen. In natural systems, some carbon dioxide comes from groundwater, air, and rain. Carbon dioxide is an important contributor to the acid content of freshwater systems, as it combines with water to form weak carbonic acid (H_2CO_3). High concentrations of carbon dioxide are often associated with low concentrations of oxygen, a situation that can be lethal to certain organisms, especially fish.

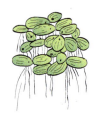

Both plants and animals use oxygen and give off carbon dioxide through respiration. But only plants use carbon dioxide and produce oxygen through the process of photosynthesis. During daylight hours aquatic plants produce more oxygen than plants and animals consume. A reserve of oxygen accumulates. At night, when photosynthesis has stopped, both plants and animals use the accumulated oxygen and produce carbon dioxide. Thus these gases fluctuate daily in both natural systems and managed aquaria.

Background for the Teacher

A new idea developed in this investigation is that of an **ecosystem**. An ecosystem is a community of organisms, all the nonliving factors that surround it, *and* the **interactions** between the living and nonliving components. Ecology is the study of the interactions between a community of organisms and the environment in which the community lives. It is the largest view of life on Earth.

What Are the Roles of Organisms in a Food Chain?

One of the major ideas in ecology is the mechanisms by which organisms acquire the material resources, or matter (chemicals in the form of minerals, gases, and organic compounds), and **energy** to conduct their lives. In most ecosystems on Earth, energy enters the ecosystem as light from the Sun and is transferred to chemical energy (food) through the complex process of photosynthesis performed by plants, algae, some bacteria, and some protists.

In the study of ecology, organisms that make food are called **producers**. Organisms that do *not* make their own food must eat other organisms to get the matter and energy they need. Organisms that eat other organisms are called **consumers**. Consumers include animals, fungi, bacteria, and protists.

This is a big idea: Energy captured by plants is the energy needed to run the whole ecosystem. The consumers can acquire the energy they need only by consuming other organisms.

The chain of feeding relationships between a series of organisms is called a **food chain**. Grass plants synthesize food, chipmunks (**herbivores**) eat grass, and hawks (**carnivores**) eat chipmunks. Animals that eat both plants and animals are **omnivores**. Waste and uneaten bodies of dead organisms fall to the base of the ecosystem, where they are reduced to component elements—minerals, gases, and water— by **decomposers** like bacteria, fungi, and other **microorganisms**. Decomposition moves the remaining energy from organic materials to raw materials. Energy and matter are again available for the next generation of producer organisms in the ecosystem.

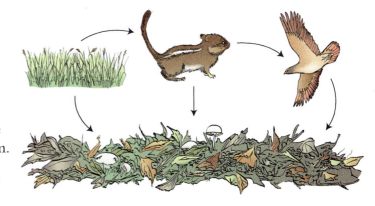

Food chains are one way to describe a series of feeding relationships in an ecosystem. But a simple food chain describes only one of the paths that matter and energy travel. In most ecosystems, there are many types of producers and consumers. Often the consumers eat more than one type of plant or animal. **Food webs** are more complex than food chains and describe all the feeding interactions in an ecosystem.

Environments Module—FOSS Next Generation

INVESTIGATION 2 – Ecosystems

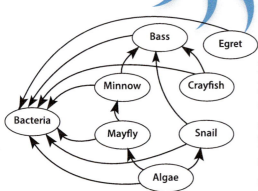

In a river ecosystem, crayfish might be food for both egrets and bass. If the river has a lot of crayfish, egrets and bass will both have plenty to eat. But if there are few crayfish, the egrets and bass will have to compete with each other for food. In this example, the egrets and bass are **predators** and the crayfish are **prey**. Predators are animals that capture and eat live animals for food; prey are the animals that are taken for food.

Animals get energy to drive their body processes from food. In the process of breaking down food and having it rearrange into new products, energy is released and made available to the cells of the body. Oxygen is essential to this process. Animals have different structures to acquire oxygen from the environment—lungs, gills, skin, and spiracles are four of the types of structures. It is not expected that students know the details of the mechanisms by which these various structures work, but rather how the diverse structures of different terrestrial and aquatic organisms deliver a similar function.

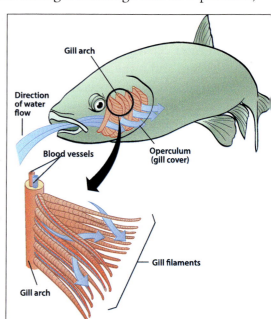

How Does Food Affect a Population in Its Home Range?

A **population** is a group of organisms of one kind that live in the same area. The size of a population (that is, the number of individuals in the group) may change greatly over a period of time. Among animals such as deer or field mice, populations may double, fluctuate slightly, decrease slowly, or "crash" to a very low number from one year to the next.

Many factors influence the size and distribution of a population. Three of the most important are (1) **competition** for available food, (2) the size of the food supply, and (3) dispersal or migration to new feeding territories. The amount of food available to a population is probably the most crucial factor in determining an area's **carrying capacity**. The carrying capacity is the greatest number of organisms that can be supported (carried) by an area without damaging it.

Background for the Teacher

Usually a population of deer or mice will forage for food in the same area year after year. This well-established territory is known as the population's **home range**. Usually only extreme conditions, such as lack of food, loss of habitat, or severe weather, will force a population out of its home range. When a population's home range no longer contains enough food, the population may disperse in search of a new range.

How Do Animals Use Their Sense of Hearing?

Bright colors and striking patterns are useless in the dark or when individuals are visually remote from one another. Other mechanisms are needed to communicate. Animals that are active at night (frogs, crickets, owls, wolves, and coyotes) make distinctive sounds that convey information to others of their kind. Sound is an effective source of stimulus. An ear is usually a funnel-shaped structure designed to intercept and direct pulses in a medium to sound receptors. Complex hearing systems (like that of the human) may have an array of sound receptors, each designed to respond to a different frequency of vibration in a medium (air or water). When the receptor is stimulated, a signal is sent along the sensory nerve to the brain, where it is interpreted. Then a response signal is initiated, which sends a message along a motor neuron to a muscle or other response target to effect a response to the stimulus. The use of auditory stimuli has its liabilities. Sound is general, and all organisms with the appropriate receptors can hear it. This is fine when it is the intended recipient who hears, but just as likely, there may be a predator listening to the announcement. To the predator, the auditory announcement is a dinner bell. A cricket hiding in the grass advertising for a mate may betray his presence to a shrew or toad. The amorous appeal may turn out to be the last sound the cricket makes. Vocalizations and other sounds are used by animals, such as wolves and coyotes, to locate and maintain cohesion among others of their pack or to locate a potential mate. Red-winged blackbirds use vocalizations to assert their "ownership" of a territory, indicating a mating opportunity to available females and a warning to other males to move along. And the various cooings and chatterings of many animals can alert a predator to the presence of a meal lurking in the dark or hiding in the foliage.

"Sound is an effective source of stimulus."

Environments Module—FOSS Next Generation

INVESTIGATION 2 – Ecosystems

NGSS Foundation Box for DCI

LS1.A: Structure and function
- Plants and animals have both internal and external structures that serve various functions in growth, survival, behavior, and reproduction. (4-LS1-1)

LS1.D: Information processing
- Different sense receptors are specialized for particular kinds of information, which may then be processed by an animal's brain. Animals are able to use their perceptions and memories to guide their actions. (4-LS1-2)

LS2.C: Ecosystem dynamics, functioning, and resilience
- When the environment changes in ways that affect a place's physical characteristics, temperature, or availability of resources, some organisms survive and reproduce, others move to new locations, yet others move into the transformed environment, and some die. (3–LS4-4, extended from grade 3)

LS4.D: Biodiversity and humans
- Populations live in a variety of habitats, and change in those habitats affects the organisms living there. (3–LS4-4, extended from grade 3)

TEACHING CHILDREN about *Ecosystems*

Developing Disciplinary Core Ideas (DCI)

The freshwater environment set up by students in this investigation serves as a representative lake or pond. As students study their aquatic systems, occasionally invite them to think about the similarities and differences between this water environment and the terrestrial environments they have been studying. It is through such comparisons that students come to understand complex ideas.

To study the freshwater environment, students introduce organisms systematically into the aquaria and are shown how to monitor environmental factors such as temperature. Students begin to see changes in the aquaria. Their observations will lead to the concept of a system of interacting factors—an ecosystem. The ecosystems that students will be observing are simple, but they will introduce some fundamental concepts that will help students understand more complex natural ecosystems.

Some of the changes students observe include plants growing roots (*Elodea*), snails laying eggs on plants or on the sides of the aquarium, and the aquarium turning green from the natural invasion of algae. Along with growth and reproduction, students might also observe death as scuds are eaten by goldfish and as goldfish die.

Students know that organisms eat other organisms to survive. This investigation will give them the opportunity to delve more deeply into feeding relationships so they can begin to see roles that each organism plays in the ecosystem—the producers, the consumers, the decomposers.

Students will also engage in an active simulation to explore animal sensory systems using the sense of hearing as a model for how animals gather information from their environment to survive.

The experiences students have in this investigation contribute to the disciplinary core ideas **LS1.A: Structure and function; LS1.D: Information processing; LS2.C: Ecosystem dynamics, functioning, and resilience;** and **LS4.D: Biodiversity and humans.**

Teaching Children about Ecosystems

Engaging in Science and Engineering Practices (SEP)

In this investigation, students engage in these practices.

- **Developing and using models** of ecosystems to investigate the interactions in natural systems.
- **Planning and carrying out investigations** with fresh water organisms in an aquarium to make observations to serve as the basis for evidence of interactions of structures and behaviors that serve the organism in survival.
- **Analyzing and interpreting data** collected from simulations and models on how organisms sense their environment and the behaviors that allow them to respond and survive.
- **Constructing explanations** about feeding relationships in terrestrial and aquatic environments and the factors that contribute to change in populations in an area.
- **Engaging in argument from evidence** about the source of energy in an environment.
- **Obtaining, evaluating, and communicating information** from books and media and integrating that with their firsthand experiences to construct explanations about survival of organisms in specific environments.

NGSS Foundation Box for SEP

- **Use a model** to test cause-and-effect relationships to interactions concerning the functioning of a natural system.
- **Make observations and/or measurements to produce data** to serve as the basis for evidence for an explanation of a phenomenon or test a design solution.
- **Represent data in tables and/or various graphical displays** to reveal patterns that indicate relationships.
- **Analyze and interpret data** to make sense of phenomena using logical reasoning.
- **Organize simple data sets** to reveal patterns that suggest relationships.
- **Use evidence** (e.g., observations, patterns) to support an explanation.
- **Identify the evidence** that supports particular points in an explanation.
- **Construct an argument** with evidence, data, and/or a model.
- **Read and comprehend grade-appropriate complex texts** and/or other reliable media to summarize and obtain scientific and technical ideas and describe how they are supported by evidence.
- **Communicate scientific and/or technical information** orally and/or in written formats, including various forms of media and may include tables, diagrams, and charts.

Environments Module—FOSS Next Generation

INVESTIGATION 2 – Ecosystems

NGSS Foundation Box for CC

- **Systems and system models:** A system can be described in terms of its components and their interactions.
- **Energy and matter:** Matter is transported into, out of, and within systems. Energy can be transferred in various ways and between objects.
- **Structure and function:** Substructures have shapes and parts that serve functions.
- **Stability and change:** Change is measured in terms of differences over time and may occur at different rates.

Exposing Crosscutting Concepts (CC)

In this investigation, the focus is on these crosscutting concepts.

- **Systems and system models.** An aquarium is a system and can be described and studied in terms of its parts and their interactions; a population of deer living in a specific environment can be described and studied in terms of its parts and their interactions.
- **Energy and matter.** Energy and matter in an ecosystem can be transferred from the Sun to producers, to consumers, to decomposers.
- **Structure and function.** The forms of an organism's structures relate to their functions.
- **Stability and change.** A population of organisms can be stable or can change depending on the environmental needs and conditions; change may occur at different rates.

Connections to the Nature of Science

- **Scientific investigations use a variety of methods.** Scientific methods are determined by questions. Scientific investigations use a variety of methods, tools, and techniques.
- **Scientific knowledge is based on empirical evidence.** Science findings are based on recognizing patterns. Scientists use tools and technologies to make accurate measurements and observations.
- **Science is a way of knowing.** Science is both a body of knowledge and processes that add new knowledge. Science is a way of knowing that is used by many people.

Algae
Aquarium
Aquatic environment
Carnivore
Carrying capacity
Competition
Consumer
Decomposer
Ecosystem
Elodea
Energy
Food chain
Food web
Freshwater environment
Herbivore
Home range
Interaction
Microorganism
Omnivore
Phytoplankton
Population
Predator
Prey
Producer
Zooplankton

162

Full Option Science System

Teaching Children about Ecosystems

Conceptual Flow

Students investigate the phenomenon of life in water and how organisms' needs are the same and different from life on land. The guiding question is how are the structures of aquatic organisms similar and different from land animals? Then, students investigate the phenomenon of animal communication by making sounds (sound source) and receiving and interpreting sounds (sound receiver). The second guiding question is how do organisms sense and interact with their environment?

This exploration comes not only through aquarium observations but also through organism cards with expository text that students manipulate.

The **conceptual flow** starts in Part 1 with students reviewing the living and nonliving factors in a terrestrial environment. The class sets up indoor **freshwater environments—aquaria—** and populates them with two kinds of fish. Later they add other organisms, including snails, plants, and small invertebrate organisms. Students monitor the environmental factors over time (**temperature**, **water level**, **organisms**). Students are introduced to the concept of **ecosystem**, the **interaction of a community of organisms with its environment**. The important word is *interaction*.

In Part 2, students look more closely at the **feeding interactions or relationships in an ecosystem**. They get to know 20 organisms and their structures in a woodland ecosystem, and develop simple **food chains** involving **producers**, **consumers**, and **decomposers**. Students practice diagramming food chains and get a glimpse into the more complex feeding model, a **food web**. Students compare the structures and their functions of land and terrestrial organisms and how these structures (both external and internal) form systems to meet basic needs.

In Part 3, students go outdoors to engage in a simulation activity involving a **population of deer in a specified home range**. Students explore how a food source affects population size. The concept of **carrying capacity** is introduced.

In Part 4, students investigate structures and functions (both internal and external) that animals use to gather information about their environment. Students explore how animals' responses to this information assists in survival. The active investigation focuses on auditory signals, which can allow predators to find prey and allow prey to avoid predators, and allow other animals to locate potential mates and/or maintain connections between individuals of a pack or family.

Complex Systems

- An aquarium is a small freshwater environment.
- Aquatic and land organisms have different structures that function in survival.
- An environment with a community of interacting organisms is an ecosystem.
- One class of interactions between organisms is the food chain.
- Organisms that synthesize their own food are called producers.
- Consumers eat other organisms for their nutrition.
- Decomposers consume organic waste and dead organisms for nutrition.
- Herbivores eat producers; carnivores eat other animals.
- Animals need oxygen to transform food into energy; they have diverse structures to do this.
- Predators capture and eat prey animals.
- Animals on land and in water have sensory systems to gather information and act on it.
- Carrying capacity is the ability of an ecosystem to produce resources to sustain population.

Environments Module—FOSS Next Generation

INVESTIGATION 2 – *Ecosystems*

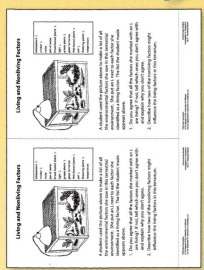

No. 8—Notebook Master

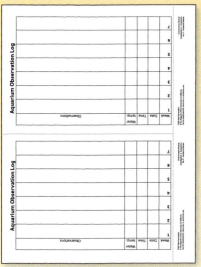

No. 9—Notebook Master

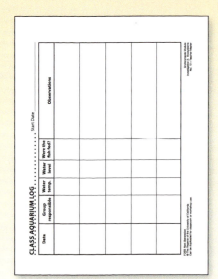

No. 12—Teacher Master

MATERIALS for
Part 1: *Designing an Aquarium*

For each student
- 1 Notebook sheet 8, *Living and Nonliving Factors*
- 1 Notebook sheet 9, *Aquarium Observation Log*
- 1 *FOSS Science Resources: Environments*
  - "Freshwater Environments"

For the class
- 3 Basins, 6 L (aquaria)
- 1 Container of gravel
- 1 Plastic cup
- 2 Fish tunnels, plastic square pipe
- 2 Goldfish ★
- 6 Guppies ★
- 10–12 Pond snails, small ★
- 4–6 Sprigs of elodea ★
- 30 Crustaceans (amphipods, especially *Gammarus*) ★
- 1 Package of fish food
- 1 Collecting net, small
- 1 Baster
- 1 Bottle of water conditioner
- 2 Thermometers, Celsius
- 2 Self-stick notes
- 8 Containers, 1/2 L
- Aged or treated tap water (See Step 4 of Getting Ready.) ★
- 1 Plastic bottle, recycled, 2–4 L ★
- 1 Teacher master 12, *Class Aquarium Log*

For embedded assessment
- *Embedded Assessment Notes*

★ Supplied by the teacher.   ❏ Use the duplication master to make copies.

164   Full Option Science System

Part 1: Designing an Aquarium

GETTING READY for
Part 1: Designing an Aquarium

1. **Schedule the investigation**
   This part will take two sessions for the active investigation and one session for the reading. Students will make periodic observations for several weeks.

2. **Preview Part 1**
   Students review the environmental factors in a terrestrial environment and compare them to environmental factors in aquatic environments. They observe guppies and goldfish and add them separately to two class aquaria. They add other organisms to both aquaria and monitor the living and nonliving factors in each environment over time. The focus question is **What are the environmental factors in an aquatic system?**

3. **Plan for aquarium locations**
   Find a convenient location for the two class aquaria (6 L basins). They should be away from any heat source, out of direct sunlight, and in a location where students can easily observe and monitor them.

4. **Treat tap water**
   Chlorine used in water treatment is toxic to fish. Chlorine will dissipate out of water when it is exposed to air for one day. This is called aging. However, chloramine, a newer additive used in place of chlorine, will *not* leave the water when exposed to air. You must use water-conditioner chemicals that specifically say "removes chloramine."

   The bottle of water conditioner in the kit will treat for both chlorine and chloramine. Follow the directions for use on the bottle. The capacity of the aquarium in the kit is 6 L.

   Two days before conducting Part 1, fill the three 6 L basins with water and let them age. Add conditioner to the water. At the same time, age some tap water in a plastic jug.

   You or your students might also check with a local aquarium store to find out how best to prepare your local water for the fish aquarium.

5. **Rinse the gravel**
   Measure a plastic cup of gravel, and rinse it with the aged and treated tap water. Put about half a cup of gravel into each aquarium.

6. **Rinse the containers**
   Wash eight 1/2 L containers with tap water to make sure that they are clean. Do not use soap. Rinse them with the aged tap water from the plastic jug, and then fill each container with conditioned tap

▶ **NOTE**
Refer to the section Planning for Live Organisms in the Materials chapter for information on setting up and monitoring the aquaria and organisms.

▶ **SAFETY NOTE**
The kit contains a bottle of water conditioner to remove chlorine and chloramine from your local water.

Environments Module—FOSS Next Generation **165**

INVESTIGATION 2 – Ecosystems

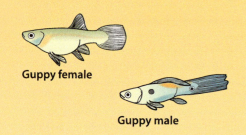
Guppy female

Guppy male

Goldfish

Elodea

Pond snail

Gammarus (scud)

water (about 3/4 full). While students are working on notebook sheet 8, *Living and Nonliving Factors*, during Step 1 of Guiding the Investigation, you will use the collecting net to place one kind of fish into each container (two goldfish and six guppies) so they are ready for students in Step 4.

7. **Purchase goldfish and guppies**
   Buy two small goldfish and six guppies from a local pet or aquarium store. Immediately return to school and float each bag in its designated aquarium for 10–15 minutes to avoid shocking the fish with an abrupt change in water temperature. It would be best if you could put the fish into the aquaria (6 L basins) when the whole class can watch but that might not be possible.

8. **Purchase other organisms for second session**
   While at the aquarium store, purchase 10–12 pond snails and 4–6 sprigs of elodea. Put all the plants in one 6 L basin for the class plant aquarium. The snails can stay in the plant aquarium.

   Order 30 or more crustaceans, such as *Gammarus* (scuds), from Delta Education. These will be used in Part 2. When they arrive, keep them in a separate container or in the plant aquarium.

9. **Plan to distribute organisms**
   The kit contains a fine-mesh collecting net to transfer animals from one place to another. A baster is provided for transferring scuds.

10. **Plan for observing and monitoring aquaria**
    In Step 7 of Guiding the Investigation, you will be introducing the class observation/recording sheets for monitoring the aquaria over the next 6-8 weeks. Devise a plan to fairly distribute the responsibility among the groups for recording weekly observations and daily feeding in the two aquaria. You might set up an early morning and a late afternoon observation schedule.

11. **Plan for the organisms' future**
    Think ahead about what will happen to the organisms at the end of the module. They can be put in a permanent class aquarium, or given to another class. They should not be released outdoors.

12. **Plan to read** *Science Resources*: **"Freshwater Environments"**
    Plan to read the article "Freshwater Environments" during a reading period after the active investigation for this part.

13. **Plan assessment: notebook entry**
    In Step 9 of Guiding the Investigation, students answer the focus question by describing the environmental factors in an aquatic system. After class, check students' notebook entries.

166 Full Option Science System

Part 1: Designing an Aquarium

GUIDING *the Investigation*
Part 1: *Designing an Aquarium*

1. **Review living and nonliving environmental factors**
   Distribute a copy of notebook sheet 8, *Living and Nonliving Factors*. Have students discuss the sheet in their groups and individually record responses in their notebooks. Then discuss the sheet as a class.

2. **Discuss aquatic environments**
   Ask students to name some **aquatic environments** they know of. Make a list on the board. Ask them to describe how those environments are similar to and different from one another and from terrestrial environments.

3. **Introduce the aquaria**
   Point out the two class **aquaria**. Tell students that each has gravel on the bottom and is filled with water. One is for the goldfish and the other one is for the guppies. Explain to students that the water was aged for 24 hours and chemically conditioned to take out the chlorine in order to make the water safe for fish.

4. **Observe the fish**
   Distribute a 1/2 L container with one fish to each group. Allow a few minutes for students to observe the fish in the containers. Ask them to describe the structures, movements, and behaviors of the fish in the same way they did with the mealworms. Two groups will have goldfish, and the rest of the groups will have guppies. If students don't recognize the fish, provide names for the fish.

5. **Place fish into their environments**
   Have groups come, one at a time, to the appropriate aquaria and gently pour in the fish. If either water level gets too high, have a student remove some water with a 1/2 L container and put it in the plastic jug or basin. Add the fish tunnels to the aquaria.

6. **Focus question: What are the environmental factors in an aquatic system?**
   Write or project the focus question as you say it aloud.

   ➤ *What are the environmental factors in an aquatic system?*

   Have students write the focus question in their notebooks. Ask students to name some of the environmental factors in the fish environments.
   - Aged or conditioned tap water
   - Gravel on the bottom

FOCUS QUESTION
What are the environmental factors in an aquatic system?

TEACHING NOTE
The discussion should include these ideas.
1. Soil and light are nonliving factors. Soil is sand and pebbles—it may have living things in it, but it is not alive.

2a. Water is necessary for plants to grow. Lizards and crickets need water.

2b. Temperature: Plants and animals need to be warm.

Students might also discuss air, soil, light, and space as nonliving.

Materials for Steps 3–6
- *Goldfish and guppies in 1/2 L containers*
- *Aquaria with gravel*
- *Fish food*
- *Fish tunnels*

EL NOTE
Discuss the root word aqua *and the suffix* tic.

Environments Module—FOSS Next Generation 167

INVESTIGATION 2 – Ecosystems

EL NOTE

Draw a diagram of the fish environment and label the factors as students name them.

TEACHING NOTE

You could have students use an electronic spreadsheet to keep track of the changes in the class aquaria.

Materials for Step 7
- Thermometers
- Self-stick notes
- *Class Aquarium Logs*
- *Aquarium Observation Logs*

aquarium
aquatic environment

- Light and heat
- Aquarium walls, plastic
- Gases and other materials in the water

Ask students to think of environmental factors that are missing from the fish environments. [Food.] In a natural environment, the food source would be other organisms—bits of plants and small animals. In the aquaria, the food source will be dry flakes of fish food. Show the container of food. A tiny, tiny pinch of food each day is enough (or twice a day). The large pieces can be ground up, so small guppies can easily consume the flakes.

7. Introduce the aquarium log

Distribute notebook sheet 9, *Aquarium Observation Log*, to each student. Project or display teacher master 12, *Class Aquarium Log*, which is similar to but not exactly the same as the student version. Explain that groups will take turns and share the responsibility of caring for the aquaria: daily feeding, checking the temperature, and monitoring the water conditions. A *Class Aquarium Log* will stay by each aquarium for easy reference. Two students in each group can attend to the goldfish and the other two students attend to the guppies. The group members can share their observations with each other and with the rest of the class through the class log.

Each individual student will keep his or her own log of observations in addition to the class log kept by each group. Students can record in their notebooks when it is their group's turn to attend to the fish. They could record at other times as well.

Have two students use thermometers to measure and report the water temperature. Discuss ways to describe the water conditions (clear, green, murky), and demonstrate the tiny pinch of fish flakes that should be given so fish will not be overfed. Have students make their first *Aquarium Observation Log* entry. Use self-stick notes to mark the water levels in each aquarium.

8. Review vocabulary

Add new words introduced in this part to the class word wall. Students might add the words to their notebooks.

9. Answer the focus question

Have students answer the focus question.

▶ *What are the environmental factors in an aquatic system?*

Students should include both living and nonliving factors and may add to their responses over time as they read about aquatic environments and monitor the class aquaria.

168 Full Option Science System

Part 1: Designing an Aquarium

10. **Assess progress: notebook entry**
    After class, look at students' answers to the focus question to check their understanding of factors in an aquatic environment.

    What to Look For

    - Living factors are organisms.
    - Nonliving factors include water, things dissolved in the water (gases and solid substances), temperature of the water, light, surfaces such as gravel or soil, and air temperature on the surface.

11. **Store the aquaria**
    Discuss the conditions surrounding the aquaria. Confirm with students that the aquaria are away from fluctuating temperatures and out of direct sunlight most of the time.

    > **TEACHING NOTE**
    > If some sunlight falls on the aquarium, that's OK. Sunlight will speed up the growth of algae, which turns the water green. This is a good thing and should stimulate student conversation.

    ### POSSIBLE BREAKPOINT

12. **Discuss other organisms**
    Ask,

    ➤ *What could we do with the aquaria to make them more like a natural water environment?*

    Materials for Steps 13–15
    - *Sprigs of elodea*
    - *Snails*
    - *Gammarus*
    - *Baster*

    Listen to students' responses to see if they suggest putting additional aquatic organisms into the aquaria to make a more complex environment for the fish. Students might add to their focus question answers over time as they read about aquatic environments and monitor the class aquaria.

13. **Enrich the aquatic environment**
    Have available the container with small pond snails and **elodea**. Have several students come to each aquarium and put in 5–6 snails and a few sprigs of elodea.

    > **EL NOTE**
    > Provide question frames for students to use in making entries. Here are some examples.
    > What happens when _____?
    > What would happen if _____?
    > How does _____ affect _____?

14. **Record notebook entries**
    Have students make entries in their science notebooks. Over the next weeks, they can describe what happens to the aquatic environment. Ask students to formulate questions about the new additions to the aquarium. Have them seek answers through observations. Have students write in their notebooks once a week.

15. **Add crustaceans to the aquarium**
    Many fascinating freshwater crustaceans will thrive in a classroom aquarium. One hardy crustacean that is big enough to observe is an amphipod, *Gammarus*. Point them out in the plant aquarium.

    > **TEACHING NOTE**
    > Over time, students will gain experience with the aquarium so they can formulate and justify predictions about what will happen in the systems.

    Students can use the baster to add *Gammarus* to the aquaria but be aware that goldfish love to eat tiny crustaceans. Adding crustaceans, however, will bring the organism interactions to life.

Environments Module—FOSS Next Generation

INVESTIGATION 2 – Ecosystems

EL NOTE

See the Science-Centered Language Development chapter for discussion protocols.

CROSSCUTTING CONCEPTS

Systems and system models

▶ **NOTE**
Go to FOSSweb for *Teacher Resources* and look for the Crosscutting Concepts—Grade 4 chapter for details on how to engage students with the concept of systems and system models.

SCIENCE AND ENGINEERING PRACTICES

Planning and carrying out investigations

TEACHING NOTE

These observations can be revisited after introducing **food chain** *in the next part. Students can describe the food chain in each aquarium.*

16. **Have a class discussion**
    Ask,

    ▶ *What is an example of a natural* **freshwater environment**? [A lake, pond, river, or stream.]

    ▶ *How is an aquarium similar to a natural freshwater environment? How is it different?* [Both have water, earth material on the bottom, plants, and animals.]

    ▶ *If you were to set up an aquarium again, what would you do differently and why?* [Set up an invertebrate community so it could support lots of crustaceans.]

17. **Introduce** *ecosystem*

    Ask students what the word **ecosystem** means to them. Call on several volunteers to offer their ideas. Build on what students say to define the word. Tell students,

    The root, eco, means home or habitat. System *refers to a group of factors or parts that work together as a whole. Putting the two parts of the word together, we have a habitat system.*

    Write the new word on the word wall. Tell students,

    An ecosystem is formed by the interaction of a community of organisms with its environment. The important word here is **interaction**. *An ecosystem is more than the nonliving and living factors that make up the environment. An ecosystem is also the interactions (how they act together) and the relationships between the environmental factors and the consequences of those interactions over time.*

    Ask students if they think that the aquarium could be called an aquatic ecosystem. Ask how the aquarium compares to the isopod terrarium. Have them discuss these topics in their groups.

18. **Observe changes in the aquaria**
    Students will learn that some changes to the environment are beneficial and some can be harmful to some of the organisms. Students might make these discoveries over time.

    - Elodea grows and forms roots.
    - Goldfish eat scuds (*Gammarus*).
    - Snails lay clear, jellylike egg masses on plants or on the sides of the aquarium.
    - Guppies reproduce; tiny guppies are seen at the surface.
    - The aquarium becomes a green-water aquarium (from the growth of microscopic algae).
    - Fish die.

170   Full Option Science System

Part 1: Designing an Aquarium

READING *in Science Resources*

19. Read "Freshwater Environments"
Read the article "Freshwater Environments," using the strategy that is most effective for your class. Tell students that this article will provide information on different types of natural freshwater environments and some of the living and nonliving factors that define them. They will be introduced to a group of organisms called plankton, which are very important in aquatic systems. Add plankton to the word wall.

20. Use a preview the text strategy
Have students preview the text by looking at and discussing the photographs, diagrams, and charts with a reading partner.

Refer students to the questions at the end of the article. Tell students that when reading informational text, it helps to have questions in mind to focus on what's important. Write the first question on the board and model how a reader would think about it.

I'm going to be looking for two categories of information: living and nonliving factors so I think I'll make a chart with those two headings in my notebook. And it says, "that define a lake's shallow-water zone" so I need to look at a particular area of a lake, not the whole lake.

Lake's shallow-water zone

Living factors	Nonliving factors

Continue modeling or have students discuss with a partner how they would look for information about the other two questions. [Number 2 is a relationship question; Number 3 is a compare and contrast question.] Have students set up these graphic organizers in their notebooks.

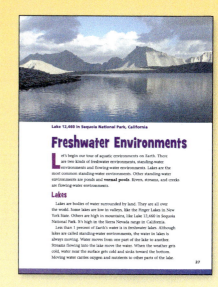

SCIENCE AND ENGINEERING PRACTICES

Obtaining, evaluating, and communicating information

ELA CONNECTION

These suggested reading strategies address the Common Core State Standards for ELA.

RI 5: Describe the overall structure of information in a text.

W 8: Gather relevant information from experiences and print, and categorize the information.

Environments Module—FOSS Next Generation

171

INVESTIGATION 2 – *Ecosystems*

ELA CONNECTION

These suggested reading strategies address the Common Core State Standards for ELA.

RI 3: Explain procedures, ideas, or concepts in a scientific text.

RI 7: Interpret information presented visually, and explain how the information contributes to an understanding of the text.

21. **Develop students' visual literacy**

    Read the first page aloud or have students read independently. Tell students to take turns summarizing the information with a partner and sharing any personal connections and/or questions they have.

    On the next page, pause and discuss the diagram of the lake. This is a good opportunity to help students develop their "visual literacy." Explain that they are looking at an analytic diagram. Ask students to discuss how this type of diagram is useful for presenting information. [It shows the relationship between an object and its parts.]

    Explain that this diagram also has a *cutaway view*. Ask students to discuss why this technique is useful. [It shows what's below the surface of the water, something we can't normally see.]

    Point out the circles in the diagram and ask students to discuss what they think they represent. Explain that the images of plankton are magnified. Explain,

    They are what we call microscopic because they are very, very tiny. What about the other images? They are also not drawn to scale. That means the organisms do not have the same size and shape relationship to one another as they would in real life.

    Have students reread the page and discuss how the diagram helps them understand the text. They can also make notes at this point about the environmental factors in their T-chart for question 1.

    Continue reading the next page aloud or have students read independently. Remind students that they are looking for the relationship between plankton and the other environmental factors in a freshwater environment. Have students discuss what they have learned about plankton with their small group and make a diagram in their notebooks that helps explain the role of plankton in a freshwater environment. (See example below.)

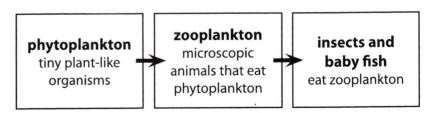

Full Option Science System

Part 1: Designing an Aquarium

22. Compare and contrast

Let students finish reading the article independently and then have them work with a partner to compare and contrast the lake and river environments using a double bubble map.

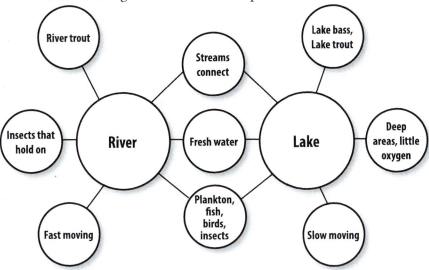

For reading strategies to support English learners and below-grade-level readers, see the Science-Centered Language Development chapter.

23. Have a sense-making discussion

Give students time to finish gathering information in their graphic organizers to answer the questions at the end of the article. Call on a Reporter from each group to share an answer to a question along with details and examples from the text. Allow other group Reporters to respond by either agreeing, disagreeing and offering alternative explanations or evidence, or adding on more information. After the discussion, ask students to refine or add to their original answers and record them in their notebooks.

➤ *What living and nonliving factors define a lake's shallow-water zone?* [Sunlight reaches the bottom; rooted plants grow in the muddy bottom; floating plants and algae; insect larvae, ducks, and small mammals may be there around the plants.]

➤ *What role do* **phytoplankton** *play in a freshwater environment?* [They are the "grass" of the lake environment and are eaten by microscopic animals.]

➤ *How are lake and river organisms different?* [Lake organisms don't have to deal with flowing water or currents. Organisms that live in flowing rivers have to be good swimmers or have structures that enable them to hold on to rocks. In the lower parts of rivers, the river water flows slowly. Plants and animals in the slower waters have structures that are more like lake organisms.]

ELA CONNECTION

These suggested strategies address the Common Core State Standards for ELA.

W 5: Strengthen writing by revising.

SL 1: Engage in collaborative discussions.

SCIENCE AND ENGINEERING PRACTICES

Constructing explanations

Environments Module—FOSS Next Generation 173

INVESTIGATION 2 – *Ecosystems*

ecosystem
elodea
freshwater environment
interaction
phytoplankton

24. Review vocabulary

Add new words introduced in this part to the word wall. Students should add the words to their notebooks. Have students add these words to their drawings of the aquarium or to a new diagram of a different freshwater environment.

WRAP-UP/WARM-UP

25. Share notebook entries

Conclude Part 1 or start Part 2 by having students share notebook entries. Ask students to open their science notebooks to the most recent entry. Read the focus question together.

➤ *What are the environmental factors in an aquatic system?*

Ask students to pair up with a partner to

- discuss their current thoughts about the answer to the focus question;
- compare the environmental factors in an aquatic environment to those in a terrestrial environment;
- discuss some of the structures that aquatic plants and animals have and how they function for the survival of the organism in the environment.

174   Full Option Science System

Part 2: Food Chains and Food Webs

MATERIALS for
Part 2: Food Chains and Food Webs

For each student
- 1 Notebook sheet 10, *Practice with Food Chains*
- 1 FOSS Science Resources: Environments
  - "What Is an Ecosystem?"
  - "Food Chains and Food Webs"

For each group
- 1 Set of food web cards for Woods Ecosystem, 20 cards per set with these organisms
  - American robin
  - Aquatic snail
  - Bacteria
  - Black bear
  - Brook trout
  - Chipmunk
  - Coyote
  - Dead plants and animals
  - Earthworm
  - Grama grass
  - Great blue heron
  - Green algae
  - Grouse
  - Hare
  - Mayfly
  - Pine trees
  - Red-tailed hawk
  - Scuds
  - *Tubifex* worm
  - Wild blueberry

For embedded assessment
- *Embedded Assessment Notes*

❏ Use the duplication master to make copies.

No. 10—Notebook Master

Environments Module—FOSS Next Generation

INVESTIGATION 2 – Ecosystems

GETTING READY for
Part 2: Food Chains and Food Webs

1. **Schedule the investigation**
   This part will take four to five sessions, two or three for active investigation and poster generation, and two for readings.

2. **Preview Part 2**
   Students work with organism cards to create food chains and food webs in a woodland ecosystem that includes terrestrial and aquatic environments. Students learn that by using the Sun's energy, plants and algae are the primary source of matter and energy entering most food chains and food webs. Students are introduced to the terms for different functional roles that organisms play in food chains. The focus question is **What are the roles of organisms in a food chain?**

3. **Check the Woods Ecosystem Food Web cards**
   Each group will need one set of 20 cards. Make sure that a complete set is in each zip bag.

4. **Plan for group work on breathing posters**
   Each group will research and design a poster showing how a particular terrestrial or aquatic organism "breathes" or gets oxygen into its body. Plan how students will work on these posters and the resources they will use.

5. **Plan to read *Science Resources*: "What Is an Ecosystem?" and "Food Chains and Food Webs"**
   Plan to read "What Is an Ecosystem?" and "Food Chains and Food Webs" during reading periods after completing this part.

6. **Plan assessment: notebook entry**
   In Step 12 of Guiding the Investigation, students answer the focus question by describing the environmental factors in an aquatic system. After class, check students' notebook entries.

Part 2: Food Chains and Food Webs

GUIDING *the Investigation*
Part 2: Food Chains and Food Webs

1. **Review *ecosystem***
   Ask students to define the term *ecosystem*. Remind students that it is more than the living and nonliving parts of the environment. It also involves how things interact in the system. An ecosystem is all about the interactions among the organisms and the nonliving factors in the environment.

2. **Focus on feeding interactions**
   Ask students to consider the aquaria. Ask,

   ➤ *How do the organisms interact in the aquaria?*

   If students don't mention feeding interactions, ask them to think about what eats what in each aquarium. Ask them to consider what eats what in the terrariums they set up in Investigation 1. Remind them that getting food is an interaction.

3. **Introduce food web cards**
   Hold up a set of cards. Tell students that each group will get a set of these 20 cards. Each card describes an organism (plant or animal) that lives in a woods or forest ecosystem. Some of the organisms live on the land and some live in the water. Each card has

   - A photograph of the organism
   - Its common name
   - Information about where it lives (natural history)
   - A list of what it eats (food)
   - A list of what eats it (**predators**)

4. **Study the organisms**
   Tell students that they will get to know the ecosystem by first getting to know the organisms that live there. They should spread out the cards, and each member of the group should take five cards to study. Once everyone has had a chance to study their five organisms, the group should look for pairs of organisms that go together because one organism eats the other. Give the groups about 15 minutes to study the organisms and find feeding pairs.

5. **Identify feeding interactions**
   After students have identified pairs of organisms, call for attention. Call on one student to identify one feeding pair—one organism that eats a second organism. Then pose another challenge.

   ➤ *<Student's name> says the hare eats the grass. Do you think some other organism eats the hare? Can you put three organisms in a row to show a larger feeding relationship?*

FOCUS QUESTION
What are the roles of organisms in a food chain?

CROSSCUTTING CONCEPTS
Structure and function

> **TEACHING NOTE**
>
> Students should focus on structures and functions of the organisms that assist in getting food and protection from predators.

Materials for Step 4
- *Woods Ecosystem Food Web cards*

SCIENCE AND ENGINEERING PRACTICES
Obtaining, evaluating, and communicating information

Environments Module—FOSS Next Generation

INVESTIGATION 2 – Ecosystems

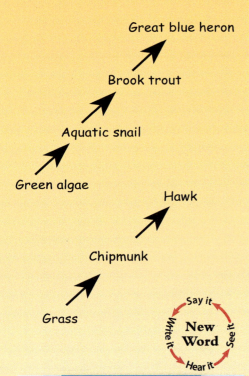

TEACHING NOTE

The direction of the arrow goes from the organism that is eaten to the organism that eats it. The arrow shows the direction of the flow of matter and energy (food) from organism to organism.

CROSSCUTTING CONCEPTS

Energy and matter

▶ **NOTE**
Go to FOSSweb for *Teacher Resources* and look for the Crosscutting Concepts—Grade 4 chapter for details on how to engage students with the concept of energy and matter.

Ask each group to look for three organisms that might have a feeding relationship. Students should put three cards in a row to show how each organism is eaten by the next organism in the row.

6. **Introduce** *food chain*

   Call on a few students to name the organisms they selected. Use an example of a feeding interaction that begins with a plant or algae, for example: grama grass, chipmunk, red-tailed hawk.

   Write the three organism names on the board with the grass on the left, the chipmunk in the middle, and the hawk on the right. Leave room between them to draw arrows.

   Ask students which organism eats the other. After they reply, describe the path the food takes in the ecosystem.

   The grama grass is eaten by the chipmunk. The chipmunk is eaten by the red-tailed hawk. I can draw arrows to show this feeding interaction. I will draw an arrow from the grass to the chipmunk. I'll draw a second arrow from the chipmunk to the hawk.

   *The path that food takes from one organism to another organism is called a **food chain**. The direction the arrow points shows the direction food (matter and energy) moves through a food chain.*

   Draw the arrows to illustrate the food chain.

   grama grass ⟶ chipmunk ⟶ red-tailed hawk

7. **Introduce** *producer*

   Ask,

   ➤ Which organisms in this food chain do not eat any other organisms? [The grass.]

   ➤ How can grass survive without eating? [Grasses are plants. Plants make their own food.]

   Tell students,

   *Organisms that make food are **producers**. Producers need **energy** from the Sun to make the food that is used in the ecosystem. Producers need sunlight, water, and carbon dioxide to make food. Plants, **algae**, and other organisms that make their own food are the primary source of matter and energy entering most food chains. In a food chain, energy of food moves from one organism to the next.*

   Ask students to find all the cards that are producers in the woods ecosystem. Make a list of those on the board under the heading "producers."

   Add the new words to the word wall.

Full Option Science System

Part 2: Food Chains and Food Webs

8. Introduce *consumer*
Tell students,

Producers like grass, pine trees, berry plants, and algae make their own food. Animals in the ecosystem don't make their own food.

▶ *How do animals get their food?* [They eat other organisms.]

Tell students,

Organisms that eat other organisms are **consumers**. *Consumers are dependent on other organisms for food. Animals that eat plants are called* **herbivores**. *Animals that eat other animals are called* **carnivores**. *Animals that eat plants and animals are called* **omnivores**.

Ask students to find cards that are consumers in the woods ecosystem. Make a list of some of those on the board under the heading "Consumers."

NOTE: Provide this vocabulary lesson. *Vore* means "one that eats." Combining *vore* with *herbi*, *carni*, and *omni* describes an animal that eats certain foods. *Herb* means "plant." Herbivores eat plants. *Carni* means "flesh" or "meat." Carnivores eat animals. *Omni* means "all." Omnivores eat all—plants and animals.

9. Introduce *decomposer*
Tell students,

Some organisms are not eaten by consumers, but instead die a natural death. Dead organisms are broken down and consumed by organisms called **decomposers**. *Organisms like bacteria and earthworms are decomposers. Everything that is not eaten by a consumer is eventually used for food by decomposers. It is recycled (or returned) to the environment as raw materials.*

Write *decomposer* on the board and under it write *earthworm* and *bacteria*.

10. Practice with food chains
Ask students to use the cards and the list of organisms on notebook sheet 10, *Practice with Food Chains*, to build several food chains of at least three organisms. Have students record their food chains in their notebooks, not on the sheets, and use arrows to show how the energy of food moves from organisms that are eaten to organisms that eat them. Finally, ask students to label the organisms as producers or consumers.

11. Review vocabulary
Review key vocabulary introduced in this part, and refer to the words on the word wall.

Environments Module—FOSS Next Generation

Producers   **Consumers**
grama grass   black bear
green algae   brook trout
pine trees   chipmunk
wild blueberry   coyote
                     great blue heron
                     grouse
                     hare
                     mayfly
                     red-tailed hawk

Decomposers
earthworm
bacteria

algae
carnivore
consumer
decomposer
energy
food chain
food web
herbivore
omnivore
predator
producer

INVESTIGATION 2 – Ecosystems

EL NOTE

For students who need scaffolding, provide a sentence frame such as, *In a food chain there are _____, and_____, and _____. The role of each is to _____. For example, _____.*

SCIENCE AND ENGINEERING PRACTICES

Analyzing and interpreting data

TEACHING NOTE

Have students draw this food web in their notebooks.

Students may suggest that the decomposers (bacteria) should be beside the food web with arrows from all the organisms.

SCIENCE AND ENGINEERING PRACTICES

Developing and using models

12. **Focus question: What are the roles of organisms in a food chain?**

    Write or project the focus question as you say it aloud.

    ➤ *What are the roles of organisms in a food chain?*

    Have students write the focus question in their notebooks and respond to the question. Students should describe the roles of producers, consumers, and decomposers.

13. **Introduce competition for resources**

    Ask students if they found that two different organisms eat the same food. Ask which ones they were.

    Students may find that
    - The coyote and hawk eat the same animals.
    - The chipmunk, hare, and grouse eat the same food.

    Tell students that the organisms that eat the same food are **competing** for a food resource. (Add the new word to the word wall.) There is competition for resources in ecosystems. Ask what might happen if the food source is in short supply.

14. **Draw a food web**

    Summarize by telling students that if you record all the feeding relationships in an ecosystem, drawings get much more complicated than a straight-line food chain. If you want to show there are three different animals eating the grass, and that there is more than one organism eating each of them, then you are drawing a **food web**. Add the words to the word wall.

    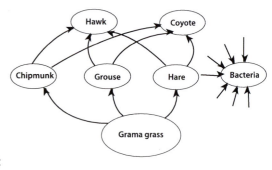

    Ask the class to help you draw a food web. The grouse, hare, and chipmunk might all eat the grass, for example.

15. **Discuss decomposers**

    Ask students what happens to dead organisms in the food web.

    ➤ *Which organisms have arrows to the decomposers? Just the top consumers or all the organisms?* [All.]

16. **Clean up**

    At the end of the session, have each group bag up the cards and have Getters return them to the materials station.

180   Full Option Science System

Part 2: Food Chains and Food Webs

READING *in Science Resources*

17. Read "What Is an Ecosystem?"

Read the article "What Is an Ecosystem?," using the strategy that is most effective for your class. This article discusses the term *ecosystem*, relates it to the terrariums and aquaria that students are observing, and introduces the idea of matter and energy in these systems. For reading strategies to support English learners and below-grade-level readers, see the Science-Centered Language Development chapter.

18. Use a concept definition map

Start by letting students preview the text by looking at and discussing the photographs and the diagram. Have them read the questions to help them focus on the main ideas of the text.

Read the title aloud and say,

We know this article will give us information about ecosystems. Let's make a concept definition map to help us better understand this concept.

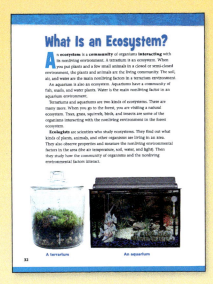

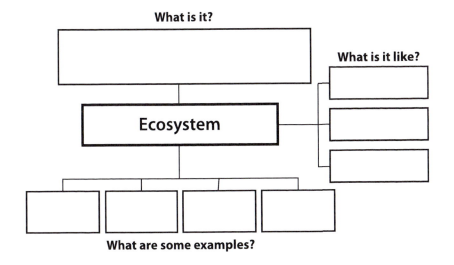

Draw the outline on the board and have students make one in their notebooks. Start with the word "ecosystem" in the middle box.

▶ *What questions should our definition answer?* [What is it? What are examples of it? What's it like? What are its characteristics?]

Tell students to look for this information as they are reading the text.

After reading the first page, have students pause and fill in what they know so far about an ecosystem on their maps. Read the rest of the article aloud or have students read independently.

ELA CONNECTION

These suggested reading strategies address the Common Core State Standards for ELA.

RI 1: Refer to details and examples in a text when explaining what the text says explicitly and when drawing inferences from text.

RI 4: Determine the meaning of general academic domain-specific words or phrases.

W 8: Gather relevant information from experiences and print, and categorize the information.

SCIENCE AND ENGINEERING PRACTICES

Obtaining, evaluating, and communicating information

Environments Module—FOSS Next Generation

INVESTIGATION 2 – Ecosystems

CROSSCUTTING CONCEPTS

Energy and matter

ELA CONNECTION

These suggested reading strategies address the Common Core State Standards for ELA.

RI 3: Explain procedures, ideas, or concepts in a scientific text.

RI 7: Interpret information presented visually, and explain how the information contributes to an understanding of the text.

---

19. Introduce a crosscutting concept

Point out that energy and matter is also a crosscutting concept—a big idea in science that helps us explain the world around us. Ask students to discuss their understanding of energy and matter with a partner using the text as evidence to support their ideas.

Refer students to the photosynthesis diagram and explain that this is a synthetic diagram, an illustration that shows connections. Ask students to discuss what they think the arrows mean and to use the diagram to take turns explaining to their partner how plants get food. If there is disagreement, students should refer back to the text to find evidence to support their ideas.

20. Discuss the reading

Give students time to finish their concept definition maps and to refine their ideas about the first two questions at the end of the article. Call on Reporters to share their group's answer to a question, along with details and examples from the text. Allow other group Reporters to respond by either agreeing, disagreeing and offering alternative explanations or evidence, or adding more information. After the discussion, let students revise or add to the original answers and record them in their notebooks.

▶ *How do plants and animals get the food they need to survive?* [Plants make their own food using sunlight, water, and carbon dioxide; animals eat organisms.]

▶ *Explain how energy from the Sun helps animals survive.* [Plants need energy from the Sun, water, and carbon dioxide to make food (sugar). They grow and become food for animals.]

▶ *What is an ecosystem?* [The interactions of the living and nonliving components in an environment.]

21. Read "Food Chains and Food Webs"

Read "Food Chains and Food Webs," using the strategy that is most effective for your class. This article provides more information about the food chains that students have been working with.

Have students preview the text by looking at and discussing the photographs and diagrams. Encourage students to ask themselves questions such as, "What do I notice? What does this image remind me of? What is this trying to show me?" Refer students to the food chains diagram on page 39. Explain that this is an example of a linear diagram. Ask students to take turns interpreting the diagram with their partner. What do the arrows represent? What is the relationship between the organisms? Refer students to the food webs diagram and ask students to discuss how this diagram is different from the food chain diagram.

182

Full Option Science System

Part 2: Food Chains and Food Webs

22. Use an anticipation guide strategy

To help facilitate text comprehension, use an anticipation guide before reading the text. Below are possible answers to the six questions at the end of the article. The possible answers will either support or challenge students' own ideas. Write or project them on the board and have students determine which ones they agree with and which ones they disagree with. Have them share their responses and their reasoning in their small groups.

1. Food is energy.
2. Plants don't need food.
3. Producers provide food for consumers.
4. Decomposers are the ecosystem's cleanup crew.
5. Animals compete with each other for food.
6. Consumers can always find food.

Next, tell students to read through the text looking for information related to each of the statements. Have them make a chart in their notebook to record evidence that either supports their ideas or refutes their ideas for each statement.

	Anticipation Guide	
	Supports my idea	**Refutes my idea**
1		
2		
3		
4		
5		
6		

23. Have a sense-making discussion

Allow time for students to share their responses and evidence from the text in their small groups and to come to a consensus. Review the statements as a whole class. Encourage students to use argumentation language structures such as, "I agree _____ because _____ ; I disagree _____ because _____ ; My evidence is _____ ; I thought _____ , but now I think _____ ."

ELA CONNECTION

These suggested reading strategies address the Common Core State Standards for ELA.

RI 1: Refer to details and examples in a text when explaining what the text says explicitly and when drawing inferences from text.

RI 7: Interpret information presented visually, and explain how the information contributes to an understanding of the text.

SCIENCE AND ENGINEERING PRACTICES

Obtaining, evaluating, and communicating information

CROSSCUTTING CONCEPTS

Systems and system models
Energy and matter

INVESTIGATION 2 – Ecosystems

Below are the correct answers students should be able to explain using details and examples from the text.

➤ *What is food? Why is it important?* [Food is a source of matter and energy, which all organisms need in order to survive. Food provides the raw materials animals need in order to grow and reproduce.]

➤ *Do plants need food? Why or why not?* [Yes, all organisms need food. But plants make their own food with energy from the Sun, carbon dioxide, and water.]

➤ *What is the role of producers in an ecosystem?* [Producers make their own food. Producers provide the energy and matter for consumers.]

➤ *Look at the food web for a freshwater river. Give three examples of animals that compete for a food source.* [Crayfish, bass, and egrets compete for minnows.]

➤ *What is the role of decomposers in an ecosystem?* [Decomposers are the ecosystem's cleanup crew. They break down dead plant and animal matter into simple chemicals. The simple chemicals are recycled in the environment.]

➤ *How might a forest fire affect the food web in a forest?* [A forest fire might destroy the producers so the consumers that eat plants would have nothing to eat. This would also starve the next level of consumers.]

24. **Read more about food chains**

    This is a good opportunity for students to learn more about food chains by reading another text. See FOSSweb for a list of recommended books. One suggestion is *Poop-Eaters: Dung Beetles in the Food Chain* by Deirdre A. Prischmann.

BREAKPOINT

ELA CONNECTION

This suggested reading strategy addresses the Common Core State Standards for ELA.

RI 9: Integrate information from two texts on the same topic.

Part 2: Food Chains and Food Webs

25. Return to concept grid for structures

In Investigation 1, Part 3, students organized the information about the structures of animals, the functions of the structures, and how the structures help the animals to survive. Now is a good time for students to return to that concept grid (or to begin another grid) to include the organisms they have studied from the woods ecosystem using the food web cards. There will be overlap between the organisms they have studied in the classroom aquaria and the cards.

Begin by reviewing the main needs of most animals.

▶ *What are the three main needs of most animals?* [Oxygen, food, and water.]

▶ *How do animals use oxygen, food, and water?* [Animals transform their food into energy, and the energy into activities that are necessary for survival. Water and oxygen are needed in those processes.]

▶ *What other things do most animals need?* [Keep their body temperature stable or within good range; space to get resources; habitat, shelter, or home; and protection.]

Have students turn in their notebooks to the concept grid they developed at the end of Investigation 1.

Suggest that students, working in their groups, add to the concept grid to keep track of the new animals studied, their structures, and how those structures help the animals survive in their environment, whether it be a terrestrial or aquatic environment. The structures can be external or internal.

Animal	Structure (external or internal)	Function of structure	How structure helps in survival
mammal, bear	claws and strong legs	running, walking, climbing trees	move to food sources

CROSSCUTTING CONCEPTS

Structure and function

▶ **NOTE**
Go to FOSSweb for *Teacher Resources* and look for the Crosscutting Concepts—Grade 4 chapter for details on how to engage students with the concept of structure and function.

Environments Module—FOSS Next Generation

INVESTIGATION 2 – Ecosystems

The woods animals can be sorted into groups. You can ask students to select one organism from each of the categories.

- Mammal: bear, rabbit, coyote, chipmunk
- Bird: robin, heron, hawk, grouse
- Fish: trout
- Mollusk: aquatic snail
- Annelid: earthworm, tubifex worm
- Insect: mayfly
- Crustacean: scuds

Students may need to get more information from other sources to answer their questions about the animals and their structures.

If a group finishes early, you could ask them to add other groups of animals such as reptiles (snakes or lizards) or amphibian (frogs or salamanders).

26. Discuss basic need for oxygen

Remind students that one of the main needs of animals is getting oxygen. Ask,

▶ *What kinds of external structures are involved in getting oxygen?* [Mouth, nose, external gills, or gill openings.]

▶ *What kinds of internal structures are involved in getting oxygen?* [Students may say tubes (trachae) and lungs in humans and mammals; gills in fish and isopods.]

Students may not know that birds, reptiles, and amphibians also have a lung system, different from mammals. In addition, amphibians may have gills in larval stages. This will give students a chance to speculate on the systems and then research one system to gather information to share.

27. Have groups research one breathing system

Make a table on the board listing eight different types of animals (mammal, bird, fish, frog, snake, earthworm, insect, and crustacean). Assign each group one type of animal and have them make a diagram with labels on chart paper (a poster) that includes:

- the external and internal parts of the system for breathing—getting oxygen into the body;
- generally how the system functions to move the oxygen to parts of the body and where waste gases are expelled.

Students need to obtain information and decide how to display (communicate) the information for others to evaluate. Depending on the resources available (books, computers, videos), this might take 20–30 minutes to gather and describe the system.

CROSSCUTTING CONCEPTS

Systems and system models

▶ **NOTE**
Go to FOSSweb for *Teacher Resources* and look for the Crosscutting Concepts—Grade 4 chapter for details on how to engage students with the concept of systems and system models.

SCIENCE AND ENGINEERING PRACTICES

Obtaining, evaluating, and communicating information

Part 2: Food Chains and Food Webs

POSSIBLE BREAKPOINT

28. Do a gallery walk
Hang the posters on the wall and conduct a gallery walk. Give students a set of self-stick notes and have them rotate from poster to poster in their groups. For each diagram they can add a comment or question with their self-stick notes. Allow time for each group to examine the questions and comments.

29. Have a sense-making discussion
As a class, review the questions and comments on the posters. Make a table of all the animals on the posters and have students classify the main structures for getting oxygen: lungs, gills, skin, openings in the hard body covering, or a combination of those structures.

30. Compare two breathing methods
Have students select two animals that get oxygen in different ways and compare the two systems. They should write their comparison in their notebooks. Students don't have to go into detail, but describe the main components of the system and their functions. Encourage them to include diagrams.

31. Assess progress: notebook entry
After class, look at students' comparison of two methods for getting oxygen. We have provided detailed illustrations for you but we don't expect the student diagrams to be this complete.

What to Look For

- *Lungs. Terrestrial organisms breathe air to get oxygen. Air comes in through mouth and nose, goes through tubes to the lungs where the oxygen goes into the blood stream and is carried to all parts of the body. Waste gas is expelled from the lungs through the mouth and nose. The heart pumps blood to and from the lungs and the body.*

TEACHING NOTE

Go to FOSSweb for Teacher Resources and look for the Science and Engineering Practices—Grade 4 chapter for details on how to engage students with the practice of obtaining, evaluating, and communicating information.

SCIENCE AND ENGINEERING PRACTICES

Obtaining, evaluating, and communicating information

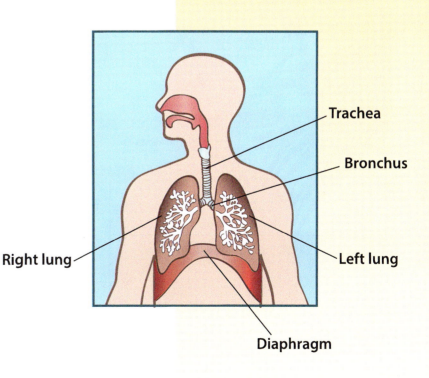

Environments Module—FOSS Next Generation

INVESTIGATION 2 – Ecosystems

- *Gills.* Many aquatic organisms (like fish) have gills, not lungs. Water comes in through the mouth and flows through the gills. The gills function like lungs to transfer oxygen to the blood to be carried to other parts of the body. The water then passes out of the fish body through the gill slits. Gills only work in water.

- *Skin.* Adult frogs and worms get oxygen into their blood right through their moist skin. The oxygen is carried to the body pumped by a heart. Adult frogs also have lungs. Frog larvae (tadpoles) have gills, not lungs, that function like fish gills.

- *Spiracles.* Insects (adults and larvae) have spiracles, holes in the body wall, where air comes in. They have an open circulatory system and a type of heart that pumps a blood-like fluid with oxygen around the body.

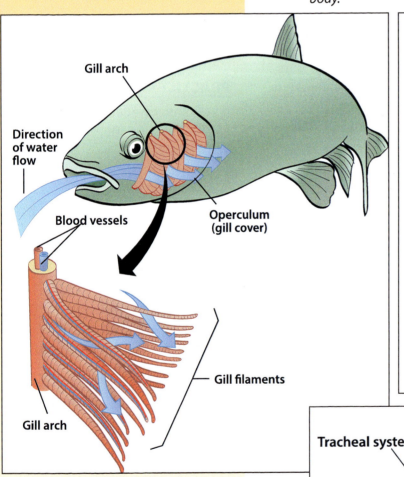

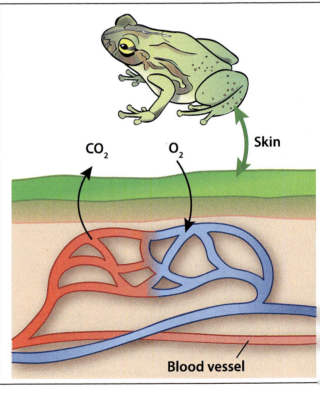

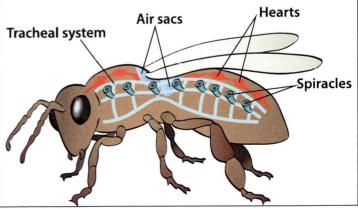

Part 2: Food Chains and Food Webs

WRAP-UP/WARM-UP

32. Share notebook entries
Conclude Part 2 or start Part 3 by having students share notebook entries. Ask students to open their science notebooks to the most recent entry. Read the focus question together.

➤ *What are the roles of organisms in a food chain?*

Ask students to pair up with a partner to
- discuss their current thoughts about the answer to the focus question;
- discuss examples of food chains.

33. Engage in argumentation
Make four word cards or pictures for: Sun, Producer, Consumer, Decomposer. Post one in each corner of the room. Ask students,

➤ *Where does a fox get its food?*
- From the Sun
- From producers
- From consumers
- From decomposers

Have students go to one of the four corners to discuss their ideas with the other students at that corner. Do they agree or disagree? What is their reasoning? Encourage students to refer to their notebooks and the text for evidence to support their ideas.

ELA CONNECTION

This suggested reading strategy addresses the Common Core State Standards for ELA.

SL 1: Engage in collaborative discussions.

SCIENCE AND ENGINEERING PRACTICES

Engaging in argument from evidence

TEACHING NOTE

See the **Home/School Connection** for Investigation 2 at the end of the Interdisciplinary Extensions section. This is a good time to send it home with students.

Environments Module—FOSS Next Generation

INVESTIGATION 2 – Ecosystems

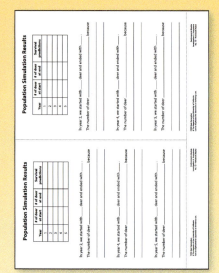

No. 11—Notebook Master

MATERIALS for
Part 3: Population Simulation

For each student
- 3 Zip bags, 1 L
- 16 Food markers, about 3 cm to 5 cm long (See Step 5 of Getting Ready.) ★
- ❑ 1 Notebook sheet 11, *Population Simulation Results*
- 1 *FOSS Science Resources: Environments*
  - "Human Activities and Aquatic Ecosystems"
  - "Comparing Aquatic and Terrestrial Ecosystems"

For the class
- Boundary marker flags
- 1 Basin
- 1 Minute timer ★
- 1 Marking pen ★
- 1 Portable whiteboard or chart paper and cardboard backing ★
- 1 Whistle ★

For embedded assessment
- ❑ • *Embedded Assessment Notes*

★ Supplied by the teacher.   ❑ Use the duplication master to make copies.

Part 3: Population Simulation

GETTING READY for
Part 3: Population Simulation

1. **Schedule the investigation**
   This part will take one active outdoor session and one reading session. Encourage students to wear sneakers to go outdoors.

2. **Preview Part 3**
   Students go to the schoolyard to simulate a population of deer foraging for food in its home range. Students are introduced to the concept of carrying capacity, the greatest number of organisms that can be supported (carried) by an area without damaging it. The focus question is **How does food affect a population in its home range?**

3. **Select your outdoor sites**
   Select a site with a large level or gently sloping natural surface (grass is ideal, but dirt will do). If you have only asphalt, encourage students to imagine they are in a natural setting. You will need one home-range area and two new-range areas. Mark the home-range area boundaries with the boundary markers. Do not visibly mark the new-range areas as the "deer" will need to discover these feeding areas during a food shortage. Try to have the new-range and home-range areas at least 10 meters (m) apart or around a corner so that the new-range area is out of sight.

4. **Check the site and set home-range area**
   Tour the outdoor site on the morning of your outdoor activity. Do a quick search for potentially distracting or unsightly items.

   Establish the home-range area (about 10 m × 10 m) with boundary marker flags at the edges.

5. **Prepare food markers**
   Each student will need 16 food markers—or about 520 food markers for a class of 32 students. You can use permanent materials for the markers, such as 5 cm lengths of white or colored plastic straws (not clear ones), or you can use organic materials, such as uncooked large elbow macaroni, rotini, or rigatoni. Based on your situation, decide which material will work best. If you are using straws, cut them in fourths.

   The food markers will need to be distributed in the home-range and new-range areas before going outdoors (see the next step). As part of the simulation, students will redistribute them after each "year." By the end of the activity, most markers will have been picked up, but students can make a final sweep to be sure that all have been retrieved.

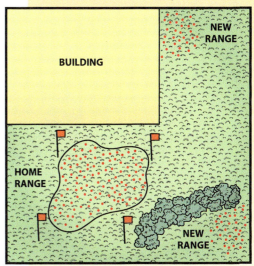

▶ **NOTE**
You could also use poker chips or bingo chips for food markers.

Environments Module—FOSS Next Generation

INVESTIGATION 2 – Ecosystems

> **NOTE**
> Do not mark the new-range areas with boundary marker flags. Only mark the boundaries of the home-range area.

Year	# of deer at start	# of deer at end	
1			
2			
3			
4			
5			

6. **Distribute food markers before going outdoors**
   Count out ten home-range food markers per student and place them in a bag. Before taking students outdoors, distribute these food markers in the home range. It is important to mark the boundaries, so that students know where to redistribute the food markers.

   Count out six more food markers per student. Scatter half of them in one new range and the other half in the second new range. These two areas will serve as new feeding areas in the fifth year of the simulation.

7. **Prepare recording table**
   During the simulation, record and display the number of deer on a piece of chart paper or portable whiteboard. If necessary, you could use a large, flat cardboard box. Set up the table indoors. Leave room for another column on the right that will be used during the third year of the simulation for predictions.

8. **Enlist additional adults**
   If possible, seek out an additional adult to join you outdoors. Remind the adult that students will need time to struggle with the challenges and that his or her job is to lend support but not to solve the problems for students.

9. **Plan to review outdoor safety rules**
   Before going outdoors, preview the rules, boundaries, and expectations for the outdoor work.

   Students will be most comfortable if you conduct this activity during mild, dry weather. They will be running around, so they should dress for active outdoor movement and wear sneakers.

10. **Plan to read *Science Resources*: "Human Activities and Aquatic Ecosystems" and "Comparing Aquatic and Terrestrial Ecosystems"**
    Plan to read these two articles during a reading period after completing the active investigation.

11. **Plan for online activities**
    Students can gain more experience comparing aquatic and terrestrial ecosystems by working with the "Virtual Aquarium" and "Virtual Terrarium" on FOSSweb. Preview the simulations by going to the Resources by Investigation section on FOSSweb.

12. **Plan to assess progress: notebook entry**
    In Step 19 of Guiding the Investigation, students answer the focus question. Check that students understand carrying capacity.

Part 3: Population Simulation

GUIDING *the Investigation*
Part 3: *Population Simulation*

1. **Review organism roles in food chain**
   Ask students to name some examples of producers and to recall where producers get their energy. [Plants get their energy from the Sun.] Ask students to name the types of organisms that eat producers. [Consumers: herbivores and omnivores.]

2. **Introduce the simulation**
   Explain to students that they are going to take part in a simulation activity that involves producers and consumers. The simulation is a kind of model to better understand food supply in a food chain. Hold up a plastic bag and explain that each bag represents one deer. Deer eat plants, which are represented by food markers (hold up the food marker). Make sure students know that a deer is a consumer and that they eat the plants, the producers.

   Say,

   The bags you will get each represent a deer in a deer population. All members of a population live and feed in the same home range. During the simulation, you will forage for food in your home range and attempt to gather at least five food markers, but not more than ten, in order to survive for a year. One minute will be equal to one year.

3. **Introduce the recording table**
   Show students the recording table and orient them to the headings. Explain that half the class will be deer in year 1 (record the number of deer in the population for the start of year 1) and the whole class will be deer in year 2 (record that number on the table).

4. **Focus question: How does food affect a population in its home range?**
   Write or project the focus question on the board as you say it aloud. Have students write the focus question in their notebooks.

   ▶ *How does food affect a population in its home range?*

5. **Review outdoor safety and expectations**
   Remind students of the rules and expectations for the outdoor work. Explain that this will be a simulation with running and excitement but that it is still science class.

6. **Go outdoors**
   Gather the materials and distribute a bag to each student. The food markers should already be distributed in all three ranges. Head outdoors, and avoid mentioning or passing the new-range areas since the deer population will need to "find" those in year 5.

FOCUS QUESTION
How does food affect a population in its home range?

▶ **NOTE**
Only half the class will be deer in year 1. The whole class will be deer in year 2.

Materials for Step 6
- *Zip bags*
- *Timer*
- *Chart*
- *Marking pen*
- *Whistle*
- *Basin*

Environments Module—FOSS Next Generation   193

INVESTIGATION 2 – Ecosystems

> **TEACHING NOTE**
>
> Tell students that collecting more than 10 pieces of food is cheating in this simulation. Each deer should collect only 5 to 10 food markers.

SCIENCE AND ENGINEERING PRACTICES

Planning and carrying out investigations.

▶ **NOTE**

All the deer in year 1 should survive as there is plenty of food for the population. If any deer didn't survive, have them join in year 2 as a deer offspring.

▶ **NOTE**

Another method to redistribute the food markers in the home range is to have students put the food in a basin and have an adult spread the food randomly in the home range. You don't want the students to form piles of the food markers.

7. **Review the rules of the simulation activity**

   Gather in a circle near the home range. Remind students that in order for each deer to survive a year, it must obtain at least five food markers, but not more than ten in their bag. Make sure students know the boundaries of the home range.

   Divide the class in half. One half will start year 1; the other half will watch during year 1 and join the simulation in year 2.

8. **Simulate year 1**

   Gather students at the edge of the home range, set the timer for 1 minute, and let the year 1 deer start foraging (half the class). If you find 1 minute is more time than necessary, you may reduce the time available for foraging. When deer have finished eating (or after 1 minute), all students should return to the gathering spot.

   Have the deer from year 1 remain separated from the other half of the class while you explain,

   a. *Count your food markers. If you have 5–10 markers, remain standing. If you have 4 or fewer markers, squat down (or fall down as if to die).*

   b. *Count how many deer did NOT get enough food markers to survive, and let's record on the table the number of survivors in year 1.*

   Allow time for students to do the math for the group and to decide what you should record on the table.

   c. *Any deer with food should redistribute the food markers in the home range, simulating the regrowth of the food resource. Be sure to spread the food out widely (don't make little piles of food).*

   Explain that the number of deer will increase through reproduction. One deer will be added to the population for every deer that survived. This is when the rest of the class joins the deer population. In year 2, all students engage in the simulation.

9. **Simulate year 2**

   Point out on the data table the recorded number of deer in the population at the start of year 2. (This was entered during Step 3.)

   Gather students at the edge of the home range, set the timer for 1 minute, and remind students that they need to collect at least five food markers in each bag. Let them forage.

   At the end of year 2, if deer have four or fewer pieces of food, they can squat down in the middle of the group. Count the casualties and record the number of surviving deer.

   Have students redistribute all the food markers in the home range. Students should check that they are being distributed randomly.

Part 3: Population Simulation

10. ## Simulate year 3
    To begin year 3, double the surviving population from the end of year 2. There won't be more students to add to the population as deer so each student should pick up a second bag and forage for two deer. There will be a lot of deer!

    Record the number of deer in the population at the start of year 3. Ask students to predict how many deer they think will survive year 3. Add a column labeled "Survival predictions." Quickly record some predictions.

    Start year 3 foraging. Stress that sharing or pooling of food markers by different players is not permitted—deer don't share their food.

    At the end of a minute of foraging for food in year 3, gather students together. Ask students to squat down if they are holding bags for one or more deer that didn't survive. Ask these students to hold up the number of bags that don't have at least five pieces of food. Count casualties and record survivors on the data board. Do not redistribute the food markers in the home range yet.

Year	# of deer at start	# of deer at end	Survival predictions
1	17	17	
2	34	34	
3	68		60, 65, 68
4			
5			

▶ **NOTE**
Do NOT redistribute the food markers in the home range yet. Introduce overbrowsing and redistribute at the end of Step 11.

11. ## Introduce overbrowsing during year 3
    Announce that during year 3, the large number of deer in the population meant that they overbrowsed many of the food plants. Explain that you will remove approximately one-fourth of the food markers. Do this by collecting one-quarter of the bags filled with food markers and place the food (only the food, not the bags) into a basin. This food is no longer available to deer in the home range.

    Now, have students redistribute the remaining food markers in the home range.

    TEACHING NOTE
    Clarify that overbrowsing means the deer harmed the plants by eating too much of the plant or too many young plants so the plants could not regrow for the next year.

12. ## Simulate year 4
    To begin year 4, double the ending population for year 3 and distribute bags appropriately. Remember, you always know how many bags are already in the hands of students—it is always the same as the previous year's starting deer population.

    Record the number of deer in the population that are entering year 4 and remind students of the damaged (reduced) food supply. Ask them how many deer they think will survive now. Record a few predictions on the table.

    Set the timer and let year 4 begin. At the end of year 4, count the casualties, record the number of surviving deer, and briefly compare the number of survivors with the group's predictions.

    Have students redistribute the same number of food markers as they had in year 4.

▶ **NOTE**
All students should participate with at least one bag in year 4 even if they don't have two surviving deer. Redistribute bags as necessary.

Environments Module—FOSS Next Generation

INVESTIGATION 2 – Ecosystems

13. Introduce new ranges for year 5

Ask students what they think a herd of deer might do when its home range no longer produces enough food for the herd to survive. If necessary, introduce the idea of expanding the feeding range beyond the home range.

Tell students that in this year, after all the available food in their home range has been eaten, they may disperse and search for new feeding areas. Do not tell them where the additional food sources are. Once one student finds the new food source, the rest of the students will quickly follow.

14. Simulate year 5

To begin year 5, double the ending population for year 4 and distribute bags appropriately. You will only need to add or subtract a few bags. Record the number of deer in the population that are entering year 5. Ask them how many deer they think will survive now. Record a few predictions on the table.

Set the timer and let year 5 begin. At the end of year 5, count the casualties, record the number of surviving deer, and briefly compare the number of survivors with the group's predictions.

15. Return to class

Have students collect all the food markers, bags, boundary markers, and any other materials. Then head back to the classroom.

16. Organize the data

Display the chart paper with the recorded results and have students fill in the data in their table at the top of notebook sheet 11, *Population Simulation Results*. Have students glue the sheet into their notebooks.

Year	# of deer at start	# of deer at end	Survival predictions
1	17	17	
2	34	34	
3	68	46	60, 65, 68
4	92	36	46, 68, 82
5	72	43	36, 40, 46

▶ **NOTE**
You always know how many bags are already in the hands of students—it is the same numbers as the previous year's starting deer population.

SCIENCE AND ENGINEERING PRACTICES

Developing and using models

Analyzing and interpreting data

Part 3: Population Simulation

17. Have a sense-making discussion
Tell students,

*Our deer population was greatly impacted by the amount of available food. Food is probably the most critical factor in determining an area's **carrying capacity** for deer. The carrying capacity is the greatest number of deer that can be supported by the food plants in an area without damaging the food source.*

Ask,

- *Which year was the easiest for each of your deer to survive? Which year was the hardest? Why?* [Easiest in year 1, plenty of food for all deer; year 4 was the hardest because the food was limited.]
- *Can we say in what year the number of deer reached carrying capacity for the home range? How do we know?* [Probably year 3. That was the year that the deer caused damage to the food source.]
- *In this simulation, which factors determined the carrying capacity of the home range?* [Amount of food and size of population.]
- *What happens to the deer population when they exceed the carrying capacity?* [The population changes in size. Some deer die, others move to new locations. In year 5, deer spread out to look for new food sources.]
- *What happens to the deer population when the size doesn't exceed the carry capacity?* [Population remains stable.]
- *How does this model help us understand what might happen to populations of other organisms, such as mealworms or fish?*

Ask students to complete the sentence frames on sheet 11 to describe the results of years 2, 4, and 5.

18. Review vocabulary
Review key vocabulary introduced in this part, and refer to the words on the word wall.

19. Answer the focus question
Have students answer the focus question in their notebooks.

- *How does food affect a population in its home range?*

Students should use the terms *carrying capacity*, *food source*, and *population* in their answers.

Help students understand and communicate the cause-and-effect relationship between food source and population size. "If there is ____ , then ____ . When there is ____ , then ____ ."

TEACHING NOTE

Refer to the Sense-Making Discussions for Three-Dimensional Learning chapter in Teacher Resources on FOSSweb for more information about how to facilitate this with students.

CROSSCUTTING CONCEPTS
Systems and system models
Stability and change

carrying capacity
home range
population

SCIENCE AND ENGINEERING PRACTICES
Constructing explanations

Environments Module—FOSS Next Generation

INVESTIGATION 2 – Ecosystems

20. Assess progress: notebook entry

Collect the notebooks after class to see what students learned from the simulation activity.

What to Look For

- Students write that the quantity of the food source in the home range limits the size of the population.

- Students write that if the population numbers go over the carrying capacity, the organisms might use up or damage the food source and the area will support fewer animals.

- Students write that the animals can leave the home range and search for new sources of food.

21. Discuss sustainability

Explain that wildlife biologists work to maintain balance in ecosystems—sustainability. Their work includes making decisions about how humans might impact the ecosystem. Have students discuss this question in their groups and then write a short response in their notebooks.

▶ *Why is it important for wildlife biologists to understand the carrying capacity of different organisms in an ecosystem?* [In order to have a sustainable ecosystem, the organisms must be able to meet their needs (for oxygen, water, food, shelter to avoid predators and extremes in temperature, and to reproduce and protect young). To make management decisions, biologists need to know what organism is causing problems in the balance and develop guidelines to reduce the impact of that organism. That might involve reducing the numbers of certain animals, particularly invasive or non-native species, protecting certain species by reducing hunting, fishing, or any impact by humans, protecting habitat by restricting use, and so on.]

> **TEACHING NOTE**
>
> Students will learn more about the work of wildlife biologists in a reading in the next part of this investigation.

Part 3: Population Simulation

READING *in Science Resources*

22. Read "Human Activities and Aquatic Ecosystems"

Read "Human Activities and Aquatic Ecosystems," using the strategy that is most effective for your class.

Let students preview the text by looking at and discussing the photographs. Tell students that this article will describe an ecological problem that happened to Lake Erie. Suggest that students use a problem/solution graphic organizer to help them focus on the main ideas and key details of the article.

Read the story aloud or have students read independently.

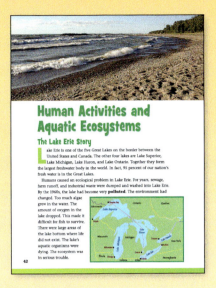

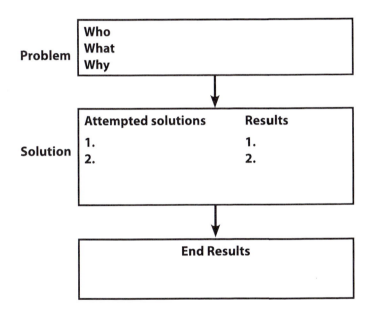

23. Discuss the reading

Ask students to review the questions at the end of the article using the information from their graphic organizers. Call on Reporters to share their group's answer to a question, along with details and examples from the text. Allow other group Reporters to respond by either agreeing, disagreeing and offering alternative explanations or evidence, or adding on more information. After the discussion, let students revise or add to the original answers and record them in their notebooks.

➤ *What organism causes most of the pollution in Lake Erie? Give examples of why you think so.* [Humans. The sources of pollution result from human activities.]

ELA CONNECTION

These suggested reading strategies address the Common Core State Standards for ELA.

RI 1: Refer to details and examples in a text when explaining what the text says explicitly and when drawing inferences from text.

W 5: Strengthen writing by revising.

SL 1: Engage in collaborative discussions.

SCIENCE AND ENGINEERING PRACTICES

Obtaining, evaluating, and communicating information

Environments Module—FOSS Next Generation

INVESTIGATION 2 – Ecosystems

➤ *What is the effect of climate change on large freshwater lakes?* [Over the past 25 years, surface temperatures have risen in all of the world's largest lakes. The change of 0.05 degrees Celsius (°C) in the temperature can change where lakes freeze and when the ice melts. Invasive species can get established and outcompete native species.]

Ask the class,

➤ *What is the effect of water pollution on an aquatic ecosystem like Lake Erie?* [Some organisms can't survive in the environment. They might be harmed by pesticides, acid water, and oil spills, or sediments might destroy their habitats. Some organisms grow too much because of the pollution. An example is the rapid growth of algae and plants when fertilizers flow into water.]

24. Read more about human impact

This is a good opportunity for students to learn more about human impact on the environment by reading another text. See FOSSweb for a list of recommended books. One suggestion is *Pollution and Conservation* by Rebecca M. Hunter.

BREAKPOINT

25. Read "Comparing Aquatic and Terrestrial Ecosystems"

Have students read the article "Comparing Aquatic and Terrestrial Ecosystems." This article succinctly reviews the concepts students have been learning. Start by having students compare and contrast the two environments they've been observing over time—terrariums and aquaria. Read the title and the first paragraph. Tell students that it is very clear that the author is making the point that aquatic and terrestrial environments are different and the same. Ask students to look for evidence and reasons the author uses to support this point.

26. Discuss the reading

Use these questions for discussion. Encourage students to use details and examples from the text to support their answers.

➤ *What are some of the ways all ecosystems are the same?* [They all have living and nonliving components; they all have complex food webs involving producers, consumers, and decomposers; they all are systems based on interactions between the organisms and the nonliving environment.]

ELA CONNECTION

This suggested reading strategy addresses the Common Core State Standards for ELA.

RI 9: Integrate information from two texts on the same topic.

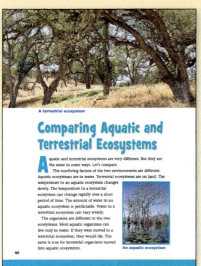

ELA CONNECTION

This suggested reading strategy addresses the Common Core State Standards for ELA.

RI 8: Explain how an author uses reasons and evidence to support particular points in a text.

200   Full Option Science System

Part 3: Population Simulation

➤ *Where do terrestrial and aquatic ecosystems get their energy?* [Ultimately from the Sun.]

➤ *How do organisms in an ecosystem get the matter and energy they need to survive?* [Plants and plantlike organisms need sunlight to make their own food (sugar), using simple chemicals from the environment (water and carbon dioxide). Plants need the energy of food to grow and reproduce. Animals get their matter and energy from plants or other animals that they eat.]

27. View online activities

Students can engage in the online activities "Virtual Aquarium" and "Virtual Terrarium" to compare the environmental factors in the two different kinds of ecosystems.

ELA CONNECTION

This suggested reading strategy addresses the Common Core State Standards for ELA.

RI 1: Refer to details and examples in a text when explaining what the text says explicitly and when drawing inferences from text.

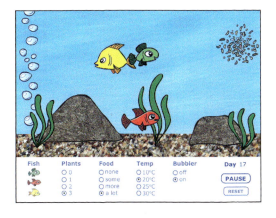

WRAP-UP/WARM-UP

28. Share notebook entries

Conclude Part 3 or start Part 4 by sharing notebook entries. Ask students to open their science notebooks to the *Population Simulation Results* sheet. Read the focus question together.

➤ *How does food affect a population in its home range?*

Make new groups of four and have students work together to come up with a model to explain their answer. Have each group draw their model on chart paper. Choose one or two to present to the class and have students discuss what they agree with and what they don't agree with in the models. Remind students to support their claims with evidence and reasoning.

➤ *What might happen if the number of guppies doubled in the aquarium?*

CROSSCUTTING CONCEPTS

Energy and matter

ELA CONNECTION

This suggested reading strategy addresses the Common Core State Standards for ELA.

SL 5: Add visual displays to presentations.

SCIENCE AND ENGINEERING PRACTICES

Developing and using models
Constructing explanations

Environments Module—FOSS Next Generation **201**

INVESTIGATION 2 – Ecosystems

MATERIALS for
Part 4: Sound Off

For each student
1 Paper-bag blindfold (See Step 8 of Getting Ready.)
1 *FOSS Science Resources: Environments*
- "Animal Sensory Systems"
- "Saving Murrelets through Mimicry"

For the class
- Noisemakers (See Step 6 of Getting Ready.)
  - 4 Clackers
  - 4 Clickers
  - 2-4 Cow bells with rubber bands
  - 4 Finger cymbals
  - 4 Gear clickers
  - 4 Hand clappers
  - 4 Jingle bells
- 8 Vials, with caps (optional)
- Rice or beans ★ (optional; see Step 6 of Getting Ready.) ★
- Rubber bands (optional) ★
- Plastic cups (optional)
- 1 Hole punch ★
- Scissors ★
- 4 Strings, 1 m long ★
- 2 Index cards
- Large brown paper bags (optional; see Step 8 of Getting Ready) ★
- 1 Carrying bag ★
- 1 Computer with Internet access ★
- 1 Projection system ★

For embedded assessment
- ❑ Notebook sheets 12–13, *Response Sheet—Investigation 2*
- ❑ *Embedded Assessment Notes*

★ Supplied by the teacher.   ❑ Use the duplication master to make copies.

No. 12—Notebook Master

No. 13—Notebook Master

Part 4: Sound Off

GETTING READY for
Part 4: *Sound Off*

1. **Schedule the investigation**
   This part will take one or two sessions for active investigation, one or two sessions for reading, and two sessions for assessment.

2. **Preview Part 4**
   Students go to the schoolyard and pretend to be animals who have poor vision or are active at night. The animals communicate with one unique sound and try to find others of their kind before being "captured" by a predator. After three rounds of this activity, students sit silently to listen to animals in the schoolyard. The focus question is **How do animals use their sense of hearing?**

3. **Select your outdoor site**
   Choose a large open area that is free of holes, ruts, or obstacles, with a uniform surface throughout. Grass or asphalt spaces are excellent choices. Ideally, the site is in a fairly quiet part of the schoolyard. If a fence or wall borders one or two sides of the site, it will help keep students in a herd. You will want a space that is about 20 meters square.

   Look for two places on the periphery of the area that could be the designated "safe" and "captured" locations. During the game, students who find their group members or are captured by the predator will remove their blindfolds and move to these designated areas to observe the remainder of the activity.

4. **Check the site**
   The morning before taking students to your selected site, do a quick search for potentially distracting or unsightly items.

5. **Plan for group size**
   For more than 20 students, we recommend having another adult with you. The size of your class will also help you decide how many predators to have. If you have 20 students or fewer, plan on one predator. If you have more than 20, plan on two predators.

6. **Plan for noisemakers**
   All students will pretend to be an animal with a specific sound. Groups of up to four animals will have the same sound. The size of your class will help determine if you will have two, three, or four animals in each group.

▶ **SAFETY NOTE**
Check to see if any student has an allergy to latex, as rubber bands are made of latex.

Environments Module—FOSS Next Generation

INVESTIGATION 2 – *Ecosystems*

The kit includes four sets of six noisemakers for prey and cowbells for predators. Empty vials in the kit can be filled with rice, sand, pebbles, coins, or anything that you have readily available, to make more noisemakers. Test these noisemakers to make sure students will be able to find their group. If you need more noisemakers, wrap a rubber band around a cup for students to twang.

Put the prey noisemakers in a bag to be carried outdoors.

7. Plan for predators

Create two or four predator ankle bracelets. Secure a large rubber band in the handle of each cowbell. For groups larger than 20, assemble four cowbell ankle bracelets.

Write "predator" on two index cards, punch two holes at the top, and tie the ends of a string (about one 1 m long) to each hole. Predators will wear the cowbells on their ankles and the predator sign around their necks. It is best if the bell is on the back of the ankle so that it makes a noise when the leg is moved.

8. Prepare paper-bag blindfolds

Every student, including the predators, will wear a paper-bag blindfold. The bags are supplied in the kit but must be cut before using them for the first time. If you are the first person to use the kit, you will need to prepare the blindfolds. Modify these instructions so the bags fit the students in your class. Provide larger brown paper grocery bags for students that might need a larger bag.

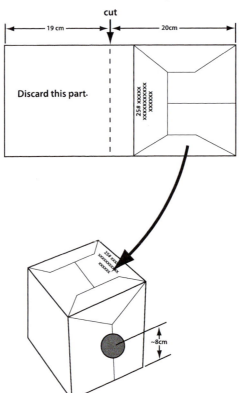

- With the bag flat, cut the top 19 cm off and recycle that part of the paper bag. The bag will now be shorter, about 20 cm in length. Ideally, the edge of the bag just covers the students' noses so that students can see their feet and their own noisemaker. You may need to cut the bag shorter.

- Cut ear holes on two sides of the bag. The center of the hole should be 8 cm from the open edge of the bag.

▶ **NOTE**
The students should adjust the fit of the bag so it is easy for them to see their feet and where they are stepping. Have them practice walking in class using the blindfold.

204   Full Option Science System

Part 4: Sound Off

9. **Plan for additional help**
   If possible, plan for additional helpers to help distribute instruments and to keep the "blind" animals from traveling out of bounds.

10. **Plan to review outdoor safety rules**
    Before going outdoors, remind students of the rules and expectations for the outdoor work. Emphasize that there is no running in this simulation activity.

11. **Plan to view videos**
    Preview the video, *Animal Language and Communication* (duration 5 minutes). Also preview chapter 6, "Sense of Hearing," duration 2 minutes 11 seconds. This chapter is from the video *All about the Senses*. Links to the videos are in the Resources by Investigation section on FOSSweb.

12. **Plan to preview image gallery**
    Plan to preview the image gallery "Animal Ears and Hearing" on FOSSweb. Show the images in Guiding the Investigation, Step 15, as you discuss sound sources and receivers.

13. **Plan to read *Science Resources*: "Animal Sensory Systems" and "Saving Murrelets through Mimicry"**
    Plan to read "Animal Sensory Systems" and "Saving Murrelets through Mimicry" during reading periods after completing the active investigation.

14. **Plan assessment: response sheet**
    Use notebook sheets 12–13, *Response Sheet—Investigation 2*, to assess students' ability to describe an animal's structures, including sensory structures, their functions, and their role in survival.

15. **Plan benchmark assessment: I-Check**
    When students have completed all the activities and readings for this investigation, plan a review session, and then give them *Investigation 2 I-Check*. Print or copy the assessment masters or schedule students on FOSSmap. Students should complete their response to the items independently.

    The Assessment Coding Guides are on FOSSweb. After you review and code students' work, plan which items you will bring back for additional discussion and self-assessment. See the Assessment chapter for guidance and information about online tools.

INVESTIGATION 2 – *Ecosystems*

FOCUS QUESTION

How do animals use their sense of hearing?

EL NOTE

If necessary provide pictures of animals that use a strong sense of hearing or play recordings of the sounds they make. Check FOSSweb for a link to the Macaulay Library of Sounds, the world's largest archive of animal sounds.

Materials for Step 5
- *Paper-bag blindfolds*

GUIDING *the Investigation*
Part 4: *Sound Off*

1. **Review animal senses/responses**
   Conduct a quick review of the ways humans respond to their environment. Ask,

   ➤ *If you didn't have your sight, what other senses could you rely on to know what's going on around you?* [Hearing, smell, touch.]

   ➤ *What animals use sound to communicate?* [Crickets, frogs, squirrels, rattlesnakes, birds, wolves, whales, dolphins, etc.]

   ➤ *What do sounds communicate?* [Mating calls, location of clan members, presence of food or danger, disposition, etc.]

2. **Set the context**
   Ask students to imagine they are animals who either have poor vision or are active at night or where visibility is limited. Ask what senses these animals might rely on. Ask the class to name some animals that might not see well and use a strong sense of hearing to help them survive [bats, crickets, whales, etc.].

3. **Introduce the activity**
   Tell students that they will go to the schoolyard and pretend to be animals that communicate by sound. Each animal will try to find a group communicating with the same sound before being "captured" by a predator. Explain that all students will be wearing blindfolds so that they will rely on their sense of hearing. The goal of the activity is to find their group (other students making the same sound) before being caught by a predator.

4. **Focus question: How do animals use their sense of hearing?**
   State the focus question while you write or project it on the board.

   ➤ *How do animals use their sense of hearing?*

5. **Introduce blindfolds and noisemakers**
   Model how students will wear the paper-bag blindfolds and explain that they will be able to see down to their feet, but not their surroundings, when they are walking. Distribute bags and have all students try on the blindfolds indoors.

   Explain that all players, including predators, will wear blindfolds and carry noisemakers. Show students all the noisemakers, paying particular attention to the cowbells that identify the predators. Whenever the predator is moving, he or she will be making noise. **Prey** are allowed to move without making noise.

Full Option Science System

Part 4: Sound Off

6. Review outdoor safety

Remind students of the rules and expectations for the outdoor work. Refer to the *FOSS Outdoor Safety* poster.

7. Go outdoors

Have students carry their blindfolds and move to your outdoor gathering spot in the usual orderly manner. Have students form a tight sharing circle. Point out the boundaries for the activity. Whenever the class needs to have a discussion, meet in the circle.

Show students where the "captured" and "safe" areas will be. Go over the rules.

- Everyone may move freely, but no running is allowed.
- No peeking!
- If the predator tags you, take off your mask and move to the "captured" area. Predators must be sure the prey know that they have been tagged.
- Prey that successfully find all of their partners should take off their masks and go to the "safe" area to watch the rest of the activity.
- Predators must wear the predator sign around their neck, and cowbells on both ankles so that they will be making noise whenever they are moving.

Make sure that all students understand the rules and the point of the activity. Be sure students know how many predators there will be and how many of each instrument there will be. (For example there are three of each kind of noisemaker except for the clappers—there will only be two clappers.)

8. Distribute noisemakers and predator signs

Have students spread out within the boundaries and put their masks on. Position two students a short distance from the other students and provide them with the predator ankle bracelets and predator signs. Quickly distribute noisemakers to the other students. Put them directly into students' hands.

Materials for Step 8
- *Predator signs*
- *Noisemakers*

9. Start the activity

Once everyone has a noisemaker, shout "Sound off!" to start the first round. Allow several minutes for the first round. The round ends when the only remaining players have lost their partners to the "captured" zone.

Environments Module—FOSS Next Generation

207

INVESTIGATION 2 – Ecosystems

SCIENCE AND ENGINEERING PRACTICES

Planning and carrying out investigations

Constructing explanations

> **TEACHING NOTE**
>
> *Refer to the Sense-Making Discussions for Three-Dimensional Learning chapter in* Teacher Resources *on FOSSweb for more information about how to facilitate this with students.*

10. **Have a sense-making discussion**

    Conduct the activity a few more times. After each round, have students remove the blindfolds and ask students a few of these questions.

    ➤ *What was it like to have restricted vision and to depend on your sense of hearing to find a partner?*

    ➤ *What happened to prey who made lots of noise all the time?*

    ➤ *Which noisemakers attracted the most attention?*

    ➤ *What listening techniques worked best to locate a partner?*

    ➤ *Was it easy or hard to walk toward a partner's sound? Why?*

    ➤ *Did any other noises in the environment make the activity more challenging? As prey? As predator?*

    ➤ *How does this simulation help you understand how animals use their sense of hearing?*

    You may want to give students a chance to modify the rules of the activity. Here are some suggestions.

    - Increase or decrease the number of predators.
    - Give prey that successfully find their group the ability to free another prey of the same kind that has been caught.

11. **Listen to schoolyard animals**

    After the last round, collect all noisemakers and blindfolds and gather students in a sharing circle. Tell them that you are going to ask them to sit silently for 2–3 minutes to see if they can hear any animals in the schoolyard environment. Invite students to continue to observe and listen to animals on their own time at home.

12. **Return to class**

    Have students return to the classroom with masks in hand.

13. **View the video**

    Prepare the classroom for observing the short video *Animal Language and Communication* (duration 5 minutes). A link to the video is in the Resources by Investigation section on FOSSweb.

    Here's what the students will see in this video.

    - Some animals cannot speak; others have found ways to communicate. (19 sec)
    - A yelping sea lion pup makes a specific sound its mother can recognize among the calls of many pups in a colony. (36 sec)

208   Full Option Science System

Part 4: Sound Off

- Male and female penguins communicate with each other through a series of honking sounds. (52 sec)
- Tasmanian devils use fearsome, screeching growls as warnings or admonitions. (37 sec)
- As an unwanted guest approaches, the vocalizations of a dominant male macaque are mimicked by the rest of the group, which quickly makes the visitor leave the area. (27 sec)
- Wolves howl to alert the rest of the pack to the presence of prey or to mourn the death of one of their own. (31 sec)
- During their nesting period, flamingos prattle constantly to comfort fledglings and to warn the rest of the flock of approaching danger. (20 sec)
- The bleating of sheep tells the flock to stick together and reminds the lambs to stay close to their mothers. (18 sec)
- Of any animal in the wild, dolphins possess the most elaborate language, a system of nearly 100 different sounds and ultrasounds that are audible at great distances. (18 sec)

POSSIBLE BREAKPOINT

14. Discuss sound sources and receivers

Have students generate possible mechanisms (external and internal structures and systems) for receiving and interpreting sounds. Have them discuss these questions in their group before having a class discussion.

> *How do animals receive sounds from a sound source?* [Sounds come from sources that vibrate. The vibrations travel through the air in all directions until they reach a receiver.]

> *What are the external structures that animals have to receive sound vibrations?* [Many land animals, such as mammals, have external ears that gather the vibrations. Some animals have big ears that can move around to point in the direction of the sound. Others have heads that can turn so the ear can better gather the sound vibrations.]

TEACHING NOTE

The discussion of sound sources and receivers should build on what students learned in grade 1.

15. Project images of animal ears

Project the images from the Image Gallery on FOSSweb called "Animal Ears and Hearing." The images include a human, a horse, a red fox, a kudu, an elephant, a seal, a sea lion, and a fish (no ears, but it has a lateral line). Ask students to provide information or questions as each animal is displayed.

Environments Module—FOSS Next Generation

INVESTIGATION 2 – *Ecosystems*

16. Discuss internal structures

Have students generate possible mechanisms for the internal structures and systems for interpreting sounds. Have them discuss this question in their group before having a class discussion.

➤ *We have looked at some external structures of animals. What might the internal structures be like that interpret sound?* [The vibrations can make internal ear structures vibrate, but somehow those vibrations need to get to the brain to be interpreted.]

17. View video

Tell students that you have a short video about how the human ear works. The ears detect sound waves; the ears turn vibrations into nerve pulses that the brain understands as sound.

Show chapter 6, "Sense of Hearing," duration 2 minutes 11 seconds. A link to the video is in the Resources by Investigation section on FOSSweb. (This is from the video *All about the Senses*.)

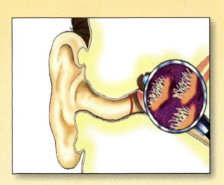

Human Ear

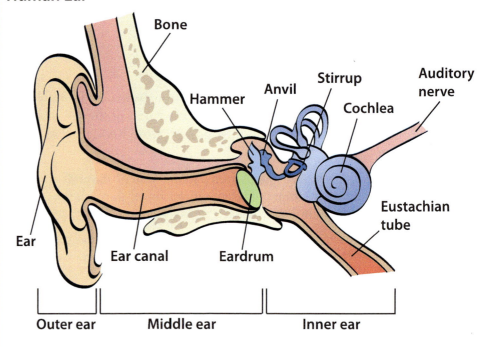

You might show the video a second time and have students record the sequence of actions that is described. After the video, have them discuss the parts of the system that make up the human auditory sensory system. Students should describe the general sequence about how ears collect and detect sound waves. The vibrations move from the outer ear through the ear canal to the

SCIENCE AND ENGINEERING PRACTICES
Developing and using models

210   Full Option Science System

Part 4: Sound Off

eardrum that begins to vibrate. The vibrations move to the inner ear (tiny bones to the cochlea), where they are converted to nerve impulses and travel to the brain through the auditory nerve.

18. Answer the focus question

Restate the focus question. Have students work independently to answer the focus question in their notebooks.

➤ *How do animals use their sense of hearing?*

Students' answers might include examples of real-life animals and the ideas of predator, prey, and mating. Students should be encouraged to share the things they personally experienced in the activity and describe how this experience relates to the real world.

19. Assess progress: response sheet

Distribute copies of notebook sheets 12–13, *Response Sheet— Investigation 2*. Ask students to work alone to respond to the tasks on the sheet.

At the end of class, review student responses.

What to Look For

- Students complete the table correctly:
  Row 1: D for structure
  Row 2: E for function
  Row 3: C for how it survives
  Row 4: A for structure
  Row 5: B for function

- Students describe another structure on the okapi, how it functions, and how it helps the animal survive. One possible structure is skin pattern—protective pattern on only part of the okapi might make it harder for predators to see them or it might be confusing to predators. Another might be strong, long legs with hooves—structures that help the okapi run fast to escape predators or to kick to defend itself or young.

EL NOTE

For students who need scaffolding provide sentence frames such as, *Animals use their sense of hearing to _____. For example, _____.*

SCIENCE AND ENGINEERING PRACTICES

Constructing explanations

Environments Module—FOSS Next Generation

INVESTIGATION 2 – Ecosystems

READING *in Science Resources*

20. Read "Animal Sensory Systems"

Read "Animal Sensory Systems," using the strategy that is most effective for your class.

One way to engage students before reading is to have students do a quick write in their notebooks. Ask them to describe three things they learned from the video (or from their own experiences) about how animals use their sense of hearing to survive. Give students a few minutes to share one example with their group.

Then, have them preview the text by looking at the photographs and reading the captions. Tell them to choose one of the photographs that interests them and write down a question or thought in their notebooks that they think will be answered in the article.

Explain that this article will discuss how animals use other senses for survival. Tell students to turn and tell their neighbor what other senses they think animals use.

21. Use a turn and talk strategy

Read the article aloud or have students read in pairs. Have them take turns reading a page and then pausing for the other to summarize. They should also keep track of any unknown words or phrases and questions they have as they read.

22. Discuss the reading

Review any questions students have and discuss the ways to determine the meaning of any unknown words or phrases. For example, students could look up the meaning of the words, *infrasound* and *ultrasound* in the glossary or they could use the prefixes (*infra* meaning below; *ultra* meaning beyond the norm).

Ask students to discuss the question,

> *How is our sense of vision like our sense of smell? How are they different?* [Both involve receptor neurons that receive information from the environment and pass it on to the brain. They differ in the type of receptor cell and the kind of stimulus the receptor responds to. Visual messages go to one part of the brain, and smell messages go to a different part of the brain. The brain interprets the information and sends a response.]

ELA CONNECTION

These suggested reading strategies address the Common Core State Standards for ELA.

RI 1: Refer to details and examples in a text when explaining what the text says explicitly and when drawing inferences from text.

RI 2: Determine the main idea of a text and explain how it is supported by key details; summarize the text.

RI 3: Explain procedures, ideas, or concepts in a scientific text.

RI 4: Determine the meaning of general academic domain-specific words or phrases.

W 7: Conduct short research projects.

W 8: Gather relevant information from experiences and print, and categorize the information.

L 4: Determine or clarify the meaning of unknown and multiple-meaning words.

Part 4: Sound Off

Assign each group an animal discussed in the article and ask them to become experts on that particular animal. If there is anything in the reading that students in the group don't fully understand, ask them to go to other sources of information for clarification.

After they have had time to study and discuss the specifics of how their animal gets sensory information from the environment, ask each group to make a diagram on chart paper that includes:

- the animal's strongest sensory receptor
- how the animal uses the information it perceives

Hang the group posters on the wall and conduct a gallery walk. Give students a set of self-stick notes and in their groups have them rotate from poster to poster. For each diagram they can add a comment or question with their self-stick notes. Allow time for each group to examine the questions and comments and to revise or add to their posters.

As an extension, individual students could do more research on a particular animal and how it uses its sensory receptors or they could choose a type of sensory receptor to investigate further.

23. Read "Saving Murrelets through Mimicry"

Read "Saving Murrelets through Mimicry," using the strategy that is most effective for your class. Tell students that this article shows how scientists can use what they know about how animals communicate to solve a problem. Have them read the title and look over the photographs. (If students are unfamiliar with the term *mimicry*, discuss the definition and a few examples.) Ask,

➤ *Without reading the text, what do you think is the problem?*

Confirm that it has something to do with the survival of the marbled murrelets, rare seabirds that nest only in old-growth redwood trees on the West Coast. (Show the photograph of the coastal redwoods and read the caption.) Tell students that these birds are endangered. Explain,

Your job is to find out why and what scientists are doing to solve the problem. You may be surprised by the solution!

Give students self-stick notes to jot down any unknown words or phrases or questions they have as they read independently. For students who need support, conduct a guided reading with a small group.

SCIENCE AND ENGINEERING PRACTICES

Obtaining, evaluating, and communicating information

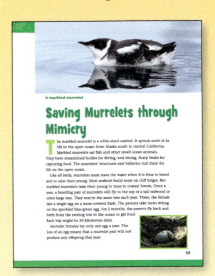

ELA CONNECTION

These suggested reading strategies address the Common Core State Standards for ELA.

RI 1: Refer to details and examples in a text when explaining what the text says explicitly and when drawing inferences from text.

RI 2: Determine the main idea of a text and explain how it is supported by key details; summarize the text.

RI 3: Explain procedures, ideas, or concepts in a scientific text.

RI 4: Determine the meaning of general academic domain-specific words or phrases.

RI 5: Describe the overall structure of information in a text.

RI 8: Explain how an author uses reasons and evidence to support particular points in a text.

SL 1: Engage in collaborative discussions.

Environments Module—FOSS Next Generation

INVESTIGATION 2 – Ecosystems

24. Determine the main idea

Clarify any unknown words or phrases and give students a few minutes to share what the article made them think about. Suggest students use a problem/solution graphic organizer such as the one used in the previous part to help them determine the main ideas and key details of the article.

Have students discuss the information they wrote in their graphic organizers about the reading with their groups, then conduct a whole group discussion using the questions on the next page as a guide. Call on an individual to share his or her answer to the question, and give others the opportunity to refine or add to the original answer.

➤ *What is the problem that biologists are trying to solve?*

➤ *What are the causes of this problem?*

➤ *How are biologists using what they know about Stellar's jays to solve the problem?*

➤ *What evidence did the biologist have that jays used their memories to guide their actions?*

➤ *How can people help the murrelets?*

To further the discussion, ask students to explain how the author uses reason and evidence to support how using mimic eggs, keeping picnic areas clean, and preserving the largest forest trees will save the murrelets.

Part 4: Sound Off

WRAP-UP

25. Review Investigation 2

Distribute one or two self-stick notes to each student. Ask students to cut each note into three pieces so each piece has a sticky end.

Ask students to take a few minutes to look back through their notebook entries to find the most important things they learned in Investigation 2. Students should include at least one science and engineering practice, one disciplinary core idea, and one crosscutting concept. They should tag those pages with self-stick notes. They might use a highlighter or colored pencils to call out the key points.

Lead a short class discussion to create a list of three-dimensional statements that summarize what students have learned in this investigation. Here are examples of the big ideas that should come forward in this discussion.

- We set up aquaria, made observations, and obtained information to provide conditions that meet the needs of the organisms. Water and temperature are two important nonliving environmental factors in aquatic ecosystems. Other nonliving factors are light (very important for plants) and oxygen. (Planning and carrying out investigations; obtaining, evaluating, and communicating information; systems and system models.)
- We analyzed and interpreted data from ecosystem cards to learn that organisms have structures that allow them to interact in feeding relationships in ecosystems. Organisms' roles include producers, consumers, and decomposers. (Developing and using models; analyzing and interpreting data.)
- We gathered information to compare the functions of structures that organisms have to meet their need for getting oxygen in terrestrial and aquatic systems. (Obtaining, evaluating, and communicating information.)
- We conducted investigations and obtained information about animals and their sensory systems that allow them to gather information about the environment and act on it. Animals communicate to warn others of danger, scare predators away, or locate others of their kind, including family members. (Planning and carrying out investigations; obtaining, evaluating, and communicating information; systems and system models.)

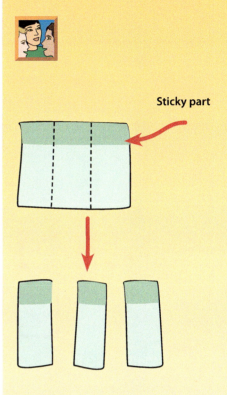

Sticky part

DISCIPLINARY CORE IDEAS

LS1.A: Structure and function

LS1.D: Information processing

LS2.C: Ecosystem dynamics, functioning, and resilience

LS4.D: Biodiversity and humans

INVESTIGATION 2 – Ecosystems

- We obtained information through readings that human activities can impact ecosystems, but that communities working together can help protect Earth's environments. (Obtaining, evaluating, and communicating information; stability and change.)

26. Discuss investigation guiding questions
Students should discuss the investigation guiding questions with a partner before responding to them in their notebooks.

▶ *How are the structures of aquatic organisms similar and different from land animals?*

▶ *How do organisms sense and interact with their environment?*

BREAKPOINT

27. Assess progress: I-Check
Give the I-Check assessment at least one day after the wrap-up review. Distribute a copy of *Investigation 2 I-Check* to each student. You can read the items aloud, but students should respond to the items independently in writing. Alternatively, you can schedule the I-Check on FOSSmap for students to take the test online.

If students take the assessment on paper, collect the I-Checks. Code the items using the coding guides on FOSSweb, and plan for a self-assessment session, identifying disciplinary core ideas, science and engineering practices, and crosscutting concepts students may need additional help with. Refer to the Assessment chapter for more information.

TEACHING NOTE

During or after these next steps with the I-Check, you might ask students to make choices for possible derivative products based on their notebooks for inclusion in a summative portfolio. See the Assessment chapter for more information about creating and evaluating portfolios.

Full Option Science System

Interdisciplinary Extensions

INTERDISCIPLINARY EXTENSIONS

Language Extension

- **Describe aquatic environments**
  Have students research and write about aquatic environments in their community (e.g., river, creek, pond, large lake).

Math Extension

- **Problem of the week**
  A student wants to set up a tropical-fish aquarium. She has $20.00 to spend on the fish. The store has four types of fish for her tank. She wants at least one of each type of fish. There is no tax charged on fish. She wants to have $2.00 or less left after she buys all of her fish.

Type of fish	Cost of fish	Length of fish
Angelfish	$2.98	7 cm
Lampeye	$1.59	3 cm
Mollies	$1.35	4 cm
Neon tetras	$1.70	2 cm

a. What combination of fish could she buy? How much money will she have left? Show your work.

b. The student's parents agreed to buy the aquarium tank for her new fish. The student remembers from her aquatic-environments project that tropical fish need 1 liter of water for every 3 centimeters of fish length in the aquarium. What size aquarium in full liters do her fish need? Show your work.

Bonus problem. Can you find another combination of fish the student could buy? What size tank does she need for these fish?

Notes on the problem. This is a multistep problem with many solutions.

- First, students need to determine that the girl will spend at least $18.00 on fish ($20.00 − $2.00 [maximum change] = $18.00).

TEACHING NOTE

Refer to the teacher resources on FOSSweb for a list of appropriate trade books that relate to this

No. 13—Teacher Master

Environments Module—FOSS Next Generation

INVESTIGATION 2 – Ecosystems

- Next they need to determine the cost of the first four fish she will buy (one of each kind):

Angelfish	$2.98
Lampeye	$1.59
Mollies	$1.35
Neon tetras	$1.70
Total for 4 fish	$7.62

   The student will have at most $12.38 to spend on additional fish ($20.00 − $7.62 = $12.38). Below are two possible solutions.

- From this point, there are many ways students will approach the problem. For example, they could guess and check for various fish combinations or estimate numbers to select combinations. She has to spend at least $10.38 ($12.38 − $2.00 = $10.38).

Solution 1. After buying one of each fish, the student has $12.38 left to spend.

She can buy at least one more of each type of fish for $7.62 and have $4.76 left ($12.38 − $7.62 = $4.76).

With the $4.76 left, she can buy an angelfish for $2.98 and have $1.78 left. This is less than $2.00.

3 angelfish	$2.98 × 3 = $8.94
2 lampeyes	$1.59 × 2 = $3.18
2 mollies	$1.35 × 2 = $2.70
2 neon tetras	$1.70 × 2 = $3.40
Total cost of fish	$18.22

Her change: $20.00 − $18.22 = $1.78

Solution 2. After buying one of each fish, the student has $12.38 left to spend. She decides to buy as many of the least expensive fish as possible.

Divide $12.38 by $1.35 (cost of one mollie) and determine that the student can buy nine mollies for $12.15. The change left is $.23 (less than $2.00). The solution is 1 angelfish, 1 lampeye, 10 mollies, and 1 neon tetra.

Her change: $20.00 − $7.62 − $12.15 = $.23

Using the results from the first part of this math problem, here are the tank sizes needed.

Interdisciplinary Extensions

Solution 1.

3 angelfish	3 × 7 cm = 21 cm
2 lampeyes	2 × 3 cm = 6 cm
2 mollies	2 × 4 cm = 8 cm
2 neon tetras	2 × 2 cm = 4 cm
Total length of fish	= 39 cm

Because these fish need 1 L per 3 cm of length, divide 39 cm by 3 cm. The student needs a 13 L tank for this combination of fish.

Solution 2.

1 angelfish	1 × 7 cm = 7 cm
1 lampeyes	1 × 3 cm = 3 cm
10 mollies	10 × 4 cm = 40 cm
1 neon tetra	1 × 2 cm = 2 cm
Total length of fish	= 52 cm

Divide 52 cm by 3 cm and round up. The students needs an 18 L tank for this combination of fish.

Science and Engineering Extensions

- **Design and build another class aquarium**
  As a class, plan and develop a large class aquarium. Make it a permanent part of the classroom and rotate responsibility for monitoring environmental conditions.

- **Add other fish**
  Introduce other fish into the aquaria, and observe which parts of the environment are used by the new fish. Have students monitor interactions between the kinds. Ask at your local aquarium store about fish that will be suitable to add to a goldfish or guppy aquarium.

TEACHING NOTE

Review the online activities for students on FOSSweb for module-specific science extensions.

TEACHING NOTE

Encourage students to use the Science and Engineering Careers Database on FOSSweb.

Environments Module—FOSS Next Generation

INVESTIGATION 2 – Ecosystems

Environmental Literacy Extension

- **Investigate water holes to mini-ponds**

    Have the class set up a water hole (a mini-pond environment) and observe it over time. This long-term outdoor activity requires regular visits to monitor the water hole for 8–10 weeks. Students will observe colonization of the water hole. Colonization is the occupation of an area by a group of organisms that did not previously live there. It involves immigration of pioneer organisms, suitable conditions for survival (sufficient nutrients, proper light and temperature), population increase (reproduction and/or immigration of more organisms), and establishing balance as populations decrease (organisms eating other organisms) and increase (as a result of emigration of organisms and reproduction).

The investigation can be made more interesting by creating two water holes and enriching one with fertilizer (compost or garden manure). Students observe and compare changes that take place in enriched and plain aquatic environments.

NOTE: Secure permission from your administration and your grounds crew to plan and conduct this investigation.

This controlled experiment will attract a number of plants and animals. These factors will increase the variety and number of organisms that immigrate:

- The time of year: warmer seasons induce more movement of organisms.
- The amount of time the water has been standing: longer is better.

Interdisciplinary Extensions

- The proximity of the water holes to a pond, lake, or stream: closer proximity will speed colonization.

Materials for two water holes

- Hoes and shovels
- 2 Children's wading pools (ideally 1 m across and 30 cm deep), or plastic sheeting to line two holes, or even two basins
- Burlap to line bottom and sides of pools
- Duct tape or string
- Water
- Fertilizer, 50 mL of composted manure
- Observation aids: hand lenses, plastic cups, spoons, dip nets, and observation containers such as modified milk cartons
- Thermometer
- Rulers or meter sticks

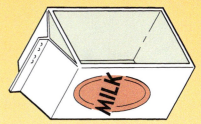

You can make a milk-carton observation container with no lid.

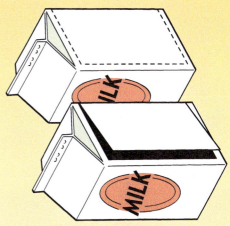

You can make a milk-carton observation container with a flap for a lid.

Site

Select a level spot that receives direct sunlight (ideally near a water tap). Avoid the base of a slope (runoff can cause problems) or near resinous trees such as pines or eucalyptus. The closer to your classroom the better!

Constructing the water holes

1. Describe the water-hole activity to the class. Explain that a water hole is a hole in the ground lined with something to retain the water. Emphasize that the two water holes will be exactly the same, except that one will have fertilizer. Tell students that the water holes will be pond environments. Divide the class into two groups.

2. Give each group a wading pool. Have each group clear an area and dig a hole for their pool deep enough so that the pool rim will be flush with the ground.

3. Have each group line their pool with burlap and place a few rocks on the burlap inside the pool to keep it down. The burlap edges can be taped or tied under the rims of the pools.

4. Have each group fill their pools with the same amount of tap water. Have one group add 50 mL of fertilizer to their water hole and stir it up. The other pool remains unfertilized.

5. Decide as a class what to record in notebooks and how to collect data. Ask students what they could measure or keep track of. Make sure students include nonliving factors. [Air temperature, water temperature, water level, water clarity.]

Environments Module—FOSS Next Generation 221

INVESTIGATION 2 – Ecosystems

Use a small collection net to transfer organisms from the pond to the water in an observation container.

Turn the collecting net inside out to make sure that all of the organisms are transferred to the container.

6. Have students record some questions or "I wonder" statements about the water holes.

Observations

1. When students observe a population of organisms in a water hole, discuss *colonization*.
2. Each group should share observations, so that the two groups can compare new changes.

Here are some ongoing questions to ask:

➤ *What changes can you see in each water hole?*

➤ *What are the major differences between the two water holes? Does the water differ in feel, smell, or color? What effect did the fertilizer have?*

➤ *What terrestrial plants or animals have established themselves around the edges of the water hole?*

➤ *What new plants and animals (colonizers) have appeared in the water?*

➤ *How do you think the appearance of organisms has affected the condition of the water?*

➤ *What organisms have remained in the water hole for a long time? What organisms have increased in numbers? Decreased in numbers?*

Home/School Connection

Students look for an aquatic environment in their neighborhoods. If no aquatic environment is close at hand, they use a local map to inventory the different kinds of aquatic environments nearby.

Print or make copies of teacher master 14, *Home/School Connection* for Investigation 2, and send it home with students after Part 2.

No. 14—Teacher Master

222

Full Option Science System

Investigation 3: Brine Shrimp Hatching

INVESTIGATION 3 – Brine Shrimp Hatching

Part 1
Setting Up the Experiment... 234

Part 2
Determining Range of Tolerance 243

Part 3
Determining Viability 256

Part 4
Variation in a Population 264

Guiding question for phenomenon:
How is optimum environment related to organism and population survival?

Science and Engineering Practices
- Developing and using models
- Planning and carrying out investigations
- Analyzing and interpreting data
- Using mathematics and computational thinking
- Constructing explanations
- Engaging in argument from evidence
- Obtaining, evaluating, and communicating information

Disciplinary Core Ideas
LS1: How do organisms live, grow, respond to their environment, and reproduce?
LS1.A: Structure and function
LS2: How and why do organisms interact with their environment and what are the effects of these interactions?
LS2.C: Ecosystem dynamics, functioning, and resilience
LS4: How can there be so many similarities among organisms yet so many different kinds of plants, animals, and microorganisms?
LS4.B: Natural selection
LS4.D: Biodiversity and humans
ESS3: How do Earth's surface processes and human activities affect each other?
ESS3.C: Human impacts on Earth systems

PURPOSE

Students investigate the environmental phenomenon of brine shrimp in Mono Lake as an important factor in the ecosystem of migratory birds by conducting controlled experiments to determine the range of tolerance to salinity changes.

Content
- Organisms have ranges of tolerance for environmental factors. Within a range of tolerance, there are optimum conditions that produce maximum growth.
- Brine shrimp eggs can hatch in a range of salt concentrations; more hatch in optimum concentrations.
- When environments change, some plants and animals survive and reproduce; others move to new locations; and some die. Human activities impact natural environments.
- Individuals of the same kind differ in their characteristics, and sometimes the differences give individuals an advantage in surviving and reproducing.

Practices
Design and conduct a scientific investigation and build explanations from evidence about the viability of brine shrimp eggs.

Crosscutting Concepts
- Cause and effect
- Systems and system models
- Structure and function

Full Option Science System

INVESTIGATION 3 – Brine Shrimp Hatching

	Investigation Summary	Time	Focus Question for Phenomenon, Practices
PART 1	**Setting Up the Experiment** Students are introduced to the phenomenon of brine shrimp living in saltwater environments. They investigate the environmental factor of salinity in hatching brine shrimp eggs. They conduct a controlled experiment to determine which one of four salt concentrations allows brine shrimp eggs to hatch. This data is used to develop evidence about the future of a salt lake.	**Active Inv.** 1 Session*	**How can we find out if salinity affects brine shrimp hatching?** **Practices** Developing and using models Planning and carrying out investigations Using mathematics and computational thinking Constructing explanations Obtaining, evaluating, and communicating information
PART 2	**Determining Range of Tolerance** Students monitor saltwater environments. They determine which environments are conducive to hatching brine shrimp eggs. Students analyze the results of a multiple-trial experiment conducted by the class and draw conclusions. They read about the Mono Lake ecosystem and create food webs using the organism cards and information in the article.	**Active Inv.** 1 Session **Reading** 2 Sessions	**How does salinity affect the hatching of brine shrimp eggs?** **Practices** Developing and using models Planning and carrying out investigations Analyzing and interpreting data Using mathematics and computational thinking Constructing explanations Obtaining, evaluating, and communicating information
PART 3	**Determining Viability** Students are challenged to manipulate the environment to see if they can get the dormant eggs to hatch and grow. They formulate and justify predictions and design an investigation to test their predictions.	**Active Inv.** 2 Sessions **Reading** 1 Session	**Does changing the salt environment allow the brine shrimp eggs to hatch?** **Practices** Planning and carrying out investigations Analyzing and interpreting data Constructing explanations Engaging in argument from evidence Obtaining, evaluating, and communicating information
PART 4	**Variation in a Population** Students go to the schoolyard in two teams, to place a population of imaginary animals in a suitable habitat based on a description of the population's natural history. Through a predator–prey simulation, students find out how variations in color and size within a population affect survival of the population.	**Active Inv.** 1 Session **Reading** 1 Session **Assessment** 2 Sessions	**What are some benefits of having variation within a population?** **Practices** Planning and carrying out investigations Analyzing and interpreting data Constructing explanations Obtaining, evaluating, and communicating information

A class session is 45–50 minutes. **Full Option Science System**

At a Glance

Content Related to DCIs	Writing/Reading	Assessment
• Brine shrimp are crustaceans that live in marine or salt-pond environments. • An environmental factor is one part of an environment. It can be living or nonliving. • Organisms have ranges of tolerance for environmental factors.	**Science Notebook Entry** *Brine Shrimp–Hatching Experiment* **Science Resources Book** "Brine Shrimp"	**Embedded Assessment** Performance assessment
• Within a range of tolerance, there are optimum conditions that produce maximum reproduction and growth. • Brine shrimp eggs can hatch in a range of salt concentrations, but more hatch in environments with optimum salt concentration. • When environments change, some plants and animals survive and reproduce; others move to new locations; and some die.	**Science Notebook Entry** *Brine Shrimp–Hatching Observations* *Brine Shrimp–Hatching Conclusions* **Science Resources Book** "The Mono Lake Story" "What Happens When Ecosystems Change?" **Online Activity** "Food Webs"	**Embedded Assessment** Science notebook entry
• Brine shrimp eggs can hatch in a range of salt concentrations, but more hatch in environments with optimum salt concentration. • When environments change, some plants and animals survive and reproduce; others move to new locations; and some die. • Humans impact natural environments.	**Science Notebook Entry** Answer the focus question **Science Resources Book** "The Shrimp Club" **Online Activity** "Virtual Investigation: Trout Range of Tolerance"	**Embedded Assessment** Response sheet
• Individuals of the same kind differ in their characteristics, and sometimes the differences give individuals an advantage in surviving and reproducing.	**Science Notebook Entry** Answer the focus question **Science Resources Book** "Variation and Selection"	**Benchmark Assessment** *Investigation 3 I-Check* **NGSS Performance Expectations addressed in this investigation** 4-LS1-1 3-LS4-2 3-LS4-4 5-ESS3-1

Environments Module—FOSS Next Generation

INVESTIGATION 3 – Brine Shrimp Hatching

BACKGROUND *for the Teacher*

Brine shrimp aren't shrimp at all. They are crustaceans, however, so they are distant cousins of the shrimp found in oceans, bays, and estuaries. The ancient brine shrimp surely evolved in the sea millions of years ago, but now they are found in only a few very specialized environments—saltwater lakes, salt or **brine** ponds, and a few ocean bays. Even though they have abandoned the open seas, the most important environmental factor influencing their survival is salt.

Under optimum conditions, populations can explode to incredible densities. For instance, the population of brine shrimp in Mono Lake in California has been estimated at 4 trillion at peak season! Because of this proliferation, they play a critical role in the life cycles of the gulls, grebes, and phalaropes that migrate through eastern central California. The future of these migrants has been the subject of a spirited environmental debate in California because the fresh water that has traditionally flowed into Mono Lake is a valuable commodity that many interests would like to divert from the lake for other uses.

How Can We Find Out if Salinity Affects Brine Shrimp Hatching?

In order to test the environmental factor of **salinity**, you need to design a **controlled experiment** where all of the factors are the same except for the salt condition in the water. The experimental design should provide extreme conditions (no salt on one end and excessive salt on the other end) and a number of salt concentrations in between. In this way, you increase the likelihood of observing differences in the hatching time and hatching success. The design needs to result in an outcome with quantifiable differences.

How Does Salinity Affect the Hatching of Brine Shrimp Eggs?

In 1840, a chemist, Justus von Liebig, contributed to the understanding of environmental factors in his book *Organic Chemistry and Its Application to Agriculture and Physiology*. In it he said, "The crops on a field diminish or increase in exact proportion to the diminution or increase of the mineral substances conveyed to it in manure." This seems like common knowledge to us today, but it was rather revolutionary for the time. He had found that each plant requires certain kinds and quantities of nutrients to **survive**. If one of these nutrients is absent, the plant will suffer. Further, if the nutrient is present in insufficient amounts, the

> *"In order to test the environmental factor of salinity, you need to design a controlled experiment where all of the factors are the same except for the salt condition in the water."*

Background for the Teacher

quality of life of the plant will be diminished. This came to be known as the law of the minimum.

Later studies of plants and animals demonstrated that not only does too little of a good thing affect the survival of an organism, but too much of a good thing can also be harmful. Each organism can survive within a certain range of conditions for any one factor, and the range is bounded by minimum and maximum limits. In 1913, V. E. Shelford expanded on the law of the minimum to describe the law of **tolerance**. Organisms can survive only within their **range of tolerance** for a particular environmental factor. The particular organisms (or species) are most abundant in areas where the environmental factor is within the optimum range for that species. The species is rare in areas where it experiences stress because the environmental factor has either too high or too low a value. Outside their range of tolerance (beyond its upper and lower limits for the environmental factor), the species fails to **reproduce**, develop, or grow.

An organism can survive when it is within its range of tolerance, but at an **optimum** place in that range, the organism **thrives**. Animals that are mobile can move to more favorable conditions and can demonstrate environmental preferences through their behavior.

Brine shrimp have a particularly good adaptation for survival in the deserts where **salt lakes** are usually found. The eggs can dry out completely and stay **viable** for a year or longer, waiting for the proper conditions of salt, water, temperature, and oxygen. When the conditions fall within their range of tolerance, they begin to hatch. Hatching is maximized when the eggs are in their optimum environment.

The truly amazing thing about brine shrimp eggs is that they will hatch only when conditions are suitable for their survival. The germ of life resting in the egg somehow senses the environment and responds to clear environmental messages. If the salt **concentration** is too low or too high, the egg keeps its isolation chamber tightly closed and waits. Soaking does not spoil the egg. If salt is added while a brine shrimp egg is immersed in plain water, the egg will hatch almost immediately. Similarly, if the egg is afloat in salt water that is too concentrated, it will remain dormant until enough fresh water dilutes the solution. Then out pops the next generation of brine shrimp!

Brine shrimp begin life as tiny larvae, the nauplii, which resemble tiny ringed triangles, making their way through the water with short jerky movements. They are about the size of the period at the end of this sentence. Consequently, they are easily overlooked by the casual observer or a farsighted investigator. Your students' eyes, however, will spot them right away, and they will create quite a stir.

Brine shrimp stages

Environments Module—FOSS Next Generation

INVESTIGATION 3 – Brine Shrimp Hatching

As the brine shrimp grow, the larvae molt (i.e., shed and re-form their outer covering) several times, becoming longer and leggier with each successive molt. In 3 to 6 weeks, the brine shrimp grow to their adult size, about 1 centimeter (cm) long.

When conditions are right and the trillions of brine shrimp explode onto the scene, they become a significant factor in the environment and a major part of the food web in salt lakes. They consume tons of minute free-floating organisms, mostly algae and bacteria. They in turn become food for thousands of **migrating** birds. If the brine shrimp were to fail because of a change in their environment, their absence would constitute a radical change in the environment of the migrating birds. This is just one example of the complex interactions that characterize ecosystems: environments nested within other environments, depending on one another for the integrity of the whole system.

> *"When conditions are right and the trillions of brine shrimp explode onto the scene, they become a significant factor in the environment and a major part of the food web in salt lakes."*

Does Changing the Salt Environment Allow the Brine Shrimp Eggs to Hatch?

Once you know the optimum salinity for hatching brine shrimp eggs, you can modify an existing environment by changing the concentration of salt. If the salt concentration is too dilute, you add an appropriate amount of salt. If the salinity is too concentrated, you dilute it by adding fresh water. Another way to deal with this problem is to mix two environments that differ in their salinity (one too dilute and one too concentrated), arriving at a median salt concentration within the range of tolerance for brine shrimp hatching.

A closing note about brine shrimp. Did you ever see that curious ad for sea monkeys in the back of a comic book or novelty magazine? You knew of course that they couldn't be monkeys, or any close kin, but what are they? Brine shrimp! An enterprising entrepreneur named Harold Von Braunhut packed brine shrimp eggs with a little package of salt and wrapped the whole bundle in a contemporary myth. Perhaps a little misleading, but certainly entertaining for those who followed the instructions carefully. Now you and your students are going to have a chance to get this sea monkey myth straightened out once and for all. And in the process you'll have more fun than a barrel of . . . brine shrimp.

Background for the Teacher

What Are Some Benefits of Having Variation within a Population?

Every living organism is a member of some population. Adult members of a population are similar in most ways, but a close look reveals that each is unique—just a little different from every other individual in that population. These small differences are called **variations**.

The rainbow trout in a mountain lake make up a population. While these fish are all rainbow trout, some are larger than others, some have more spots on their fins, some have darker coloration on their backs, and some have longer jaws. The eastern brook trout in the same lake make up another population. The brook trout also display size and color variations, but they are all members of the brook trout population. Rainbow trout and eastern brook trout are two of the many populations of organisms that live in the mountain lake.

These variations can be observed. But other variations in a population are hard to observe unless you monitor the population over time. Individuals of the same kind differ in their characteristics, and sometimes the differences give individuals an advantage in surviving and reproducing. Some will be better at getting food. Others will be better at avoiding predators. Some will have a slightly larger range of tolerance for changes in temperature. Those individuals that can thrive, given the pressures that the environment puts on them, are more likely to reproduce and have offspring that share those characteristics. Traits that are passed on to offspring are **inherited traits**. Those traits will help the offspring thrive in that same environment and pass their traits on to their offspring.

"Those individuals that can thrive, given the pressures that the environment puts on them, are more likely to reproduce and have offspring that share those characteristics."

Environments Module—FOSS Next Generation

INVESTIGATION 3 – Brine Shrimp Hatching

NGSS Foundation Box for DCI

LS1.A: Structure and function
- Plants and animals have both internal and external structures that serve various functions in growth, survival, behavior, and reproduction. (4-LS1-1)

LS2.C: Ecosystem dynamics, functioning, and resilience
- When the environment changes in ways that affect a place's physical characteristics, temperature, or availability of resources, some organisms survive and reproduce, others move to new locations, yet others move into the transformed environment, and some die. (3–LS4-4, extended from grade 3)

LS4.B: Natural selection
- Sometimes the differences in characteristics between individuals of the same species provide advantages in surviving, finding mates, and reproducing. (3-LS4-2, extended from grade 3)

LS4.D: Biodiversity and humans
- Populations live in a variety of habitats, and change in those habitats affects the organisms living there. (3–LS4-4, extended from grade 3)

ESS3.C: Human impacts on Earth systems
- Human activities in agriculture, industry, and everyday life have had major effects on land, vegetation, streams, oceans, air, and even outer space. But individuals and communities are doing things to help protect Earth's resources and environments. (Foundational for grade 5)

TEACHING CHILDREN about *Testing Environmental Factors and Variation in a Population*

Developing Disciplinary Core Ideas (DCI)

Research indicates that children generally think about organisms as individuals—the pet they have, a horse they can name, an animal in a zoo. This way of thinking about organisms carries into adulthood if the learner does not have opportunities to think about organisms differently.

This investigation involves students with a *population* of organisms rather than with individual organisms. This experience enables students to broaden their concept of organisms.

Brine shrimp live and breed in large populations, allowing students to study a large number of organisms *en masse*. The brine shrimp eggs can hatch in waters of various salt concentrations, but they hatch most reliably when provided with an optimum salt concentration. To experience this important phenomenon, students design controlled experiments. They systematically test the effect of four salt concentrations on the hatching of brine shrimp eggs. This investigation provides an opportunity for students to help plan and manage a multiple-trial experiment. They will have other opportunities to engage in such experiments in Investigation 4, working with plants.

Through discourse about data, students acquire the knowledge that brine shrimp eggs can hatch in a range of salt concentrations, hatch best under certain salt conditions, and can still hatch even after being subjected to salt concentrations outside their range of tolerance. The structure of the brine shrimp egg responds to the environment and will not hatch unless the conditions are within a range of tolerance for survival. Conditions must be right for the eggs to hatch. Understanding is enhanced by engaging students with thoughtful questions: If you were going to raise brine shrimp, what range of salt concentrations could you use? What salt concentration would be ideal for hatching eggs? What would be too much or too little salt for brine shrimp to hatch? What data can you provide to support your answers to these questions?

As students carry out their experiments, the *FOSS Science Resources* reading provides detailed information about brine shrimp—their life cycle and where they live naturally. Students are introduced to the Mono Lake ecosystem and its living and nonliving environmental components. They learn about the other organisms that live in the lake by studying organism cards, and they investigate the complex feeding interactions by building a

Teaching Children about Environmental Factors and Variation

food web. The importance of brine shrimp and microscopic algae to the Mono Lake ecosystem becomes apparent in this investigation.

The experiences students have in this investigation contribute to the disciplinary core ideas **LS1.A: Structure and function; LS2.C: Ecosystem dynamics, functioning, and resilience; LS4.B: Natural selection; LS4.D: Biodiversity and humans;** and **ESS3.C: Human impacts on Earth systems.**

Engaging in Science and Engineering Practices (SEP)

In this investigation, students engage in these practices.

- **Developing and using models** of brine shrimp environments, such as Mono Lake, to test the salt conditions that will support brine shrimp egg hatching to better understand the natural lake system.
- **Planning and carrying out investigations** collaboratively, making observations over time, and combining resulting data to serve as the basis for making a decision about the proper salt environment for brine shrimp egg hatching. Make a prediction about the viability of eggs that were in conditions outside the range of tolerance for hatching.
- **Analyzing and interpreting data** of brine shrimp hatching in different environmental conditions by representing data in tables and charts to reveal patterns, and using those patterns to make sense of phenomena using logical reasoning.
- **Using mathematics and computational thinking** by organizing simple data sets to show the relationship between salinity and brine shrimp egg hatching.
- **Constructing explanations** about the range of tolerance for salt for brine shrimp egg hatching in salt lake environments.
- **Engaging in argument from evidence** to evaluate cause-and-effect claims about brine shrimp egg hatching.
- **Obtaining, evaluating, and communicating information** from books and media and integrating that with their firsthand experiences to construct explanations and the functioning of ecosystems.

NGSS Foundation Box for SEP

- **Use a model** to test cause-and-effect relationships to interactions concerning the functioning of a natural system.
- **Plan and conduct an investigation** collaboratively to produce data to serve as the basis for evidence.
- **Make observations and/or measurements to produce data** to serve as the basis for evidence for an explanation of a phenomenon or test a design solution.
- **Make predictions** about what would happen if a variable changes.
- **Represent data in tables and/or various graphical displays** to reveal patterns that indicate relationships.
- **Analyze and interpret data** to make sense of phenomena using logical reasoning.
- **Organize simple data sets** to reveal patterns that suggest relationships.
- **Use evidence** (e.g., measurements, observations, patterns) to support an explanation.
- **Identify the evidence** that supports particular points in an explanation.
- **Construct an argument** with evidence, data, and/or a model.
- **Read and comprehend grade-appropriate complex texts** and/or other reliable media to summarize and obtain scientific and technical ideas and describe how they are supported by evidence.
- **Obtain and combine information** from books and other reliable media to explain phenomena.
- **Communicate scientific and/or technical information** orally and/or in written formats, including various forms of media and may include tables, diagrams, and charts.

Environments Module—FOSS Next Generation

INVESTIGATION 3 – Brine Shrimp Hatching

> **NGSS Foundation Box for CC**
> - **Cause and effect:** Cause-and-effect relationships are routinely identified and used to explain change.
> - **Structure and function:** Substructures have shapes and parts that serve functions.
> - **Systems and system models:** A system can be described in terms of its components and their interactions.

Exposing Crosscutting Concepts (CC)

In this investigation, the focus is on these crosscutting concepts.

- **Cause and effect.** Organisms have structures that function by responding to environmental conditions.

- **Structure and function.** The forms of an organism's structures relate to their functions in a particular environment.

- **Systems and system models.** A lake ecosystem can be described in terms of its living and nonliving components; models are useful in testing the effect of environmental factors on the functioning of the system.

Connections to the Nature of Science

- **Scientific investigations use a variety of methods.** Scientific methods are determined by questions. Scientific investigations use a variety of methods, tools, and techniques.

- **Scientific knowledge is based on empirical evidence.** Science findings are based on recognizing patterns. Scientists use tools and technologies to make accurate measurements and observations.

- **Science is a human endeavor.** Men and women from all cultures and backgrounds choose careers as scientists and engineers. Most scientists and engineers work in teams. Science affects everyday life. Creativity and imagination are important to science.

Brine
Brine shrimp
Concentration
Controlled experiment
Inherited trait
Migrate
Optimum
Range of tolerance
Reproduce
Salinity
Salt lake
Survive
Thrive
Tolerance
Variation
Viable

Teaching Children about Environmental Factors and Variation

Conceptual Flow

Students investigate the phenomenon of **brine shrimp** eggs hatching and adult shrimp living in saltwater environments. The guiding question for the investigation is how is optimum environment related to organism and population survival?

The **conceptual flow** for this module starts when students are introduced to the brine shrimp organism that lives in a **specialized environment—salt lakes**. Salt is the defining factor in the aquatic environment of the brine shrimp. Students set up an experiment to discover the concentration of salt that supports the hatching and survival of brine shrimp.

In Part 2, students analyze the brine shrimp hatching results to determine the range of salt concentrations in which eggs hatch to determine their **range of salt tolerance, and the optimum salt concentration for supporting the hatching of brine shrimp**.

In Part 3, students explore the **viability** of brine shrimp eggs after they have been exposed to saltwater environments that are outside the range of tolerance (salt solutions too concentrated or too dilute). Students find that **brine shrimp eggs maintain their viability** and respond by hatching when salt concentration is corrected to fall within the range of tolerance.

In Part 4, students engage in a simulation, which introduces the idea that the members of a population are not all identical, but actually exhibit a significant amount of variation (color and size for example). **Variation within a population** is inherent in all populations. Those variations can give individuals an advantage in surviving and reproducing in certain conditions. Adults that have the advantage might reproduce and their offspring will inherit those traits and have those same characteristics that allow them to survive in the changed environment.

Complex Systems

- *Salt lakes and ponds are specialized aquatic environments characterized by high concentrations of salts.*
- *Brine shrimp eggs hatch within a range of salt concentration.*
- *Within the range of tolerance, one salt concentration represents the optimum concentration for egg hatching.*
- *Viability is the ability of an embryo to emerge from dormancy to become active.*
- *Brine shrimp eggs remain viable even after inundation in concentrated and dilute solutions.*
- *Individuals within a population exhibit variation and that variation might provide an advantage.*
- *Offspring inherit traits from their parents; parents that thrive in an environment will produce offspring with a greater chance of surviving.*

Environments Module—FOSS Next Generation

233

INVESTIGATION 3 – Brine Shrimp Hatching

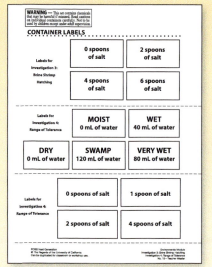

No. 14—Notebook Master

No. 15—Teacher Master

MATERIALS for
Part 1: Setting Up the Experiment

For each student
- 1 Safety goggles ★
- ❑ 1 Notebook sheet 14, *Brine Shrimp–Hatching Experiment*
- 1 *FOSS Science Resources: Environments*
  - "Brine Shrimp"

For each group
- 5 Plastic cups
- 4 Cup lids
- 4 Self-stick notes or prepared labels
- 1 Vial with cap
- 1 Minispoon
- 1 Spoon, 5 mL
- 1 Craft stick
- 1 FOSS tray
- 1 Container, 1 L
- 1 Beaker, 100 mL

For the class
- Brine shrimp eggs
- Kosher salt
- 16 Tray columns, plastic tube
- Bottled water (See Step 4 of Getting Ready.) ★
- Transparent tape
- ❑ 8 Teacher master 15, *Container Labels* (optional)

For embedded assessment
- *Performance Assessment Checklist*

★ Supplied by the teacher.   ❑ Use the duplication master to make copies.

Full Option Science System

Part 1: Setting Up the Experiment

GETTING READY for
Part 1: Setting Up the Experiment

1. **Schedule the investigation**
   This part will take one session and is referred to as day 1, since it is the first day of a multi-day experiment. The timing of subsequent, consecutive days for observation and further testing is critical. It's best to begin Part 1 on a Monday, followed by Part 2 on a Tuesday, Wednesday, and Thursday (days 2, 3, and 4), with Part 3 on a Friday.

2. **Preview Part 1**
   Students investigate the environmental factor of salinity in hatching brine shrimp eggs. They conduct a controlled experiment to determine which one of four salt concentrations allows brine shrimp eggs to hatch. The focus question is **How can we find out if salinity affects brine shrimp hatching?**

3. **Familiarize yourself with the two spoons**
   This investigation uses two spoons. The larger 5 milliliter (mL) spoon is used to measure salt. The smaller minispoon is used to measure the brine shrimp eggs. Make sure students use the appropriate spoon for each purpose.

4. **Obtain bottled water**
   Each group will need 600 mL of bottled water (rather than using aged tap water). Obtain 6 L of bottled water for this part.

5. **Test water and eggs ahead of time**
   We recommend that you test the suitability of your local bottled water and the viability of your brine shrimp eggs a week or so before you plan to do the investigation. Put 150 mL of bottled water in two cups. Add two 5 mL spoons of salt to one cup and label it "2 spoons salt" and three 5 mL spoons of salt to the other cup and label it "3 spoons salt." Transfer 1 level minispoon of brine shrimp eggs to each cup. Put the lids on and gently swirl the cups to wet the eggs.

   In 24 to 48 hours, at room temperature, you should see the tiny brine shrimp larvae swimming about. You must look closely and you might want to use a flashlight. Any movement of the water will interfere with your ability to see hatched shrimp, so leave the cup on the table and look into it from the side or down from the top (take the lid off). Eggs float, but larvae swim about.

   If no eggs have hatched after 72 hours, the viability of the eggs is suspect. Viability is certain for a year or two, so if your eggs are old, replace them. Eggs can be purchased from Delta Education or from tropical-fish stores.

▶ **NOTE**
To prepare for this investigation, view the teacher preparation video on FOSSweb.

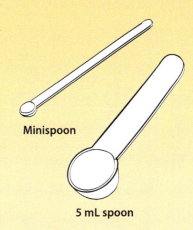

Minispoon

5 mL spoon

▶ **NOTE**
Transfer the bottled water into a 1 L container for each group before the investigation begins.

▶ **SAFETY NOTE**
Find out if any students are allergic to seafood and/or shellfish as they are likely to also be allergic to brine shrimp.

▶ **NOTE**
Temperature is a variable in brine shrimp egg hatching. In colder temperatures hatching takes longer.

Environments Module—FOSS Next Generation

INVESTIGATION 3 – Brine Shrimp Hatching

▶ **SAFETY NOTE**
Remind students to follow safety rules when working with chemicals. Safety goggles should be worn at all times when working with chemicals.

6. **Prepare cups of salt**
Each group will need a plastic cup half full of kosher salt. Kosher salt is provided because it is pure sodium chloride. Brine shrimp don't hatch well in iodized salt. Noniodized table salt contains a free-running agent, sodium aluminosilicate, that makes the water cloudy. The brine shrimp will hatch, but they will be more difficult to see.

7. **Prepare the vials of brine shrimp eggs**
Each group will need about 4 minispoons of brine shrimp eggs. Measure 4 minispoons of eggs into one vial, then pour an equivalent volume of eggs into a vial for each group. The eggs will just cover the bottom of the vial. Cap the vials.

8. **Prepare labels**
Students can prepare labels for their brine shrimp hatcheries with self-stick notes and tape, or they can tape on labels copied from teacher master 15, *Container Labels*.

9. **Plan a hatchery area**
The FOSS trays are designed to be stacked using the plastic tube columns. Stack the trays two high so that students can easily make observations.

10. **Plan to view video**
Preview the video, *The Mono Lake Story* (duration 27 minutes). After watching the entire video, you can select short video clips to use at the beginning of Part 1 to introduce students to the Mono Lake environment and to the phenomenon of brine shrimp. This is described in the EL Note by Step 1 of Guiding the Investigation and it would be good for all students.

The Mono Lake Story is listed as an extension for this investigation. Along with the main video, their are four short clips some students might be interested in viewing. Links to the videos are in the Resources by Investigation section on FOSSweb.

11. **Plan to read *Science Resources*: "Brine Shrimp"**
Plan to read "Brine Shrimp" during a reading period after completing the active investigation for this part.

12. **Plan assessment: performance assessment**
Students set up a controlled experiment to determine the effect of salinity on the hatching of brine shrimp eggs. This will provide a good opportunity for you to observe students' scientific practices. Carry the *Performance Assessment Checklist* with you as you visit the groups while they are working. For more information about what to look for during observations, see the Assessment chapter.

Part 1: Setting Up the Experiment

GUIDING *the Investigation*
Part 1: *Setting Up the Experiment*

1. **Introduce *salt lake* and *brine***
   Review the different types of environments that students have been investigating, both terrestrial and aquatic. Then explain,

   ▶ One specialized kind of environment found around the world is a **salt lake**. What do you think a salt-lake environment is like? What might live there? [This environment has salt and water, but is smaller than the ocean. Plants and animals that live there might be similar to those living in the ocean.]

   Tell them that the word **brine** means "salty water." A brine pond is smaller than a lake but, like a salt lake, it is very salty.

2. **Introduce the problem**
   Give some background information.

   There is a lake in Northern California called Mono Lake. Long ago it was quite large, but for thousands of years it has been drying up and getting smaller. As a result, Mono Lake is a salt lake today. As the level of the lake falls, the **concentration** of the salt, or **salinity**, increases. Salt concentration means the amount of salt in a given amount of water.

   Introduce the challenge. Tell them,

   Dr. Salina Bryan is an ecologist who has been interested in the Mono Lake environment for a long time. She studies the organisms that live in and around the lake. Dr. Bryan has found that thousands of **migrating** birds depend on small aquatic animals called **brine shrimp** for food. Migrating birds move from one place to another, moving from the north to the south for the winter months.

   Dr. Bryan is concerned. If the lake continues to get saltier, either because the lake is drying up or because the fresh water flowing into the lake is reduced, the salty environment might affect the population of brine shrimp.

3. **Introduce brine shrimp**
   Hold up one of the vials and show the brine shrimp eggs to the class. Tell students,

   Dr. Bryan needs to find out if the amount of salt in the water, the salinity of the water, is an environmental factor that affects the hatching of Mono Lake's brine shrimp eggs. She needs to design an experiment to find an answer to her question.

FOCUS QUESTION

How can we find out if salinity affects brine shrimp hatching?

EL NOTE

Show or project a picture of Mono Lake from the FOSS Science Resources book or show a short video clip from The Mono Lake Story video. Do the same for the brine shrimp. Start a new word wall and add the vocabulary words with a simple illustration.

Environments Module—FOSS Next Generation

237

INVESTIGATION 3 – Brine Shrimp Hatching

4. **Discuss Dr. Bryan's question**
   Ask,

   ▶ *In order to help Dr. Bryan answer her question, how would you set up an experiment to find out how the amount of salt in the water affects brine shrimp egg hatching? How can you find out the best salinity for hatching brine shrimp eggs?*

   Listen to students' ideas. Students may suggest placing eggs in a series of salinities and observing the hatching results over time.

5. **Focus question: How can we find out if salinity affects brine shrimp hatching?**
   Write or project the focus question on the board as you say it aloud.

   ▶ *How can we find out if salinity affects brine shrimp hatching?*

   Have students write the focus question in their notebooks.

6. **Discuss an experimental procedure**
   Propose the plastic cups (hold one up) as brine shrimp hatcheries. Guide students to think through a **controlled experiment** in which all the factors are the same except for the amount of salt—same kind of cup, same amount of water, same number of eggs, same temperature, but different amounts of salt. Write the elements of the experiment on the board as the procedure evolves.

7. **Confirm the procedure**
   Distribute notebook sheet 14, *Brine Shrimp–Hatching Experiment*. Carefully go through the instructions for setting up the experiment. Hold up each piece of equipment as it is mentioned on the sheet. Make sure that all students can associate the written name of the measuring tool with the actual tool.

   a. Label four cups—0 spoons salt, 2 spoons salt, 4 spoons salt, and 6 spoons salt. Write your group name on each label.

   b. Measure 150 mL of water into each of the cups, using the 100 mL beaker.

   c. Measure salt into the labeled cups, using the 5 mL spoon. Use a craft stick to level the measure. Put lids on the cups, and gently swirl them to dissolve the salt.

   d. Put 1 level minispoon of brine shrimp eggs into each cup. Put the lids on, and gently swirl the water to wet the eggs.

   e. Place all four cups on a FOSS tray.

TEACHING NOTE

After giving the focus question, let students determine how to conduct the experiment. Encourage them to look back at how they tested the isopods. If students are stumped, then continue with Steps 6 and 7.

▶ **SAFETY NOTE**

Find out if any students are allergic to seafood and/or shellfish as they are likely to also be allergic to brine shrimp.

TEACHING NOTE

Since each group will be doing exactly the same experiment, the class data will represent multiple trials. Discuss how conducting multiple trials improves the accuracy of experimental results.

Part 1: Setting Up the Experiment

8. **Set up the hatching experiment**
   Each group can decide how it would like to divide up the following tasks.
   - Get containers of water and other materials.
   - Label cups.
   - Measure water into the cups.
   - Measure and add salt.
   - Measure brine shrimp eggs into the cups.

   Let Getters get materials for their four experimental cups. Allow 15–20 minutes for students to set up the experiment.

9. **Assess progress: performance assessment**
   Circulate from group to group. Listen to the group discussions and observe how students work together. Make notes on the *Performance Assessment Checklist* while students work.

   ### What to Look For
   - *Students are engaged in and contributing to conducting a well-reasoned investigation. (Planning and carrying out investigations.)*
   - *Students can describe how the investigation models possible conditions in a saltwater environment. (Developing and using models; LS2.C: Ecosystem dynamics, functioning, and resilience; systems and system models.)*
   - *Students understand by keeping the water amount the same, but increasing the amount of salt they add to various containers that they are varying salinity. (Using mathematics and computational thinking.)*

10. **Clean up**
    When the experiments are set up, have Getters return the materials to the materials station. Return the unused eggs to the supply vial *if they are dry* and return the salt to its storage container.

11. **Set up the brine shrimp hatchery**
    When all the groups have set up the experiment, have a Getter from each group bring the FOSS tray to the hatchery area. Stack trays two high, using the tray columns.

Materials for Step 8
- *Cups with lids*
- *Labels*
- *Cups of salt*
- *Spoons, 5 mL*
- *Beakers, 100 mL*
- *Vials of eggs*
- *Minispoons*
- *Craft sticks*
- *FOSS trays*
- *Containers, 1 L*
- *Bottled water*
- *Safety goggles*
- *Transparent tape*

SCIENCE AND ENGINEERING PRACTICES

Developing and using models

Planning and carrying out investigations

Using mathematics and computational thinking

DISCIPLINARY CORE IDEAS

LS2.C: Ecosystem dynamics, functioning, and resilience

CROSSCUTTING CONCEPTS

Systems and system models

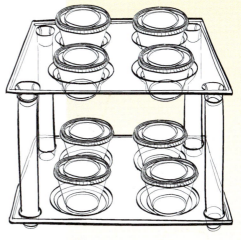

Environments Module—FOSS Next Generation

INVESTIGATION 3 – Brine Shrimp Hatching

brine
brine shrimp
concentration
controlled experiment
migrating
salinity
salt lake

E L N O T E

For students who need scaffolding, provide sentence frames such as, I predict _____ because _____. We can find out by _____.

TEACHING NOTE

Informally review some notebooks and look to see if students are writing about doing a controlled experiment in which one factor (salt in this case) is isolated and tested. All other factors are held constant (controlled).

12. Review vocabulary
Add new words to the word wall.

- *Brine* means salty water.
- *Brine shrimp* are tiny crustaceans related to crabs and lobsters. Brine shrimp are found in brine ponds and salt lakes.
- *Concentration* refers to the amount of solid material, such as salt, in a volume of water. Two spoons of salt in 150 mL of water makes a more concentrated salt environment than one spoon of salt in 150 mL of water. The higher the concentration of salt, the saltier the water.
- *Migrating* birds travel from one place to another. The birds that fly to Mono Lake are migrating, or moving south for the winter months.
- *Salinity* is how salty water is, or the concentration of salt in water.
- A *salt lake* is a body of water that contains salt.

13. Answer the focus question
Ask students to answer the focus question by writing a summary of the procedure they are using to test different salinities.

➤ *How can we find out if salinity affects brine shrimp hatching?*

Part 1: Setting Up the Experiment

READING *in Science Resources*

14. Read "Brine Shrimp"
This short one-page article describes the natural history of the brine shrimp. Tell students that this article will give them more information about the organism they are investigating in the classroom. Point out the photograph and ask students what they notice, what the purpose of the image is, and what they can learn from looking at it. Ask,

▶ *What do you notice about the brine shrimp structures?*

▶ *What do you think the functions of these structures are?*

Before reading, ask students to make a list of three things they think they already know about brine shrimp and a question they have about the photograph. Give students a few minutes to share their ideas with their groups and call on a few students to share questions. Have students read the article independently noting any words or phrases they are not sure of.

For reading strategies to support English learners and below-grade-level readers, see the Science-Centered Language Development chapter.

15. Discuss the reading
Ask students to review their lists of things they knew about brine shrimp before they read the article and to discuss any new or different ideas they learned from the text with their group. Ask,

▶ *How do the structures of the brine shrimp and their function compare to the other organisms we've been investigating?*

You might remind students of other crustaceans they have studied—isopods, *Gammarus*, and perhaps crayfish from grade 3.

Review any unknown words or phrases and questions students have about the text. Then ask them to summarize the text using the summarizing chart below. Students can work with a partner or their group to come up with a description of the brine shrimp, the food it eats, where it lives, and some interesting facts they learned.

Brine Shrimp	
Description	Food
Habitat	Interesting facts

ELA CONNECTION

These suggested reading strategies address the Common Core State Standards for ELA.

RI 2: Determine the main idea of a text and explain how it is supported by key details; summarize the text.

RI 4: Determine the meaning of general academic domain-specific words or phrases.

RI 6: Compare and contrast a firsthand and secondhand account of the same topic.

RI 7: Interpret information presented visually, and explain how the information contributes to an understanding of the text.

W 8: Gather relevant information from experiences and print, and categorize the information.

SCIENCE AND ENGINEERING PRACTICES

Obtaining, evaluating, and communicating information

Environments Module—FOSS Next Generation

INVESTIGATION 3 – Brine Shrimp Hatching

SCIENCE AND ENGINEERING PRACTICES
Constructing explanations

CROSSCUTTING CONCEPTS
Structure and function

16. Discuss survival in saltwater environments

Ask students to talk in their groups about what survival challenges brine shrimp have living in salt lakes. Have students refer back to the basic needs of organisms (oxygen, food, water, stable body temperature, safe place for young, shelter and protection from predators). Ask each Reporter to offer a challenge for a class list.

Then, have students generate ideas about structures or systems of structures brine shrimp have for survival in salt lakes. If they don't have ideas, help students generate specific questions. For example,

➤ *What brine shrimp structures take oxygen from the water?* [Aquatic organisms usually have gills, so the brine shrimp should too.]

➤ *How do the structures of the brine shrimp help them get food?* [They gather tiny microscopic bits of food with salt water. The food goes through a tube to their gut (stomach tube).]

➤ *How do the structures of the brine shrimp help them to get rid of salt from the water?* [They might have some kind of active pump that takes the salt out of the water and gets rid of it.]

➤ *How do they protect their young?* [They lay eggs that can survive in dry conditions and be dormant for up to three years, until the environmental conditions are right for hatching.]

You can provide more information about brine shrimp structures.

- Brine shrimp have an internal fluid that moves gases and nutrients through their bodies (similar to blood). This fluid contains proteins that bind strongly to oxygen (similar to the way hemoglobin in human blood carries oxygen).

- The "skin" of the shrimp is waterproof, so the only way for water to enter the shrimp is through the mouth with food.

- Water comes in through the mouth and goes through a tube to the gut, where the gut lining absorbs salt so it is not digested. Salt is moved into the gills, and actively pumped out.

- Brine shrimp have a second salt pump in a gland in their neck.

WRAP-UP/WARM-UP

16. Share notebook entries

Conclude Part 1 or start Part 2 by having students share notebook entries with a partner. Read the focus question together.

➤ *How can we find out if salinity affects brine shrimp hatching?*

Students should predict what they think might happen in each of the salinity conditions.

Part 2: Determining Range of Tolerance

MATERIALS for
Part 2: Determining Range of Tolerance

For each student
- 1 Hand lens
- ❏ 1 Notebook sheet 15, *Brine Shrimp–Hatching Observations*
- ❏ 1 Notebook sheet 16, *Brine Shrimp–Hatching Conclusions*
- 1 *FOSS Science Resources: Environments*
    - "The Mono Lake Story"
    - "What Happens When Ecosystems Change?"

For each group
- 1 FOSS tray with hatchery cups (from Part 1)
- 1 Set of Mono Lake Food Web cards, 12 cards per set
    - Bottom algae
    - Brine fly
    - Brine shrimp
    - California gull
    - Caspian tern
    - Coyote
    - Eared grebe
    - Floating algae
    - Halobacteria
    - Red-necked phalarope
    - Snowy plover
    - Wilson's phalarope
- 1 Marking pen ★

For the class
- White paper (or reuse arrow strips from Part 4 of Investigation 2) ★
- 1 Paper cutter ★
- 1 Projection system (optional) ★
- Low-power microscopes (optional) ★
- Safety goggles ★
- Computers with Internet access ★

For embedded assessment
- ❏ *Embedded Assessment Notes*

★ Supplied by the teacher.   ❏ Use the duplication master to make copies.

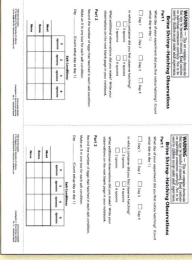

No. 15—Notebook Master

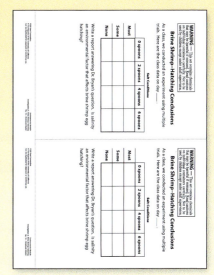

No. 16—Notebook Master

Environments Module—FOSS Next Generation

243

INVESTIGATION 3 – Brine Shrimp Hatching

> **NOTE**
> Timing is critical to the success of this investigation. Part 1 starts on Monday, Part 2 starts on Tuesday, and Part 3 starts on Friday of the same week.

GETTING READY for
Part 2: Determining Range of Tolerance

1. **Schedule the investigation**
   This part will take three sessions in the same week. The first session of Part 2 should take place the day after Part 1 (Tuesday). On Wednesday and Thursday, students make experimental observations and read articles. On Thursday, students also draw conclusions from the experiment. Part 3 of the investigation, Determining Viability, should start on Friday.

2. **Preview Part 2**
   Students monitor saltwater environments. They determine which environments are conducive to hatching brine shrimp eggs. Students analyze the results of a multiple-trial experiment conducted by the class and draw conclusions. They read about the Mono Lake ecosystem and create food webs using the organism cards and information in the article. The focus question is **How does salinity affect the hatching of brine shrimp eggs?**

3. **Check the Mono Lake Food Web cards**
   Each group will need one set of 12 Mono Lake Food Web cards. Make sure that a complete set is in each zip bag.

4. **Make arrow strips**
   Students will create a Mono Lake food web by placing organism cards in a logical array and connecting them with arrows to indicate what eats what. Prepare paper strips ahead of time, using a paper cutter. Place a standard sheet of paper in a landscape orientation on the cutter and cut strips about 2 cm wide (about 3/4"). This makes strips 2 × 21 cm. One sheet of paper will make 14 strips, enough for one group.

   Recycled paper is fine for this as long as one side of the paper is relatively clean. Students will draw the arrows on the strips and can make more strips by cutting the long strips in half to make shorter strips.

Full Option Science System

Part 2: Determining Range of Tolerance

5. **Preview the video**
   See the Interdisciplinary Extensions for information on the video *The Mono Lake Story* available from the Mono Lake Committee. The video will complement the reading.

   You will find the video on FOSSweb in Resources by Investigation and in the streaming video section.

6. **Plan to read "The Mono Lake Story" and "What Happens When Ecosystems Change?"**
   Plan to have students read two articles "The Mono Lake Story" and "What Happens When Ecosystems Change?" on consecutive days after they observe the experiments.

7. **Plan for online activities**
   Students can gain more experience with the Mono Lake food web by exploring the food web simulation on FOSSweb. Preview the simulation by going to the Resources by Investigation section on FOSSweb. There are three other ecosystems with food webs to study in this online activity (desert, woods, kelp forest).

8. **Plan assessment: notebook entry**
   In Step 27 of Guiding the Investigation, check whether students can state a claim and support it with evidence from the multiple-trial investigation to argue how salinity as an environmental factor affects brine shrimp hatching.

Environments Module—FOSS Next Generation **245**

INVESTIGATION 3 – Brine Shrimp Hatching

FOCUS QUESTION

How does salinity affect the hatching of brine shrimp eggs?

Materials for Step 2
- *FOSS trays with hatcheries*
- *Hand lenses*
- *Safety goggles*

▶ **SAFETY NOTE**

Find out if any students are allergic to seafood and/or shellfish as they are likely to also be allergic to brine shrimp.

SCIENCE AND ENGINEERING PRACTICES

Planning and carrying out investigations

EL NOTE

For students who need scaffolding, provide sentence frames such as, *I observed _____. Before _____, now _____.*

GUIDING *the Investigation*
Part 2: *Determining Range of Tolerance*

1. **Focus question: How does salinity affect the hatching of brine shrimp eggs?**
   Write or project the focus question on the board as you say it aloud.

   ▶ *How does salinity affect the hatching of brine shrimp eggs?*

   Have students write the focus question in their notebooks.

2. **Observe hatcheries on day 2**
   On the day after setting up hatcheries, have the Getters get their group's tray. Let students check each cup for changes. The first brine shrimp might hatch on day 2 (the day after setup), but will surely hatch by day 3.

   Students should let the cups sit undisturbed for a minute before making observations. Unhatched dark eggs will move around the containers with the slightest disturbance, making it hard to distinguish hatched from unhatched eggs. But once the systems are still, students will be able to look through the side of the cup or down from the top to see the minute, white larvae swimming around in different directions with jerky motions. Students may be surprised that the newly hatched brine shrimp are so tiny. Students can use hand lenses to make observations.

3. **Introduce the record sheet**
   When the groups observe hatching, have the Getter from each group get four copies of notebook sheet 15, *Brine Shrimp–Hatching Observations*. Students should glue the notebook sheet into their science notebooks. Have students answer the questions in Part 1. Ask them to write more about their observations on the next blank page in their notebooks.

4. **Store the trays**
   Have Getters return the trays to the hatchery area.

BREAKPOINT

5. **Observe hatcheries on day 3**
   On the second day after setting up hatcheries, have Getters get their group's tray. Let students check each cup for changes and record observations in their notebooks.

246   Full Option Science System

Part 2: Determining Range of Tolerance

READING *in Science Resources*

6. Read "The Mono Lake Story"

Read the article "The Mono Lake Story," using the strategy that is most effective for your class. The article situates the brine shrimp in the Mono Lake ecosystem and introduces the living and nonliving factors that influence the food web.

Have students preview the text by looking at and discussing the photographs and diagrams with a reading partner. They should also read the subtitles and captions. This is a good opportunity to have students practice summarizing and responding to text. Have students set up their notebook to take two-column notes, titled "Summary" and "Response." In the first column students write a summary of each section. In the second column they write their own personal response to the text. Model by reading the first section aloud, summarizing the section, and adding your personal response. Continue reading each section aloud and pausing for students to summarize and respond or have students work independently or in guided reading groups.

7. Focus on diagrams

Before reading the last section, have students examine and discuss the diagrams. Ask students what they notice about the structure of these diagrams. Explain that these are web diagrams. This type of diagram shows many interconnecting lines or arrows that link the parts of a system. Ask students why this type of diagram is useful. [It shows the relationship among many parts of a system.]

Ask students to discuss what the parts of this food web system are and what the arrows represent. [Producers, composers, decomposers, and how energy is transferred.] Have students interpret the Mono Lake Food Web diagram with a partner.

First look at the images of the organisms and read the captions. Determine which are the producers, consumers, and decomposers. Use your finger to trace through the food web. Start with an organism and explain its relationship to the other organisms. As you trace through the arrows, say "is eaten by" or "transfers energy to."

Refer students to the simple diagram on the next page and ask,

➤ What is different about this food web?

➤ Why is the Sun represented?

➤ What do the arrows mean? [Here the arrows mean "transfers energy to." Algae is not eaten by the Sun. Algae need energy from the Sun to make its own food.]

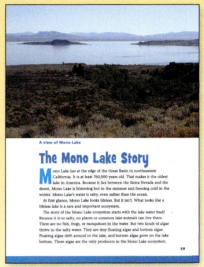

Summary	Responses

SCIENCE AND ENGINEERING PRACTICES

Constructing explanations

Obtaining, evaluating, and communicating information

ELA CONNECTION

These suggested reading strategies address the Common Core State Standards for ELA.

RI 2: Determine the main idea of a text and explain how it is supported by key details; summarize the text.

RI 7: Interpret information presented visually, and explain how the information contributes to an understanding of the text.

W 8: Gather relevant information from experiences and print, and categorize the information.

Environments Module—FOSS Next Generation

247

INVESTIGATION 3 – Brine Shrimp Hatching

8. Discuss competition for resources
Tell students,

Many of the animals in the Mono Lake ecosystem eat more than one kind of organism. And some organisms are eaten by many animals. When you show all the possible feeding interactions, you no longer have a simple food chain but rather a food web, with lines going off in different directions.

Refer to the Mono Lake food web diagram.

➤ *Describe two kinds of competition in the Mono Lake ecosystem.* [Grebes and phalaropes both eat brine shrimp; lots of birds eat brine flies.]

9. Discuss the reading
Have students share their summaries and responses with their groups. Review any unknown words or phrases and the strategies students used to determine their meaning. Call on a few students to share something they heard another student in their group say that was interesting. Give students a few minutes to discuss the questions at the end of the article in their groups. Call on reporters to share their group's answer to a question, along with details and examples from the text. Allow other group reporters to respond by either agreeing, disagreeing and offering alternative explanations or evidence, or adding on more information. After the discussion, let students revise or add to the original answers and record them in their notebooks.

➤ *What is the main environmental factor that affects the health of the Mono Lake ecosystem? Explain your answer.* [Salt concentration. Salt concentration was the only factor that changed in the cups.]

➤ *Why did the California gull chicks not survive at Mono Lake in 1982?* [There were fewer shrimp for the gulls to eat; the lower water level allowed foxes to cross the land bridge to eat the gull eggs.]

10. Make a Mono Lake food web
Distribute a set of Mono Lake Food Web cards to each group. Show students the paper strips. There are about 14 strips for each group. Explain that students can draw arrows on the strips with a marking pen and use the arrows to connect the organisms in a food web. If they want to make a shorter arrow, they can cut the strip in half and draw a shorter arrow. The arrows will show the feeding interactions in the ecosystem.

ELA CONNECTION
These suggested reading strategies address the Common Core State Standards for ELA.

RI 4: Determine the meaning of general academic domain-specific words or phrases.

SL 2: Paraphrase information presented orally.

SL 4: Report on a text in an organized manner, using appropriate facts and relevant, descriptive details to support main ideas or themes; speak clearly at an understandable pace.

Materials for Step 10
- *Mono Lake Food Web cards*
- *Paper strips*
- *Marking pens*

Part 2: Determining Range of Tolerance

Challenge each group to make a food web for the Mono Lake ecosystem, with producers, consumers, and decomposers. They should use arrows marked on the paper strips to show how the food moves through the food web. Each student should record the food web in his or her science notebook. Allow 15 minutes for the groups to make their food webs.

11. Monitor progress
Circulate from group to group. Listen to the group discussions and observe how students work together.

12. Discuss decomposers
As you monitor the group work, and see how they are incorporating the decomposers (bacteria) into the food web, ask them to explain where they think the bacteria card should go in the web. Some groups may decide that the bacteria card should be in a central location and lines should go from every other organism to the bacteria card.

13. Record the food web
Have students draw their food webs in their notebooks using the organisms described on pages 64–65 of the *FOSS Science Resources* book. The organisms are:

- coyotes
- bottom algae
- grebes
- floating algae
- phalaropes
- brine shrimp
- gulls
- bacteria
- brine flies

Have students take turns interpreting their food webs aloud. Before students put the cards away, have them "tell the story" with the cards of what happened to Mono Lake when the water was diverted from the lake. If students need a hint, have them refer back to the Mono Lake Story article. They should start by taking out the algae and brine fly cards, then the brine shrimp, California gull, etc. Discuss how changing the environmental factor of water level had a ripple effect on the entire Mono Lake ecosystem.

14. View online activity
Have students engage with the online activity "Food Webs" and use the interactive to construct a Mono Lake food web. Students can work in small groups or as individuals.

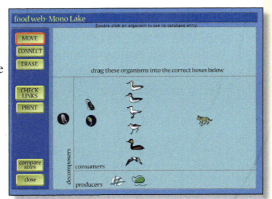

> **TEACHING NOTE**
>
> For students having difficulty getting started, you might suggest that they organize the cards into producers, herbivores, carnivores, and decomposers.

> **SCIENCE AND ENGINEERING PRACTICES**
>
> Developing and using models

> **ELA CONNECTION**
>
> This suggested reading strategy addresses the Common Core State Standards for ELA.
>
> SL 4: Report on a text in an organized manner, using appropriate facts and relevant, descriptive details to support main ideas or themes; speak clearly at an understandable pace.

Environments Module—FOSS Next Generation

INVESTIGATION 3 – Brine Shrimp Hatching

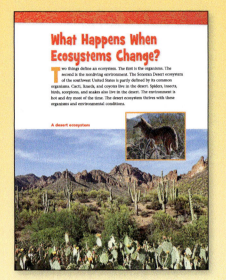

SCIENCE AND ENGINEERING PRACTICES

Obtaining, evaluating, and communicating information

CROSSCUTTING CONCEPTS

Cause and effect

ELA CONNECTION

These suggested reading strategies address the Common Core State Standards for ELA.

RI 1: Refer to details and examples in a text when explaining what the text says explicitly and when drawing inferences from text.

RI 5: Describe the overall structure of information in a text.

READING *in Science Resources*

15. Read "What Happens When Ecosystems Change?"

Review with students what they learned about changes in ecosystems from reading the Mono Lake Story and using the food web cards to make a model of what happened. Write the question and title of the article "What Happens When Ecosystems Change?" on the board or chart paper. Have students Think-Pair-Share this question and record a few of their ideas on the chart paper under a title, ("Our thinking"). Then write "Questions we have:" and repeat the process. Tell students that this article gives another example of how changing one environmental factor in an ecosystem can affect the entire system.

Refer students to the first two pages and have them describe the living and nonliving environmental factors they observe in the photographs of the desert ecosystem and the tropical marine ecosystem. Read aloud or have students read the first two pages independently.

Bring out the Environments Concept Grid from Investigation 1 and update it with any new information students found from the first two pages of the reading.

16. Have students use a graphic organizer

Tell students that the next three pages will tell the story of a major change in the Colorado River Basin ecosystem. Add that thinking about the cause-and-effect relationships between the different environmental factors will help them understand why the ecosystem changed.

Cause	Effect

Suggest that students make a cause-and-effect graphic organizer in their notebooks to help them discover what changed and why. Or, use the problem/solution graphic organizer from the previous investigation. Read the rest of the article aloud or have students read independently.

---

250 Full Option Science System

Part 2: Determining Range of Tolerance

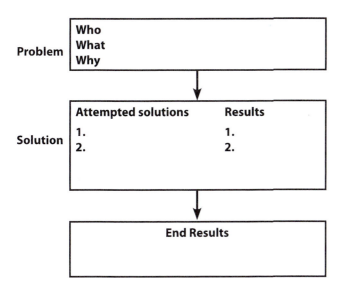

17. Discuss the reading

Review the cause-and-effect or problem/solution relationships that students recorded on their graphic organizers and any unknown words or phrases. Discuss the concepts of range of tolerance using examples from the previous investigations with mealworms and darkling beetles.

Give students a few minutes to discuss the question at the end of the article in their groups. Call on a Reporter from each group or individual to share their answer to a question along with details and examples from the text. Allow other group Reporters to respond by either agreeing, disagreeing and offering alternative explanations or evidence, or adding more information. After the discussion, let students refine, revise, or add to their original answers and record them in their notebooks.

➤ *How did the change of water temperature after the dam was completed affect the chub population? Why?* [First, the water released from the Glen Canyon dam made the river water below the dam colder. This cold water was good for the trout and the trout population increased. The increased trout population put more pressure on the chub population because trout eat chub. Second, the spring floods down river were stopped by the dam. The spring floods used to wash the willow trees and plants away, but this no longer happened. As a result, the shallow water habitat needed by baby chubs was destroyed. Both of these things caused the chub population to get so small that the chub was put on the endangered species list.]

ELA CONNECTION

These suggested reading strategies address the Common Core State Standards for ELA.

RI 3: Explain procedures, ideas, or concepts in a scientific text.

RI 4: Determine the meaning of general academic domain-specific words or phrases.

Environments Module—FOSS Next Generation

INVESTIGATION 3 – Brine Shrimp Hatching

> **TEACHING NOTE**
>
> This is an excellent opportunity for students to collect and analyze detailed information about brine shrimp using a microscope. Have them record their observations each day with drawings and compare the shrimp structures and features that can be seen as the shrimp grow.

SCIENCE AND ENGINEERING PRACTICES

Using mathematics and computational thinking

18. Make comparisons on day 4
On day 4, have students observe and compare the hatching success in each of their four containers. They should judge the hatching success of each cup based on a rough estimate of the density of larvae: none, some, or most of the eggs hatched.

19. Use a projection system (optional)
You may want to use a projection system (document camera or microscope) to help students judge how many is "some" and how many is "most." Borrow two cups with hatched brine shrimp from one group. Put them on the projector, and focus the image on a screen. Have students compare the two concentrations of hatchlings to determine which one they think has more hatching.

20. Record observations on day 4
Have students fill in the chart in Part 2 of the *Brine Shrimp–Hatching Observations* sheet. They should put an *X* in each column to indicate the hatching success in each concentration of salt for their experiment.

Day 4   Salt Conditions

	0 spoons	2 spoons	4 spoons	6 spoons
Most		X		
Some			X	
None	X			X

21. Organize class results
Draw a large replica of the chart on the board. Have one member of each group come up and transfer his or her *X*s to the chart on the board. Have all students record the class results on notebook sheet 16, *Brine Shrimp–Hatching Conclusions*.

Day 4   Salt Conditions

	0 spoons	2 spoons	4 spoons	6 spoons
Most		XXX	XXX XXX	
Some		XXX XX	XX	
None	XXXX XXXX			XXXX XXXX

252   Full Option Science System

Part 2: Determining Range of Tolerance

22. Have a sense-making discussion
Ask,

➤ *Were there cups that had conditions favorable for brine shrimp hatching? Which one(s)?* [The ones with 2 and 4 spoons of salt.]

➤ *Were there cups that had conditions that were unfavorable for brine shrimp hatching? Which one(s)?* [The ones with 0 and 6 spoons.]

➤ *What do these results tell us about the effect of salinity on brine shrimp hatching?* [There is an upper salinity limit and a lower salinity limit for brine shrimp hatching.]

You can scaffold this last question by asking,

➤ *Do brine shrimp eggs need salt in order to hatch?* [Yes.]

➤ *Can you give brine shrimp eggs too much salt?* [Yes.]

Focus on the use of models to understand natural phenomena.

➤ *How does our experiment help us understand Mono Lake?*

23. Introduce *range of tolerance*
Explain that brine shrimp eggs hatched not just in one salt condition but in a range of salt conditions. These conditions define the range of salt tolerance for hatching the eggs. **Tolerance** is the ability of an organism to survive under a given set of conditions. Every organism has a **range of tolerance** for each factor in its environment. To find the range for one factor, you can set up an experiment where all the factors are the same except for one.

➤ *What is the range of tolerance that brine shrimp eggs have for the environmental factor of salt (salinity)?* [10–20 mL of salt in 150 mL of water.]

24. Introduce *optimum*
Explain that within the range of tolerance, the condition or conditions where an organism does best is called the **optimum**.

➤ *Which salt condition(s) seemed to be optimum (best) for brine shrimp hatching?* [10 mL of salt in 150 mL of water seems good but 20 mL of salt in 150 mL of water might be better.]

Emphasize this last point: the best environmental conditions for an organism are its optimum conditions.

25. Review vocabulary
Review the new vocabulary on the word wall.

- The varying conditions of one environmental factor in which an organism can survive is called *range of tolerance*.
- *Optimum* environmental conditions are the best conditions for an organism.

TEACHING NOTE

Go to FOSSweb for Teacher Resources and look for the Science and Engineering Practices—Grade 4 chapter for details on how to engage students with the practice of analyzing and interpreting data.

SCIENCE AND ENGINEERING PRACTICES

Developing and using models

Analyzing and interpreting data

CROSSCUTTING CONCEPTS

Cause and effect

EL NOTE

Have students add these new words to their glossary or index. This is a good opportunity to do a vocabulary activity to help students understand the relationships between this new concept and the concepts developed in previous investigations. See the Science-Centered Language Development chapter for suggestions.

 optimum
range of tolerance
tolerance

Environments Module—FOSS Next Generation

INVESTIGATION 3 – Brine Shrimp Hatching

SCIENCE AND ENGINEERING PRACTICES

Constructing explanations

EL NOTE

For students who need scaffolding, provide a framework for writing a formal report:
Dear Dr. Bryan:
We are investigating _____.
Our question was _____.
We found out that _____.
Our evidence is _____.
Therefore,
I recommend _____.

26. Answer the focus question

Remind students that Dr. Bryan wanted to know if salinity is an environmental factor that affects the hatching of Mono Lake's brine shrimp eggs. Restate the focus question,

➤ *How does salinity affect the hatching of brine shrimp eggs?*

To answer the focus question, have students write a report to Dr. Bryan. Their reports should address how the water salinity affects the hatching of brine shrimp eggs. They should cite evidence from the investigation to support their conclusions. Collect and review the notebooks.

27. Assess progress: notebook entry

After class, look at students' reports to get a sense of how well they can analyze and interpret data from the experiment to provide evidence for their claims.

What to Look For

- Salinity does affect brine shrimp hatching.
- Evidence from the class investigation is used to support the students' claims regarding range of tolerance and optimum conditions of salt for the best hatching.

28. Clean up

Ask Getters to return the trays to the hatchery area.

29. View online activity

Have students engage with the online activity "Food Webs." This time, use the interactive to construct food webs of other ecosystems (wetlands, kelp forest, desert). Students can work in small groups or as individuals.

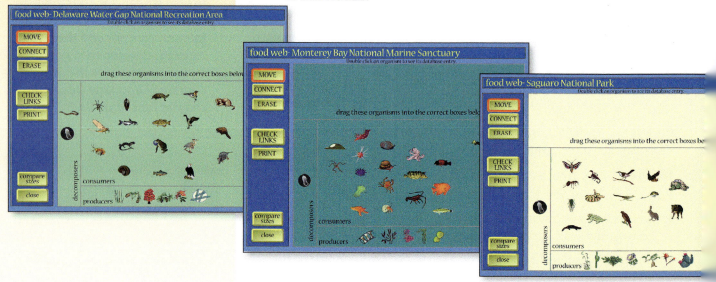

Full Option Science System

Part 2: Determining Range of Tolerance

WRAP-UP/WARM-UP

30. Share notebook entries

Conclude Part 2 or start Part 3 by having students share notebook entries. Ask students to open their science notebooks to the most recent entry. Read the focus question together.

➤ *How does salinity affect the hatching of brine shrimp eggs?*

Ask students to pair up with a partner to discuss what they determined about hatching brine shrimp eggs.

Students could also discuss what other environmental factors of the brine shrimp environment they could test to determine the optimum conditions. Or they could discuss what environmental factors they could test for goldfish, guppies, isopods, or mealworms.

31. Use the anonymous student work strategy

If student responses are not up to expectations, try the anonymous student work strategy. Make up a set of 4–5 responses ranging from not so good to fairly good. Tell students you are going to show them some responses to the focus question that students did last year. Project the responses one by one starting with the best one and have students critique them with their partner. Ask,

➤ *What is good about this response?*

➤ *What would make it better?*

After students have critiqued the responses, give them a few minutes to revise their own responses. See the assessment section or the notebook chapter for other next-step strategies.

EL NOTE

For students who need scaffolding, provide a sentence frame such as, Brine shrimp eggs will hatch when ____. We observed that ____. Therefore, I think ____.

ELA CONNECTION

This suggested reading strategy addresses the Common Core State Standards for ELA.

W 5: Strengthen writing by revising.

Environments Module—FOSS Next Generation

255

INVESTIGATION 3 – Brine Shrimp Hatching

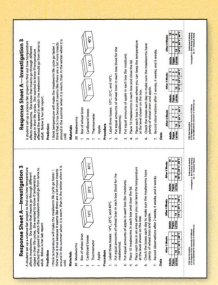

No. 17—Notebook Master

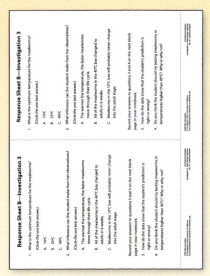

No. 18—Notebook Master

MATERIALS for
Part 3: Determining Viability

For each student
1   Hand lens
1   Safety goggles ★
1   *FOSS Science Resources: Environments*
- "The Shrimp Club"

For each group
1   FOSS tray with hatchery cups
2   Plastic cups
1   Spoon, 5 mL
1   Craft stick
1   Container, 1 L
1   Container, 1/2 L
2   Self-stick notes
1   Beaker, 100 mL

For the class
- Kosher salt
2   Pitchers
- Bottled water (1–2 L) ★
1   Collecting net
- Yeast (optional) ★
- Computers with Internet access ★

For embedded assessment
❑   • Notebook sheets 17–18, *Response Sheet A* and *B— Investigation 3*
❑   • *Embedded Assessment Notes*

★ Supplied by the teacher.   ❑ Use the duplication master to make copies.

Part 3: Determining Viability

GETTING READY *for*
Part 3: *Determining Viability*

1. **Schedule the investigation**
   This part should start on Friday, day 5 of the brine shrimp investigation, and will take two sessions for active investigation and one reading session. The weekend will provide 2 days for the eggs to hatch. The second active session should take place on Monday.

2. **Preview Part 3**
   Students are challenged to manipulate the environment to see if they can get the dormant eggs to hatch and grow. They formulate and justify predictions and design an investigation to test their predictions. The focus question is **Does changing the salt environment allow the brine shrimp eggs to hatch?**

3. **Have salt and water available**
   Students may decide to mix the hatcheries without salt and those with six spoons of salt into a 1/2 L container and then pour half the mixture into cups to observe over time. Other students will add water to the cups with six spoons of salt, and salt to the cups without salt. Make sure you have enough bottled water and salt available for the groups to implement their plans.

4. **Plan for the future of brine shrimp**
   Plan to discuss with students what to do with the brine shrimp at the end of the experiment. Without food (yeast) they will not survive after day 5. If you want to grow the brine shrimp to maturity, see teacher master 18, *Home/School Connection*.

   ▶ **SAFETY NOTE**
   Brine shrimp require little care and are fascinating to observe.

5. **Plan to read *Science Resources*: "The Shrimp Club"**
   Plan to read "The Shrimp Club" during a reading period after setting up the investigation.

6. **Plan for online activity**
   Preview "Virtual Investigation: Trout Range of Tolerance." Students analyze trout survival data and make a recommendation about the building of a dam. The link to this activity for teachers is in Resources by Investigation on FOSSweb.

7. **Plan assessment: response sheet**
   Use notebook sheets 17 and 18, *Response Sheet A and B—Investigation 3*, for a closer look at students' ability to read and interpret an experimental report. Sheet 17 describes the investigation and provides the data. The questions for students to respond to are on sheet 18. You can read sheet 17 aloud to the class or to small groups. Plan to spend time discussing the sheets with students after you have reviewed them.

Environments Module—FOSS Next Generation

INVESTIGATION 3 – Brine Shrimp Hatching

FOCUS QUESTION
Does changing the salt environment allow the brine shrimp eggs to hatch?

Materials for Step 1
- FOSS trays with hatcheries

SCIENCE AND ENGINEERING PRACTICES
Planning and carrying out investigations

Materials for Step 4
- Cups of salt
- Spoons, 5 mL
- Beakers, 100 mL
- Craft sticks
- Containers, 1/2 L
- Containers, 1 L
- Self-stick notes
- Bottled water
- Safety goggles

TEACHING NOTE
Viable eggs in the 0-spoons and 6-spoons environment should start to hatch in 2 days after the salinity comes into range. Conduct the final observations session on Monday.

GUIDING the Investigation
Part 3: Determining Viability

1. **Distribute the trays**
   Have Getters get the hatchery trays. Give students 2 minutes to check the containers for brine shrimp larvae. Ask,

   ➤ *Did any more brine shrimp eggs hatch?*

   Listen to students' observations.

2. **Question the viability of unhatched eggs**
   Tell students,

   When something is alive and able to live and grow, it is said to be viable. **Viable** *means alive or able to grow.*

   ➤ *Do you think the eggs that didn't hatch are viable or dead? How could we find out if the eggs that have not hatched in the cups are still viable?*

3. **Focus question: Does changing the salt environment allow the brine shrimp eggs to hatch?**
   Write or project the focus question on the board.

   ➤ *Does changing the environment allow the brine shrimp eggs to hatch?*

   Have students write the focus question in their notebooks.

4. **Formulate a prediction and test it**
   Students may suggest adding salt to the 0-spoons cup and more water to the 6-spoons cup. Acknowledge these as good ideas. Ask students to work in their groups to come up with a plan.

   a. *Talk about how much salt and/or water you will add to the "extreme" cups to bring the salt concentration within the range of tolerance for hatching.*

   b. *Predict what changes to the environment will promote hatching. Write out your plan in your science notebook, including the amount of salt or water you plan to add, and the reason for adding that amount.*

   c. *Get my approval for your prediction and your plan.*

   d. *Set up the investigation, making sure to relabel the cups.*

   Let Getters get materials and put their plan into action. Ask one group to leave its cups unaltered as a control.

5. **Return the trays to the hatchery area**
   Have students return their trays to the hatchery area and return the unused materials to the materials station.

Part 3: Determining Viability

BREAKPOINT

6. Observe the results of the viability test

After 2 days (on Monday) ask Getters to retrieve their group's tray and observe the results. After a few minutes of observation, ask,

➤ *Did changing the salinity in the extreme conditions allow the eggs to hatch?* [Yes.]

➤ *Did a lot of eggs hatch or did only a few hatch?*

➤ *Were the eggs in the 0-spoons and 6-spoons cups still viable?* [Yes.]

➤ *How do brine shrimp eggs respond to their environment?* [It depends. If the environmental factor of salinity is within their range of tolerance, the eggs hatch. If it isn't, they don't hatch.]

➤ *What advantage is it to the brine shrimp to delay hatching in salt solutions that are very dilute or very concentrated?* [They delay hatching until conditions are right for their survival.]

7. Review vocabulary

Add new words to the word wall.

- *Viable* means alive and able to grow.

8. Answer the focus question

Ask students to describe in their science notebook what happened in each of the brine shrimp hatcheries and discuss what that means about the viability of brine shrimp eggs.

➤ *Does changing the salt environment allow the brine shrimp eggs to hatch?*

9. Have a sense-making discussion

Help students make the connection between range of tolerance, preferred environment, and responding to environmental conditions with these questions.

➤ *How do organisms respond when they are in an environment that is not in their range of tolerance?*

➤ *What do beetles and isopods do when they aren't in their preferred environment?*

➤ *How is this similar to the hatching of brine shrimp eggs?*

➤ *What are other examples of how animals respond to conditions outside of their range of tolerance?* [Animals migrate, hibernate or go dormant, dig burrows, and so on.]

Materials for Step 6
- Hand lenses
- FOSS tray with hatcheries

TEACHING NOTE

This is an opportunity to discuss adaptation. The egg remains viable for a long time in and out of water until conditions are in the range of tolerance. This is an adaptation that brine shrimp have.

 viable

TEACHING NOTE

Refer to the Sense-Making Discussions for Three-Dimensional Learning chapter in Teacher Resources *on FOSSweb for more information about how to facilitate this with students.*

SCIENCE AND ENGINEERING PRACTICES

Constructing explanations

Environments Module—FOSS Next Generation

INVESTIGATION 3 – Brine Shrimp Hatching

SCIENCE AND ENGINEERING PRACTICES

Analyzing and interpreting data

Constructing explanations

EL NOTE

You can read the investigation description on notebook sheet 17 aloud to the class or in small groups. Then have students work alone to write a response.

TEACHING NOTE

*See the **Home/School Connection** for Investigation 3 at the end of the Interdisciplinary Extensions section. This is a good time to send it home with students.*

---

10. Assess progress: response sheet

Distribute a copy of notebook sheets 17–18, *Response Sheet—Investigation 3*, *Parts A* and *B* to each student. Tell students that sheet 17 describes an investigation and sheet 18 includes questions. They should work alone to write a response. Collect the science notebooks after class and review students' responses. Discuss the sheets with students after you review them.

What to Look For

- *The optimum temperature is 40°C.*
- *The best inference is A: Given the conditions tested, the warmer the temperature, the faster mealworms move through their life cycle.*
- *Students write that the data support the prediction because there were 10 mealworms in each condition and the mealworms overall become adults faster at 25°C than at 10°C and at 40°C than at 25°C.*
- *Students write that it might be possible to increase the temperature a little (a few degrees) but they would need to research to see what temperature might harm the mealworms.*

11. Clean up

When students have completed their brine shrimp investigations, have them decide what they would like to do with the brine shrimp that have hatched. The brine shrimp can be put in a separate aquarium and provided with yeast as food.

See the Science Extensions in the Interdisciplinary Extensions section for some suggestions to continue and extend the investigations with brine shrimp.

Any dead brine shrimp and water should be poured down the sink.

Make sure all cups and lids are rinsed with water (don't use soap) and dried thoroughly before storing them in the kit.

Part 3: Determining Viability

READING *in Science Resources*

12. Read "The Shrimp Club"

Read the article "The Shrimp Club" using the strategy that is most effective for your class. The article describes how a school became involved in an environmental action project that had lasting and positive impact on the local environment.

Begin by letting students preview the text by looking at and discussing the photographs with a partner. Ask students what they notice about how this article is structured. [Question and answer.] Confirm that the text is the write-up of an interview with Laurette Rogers, a former 4th grade teacher. Explain that Ms. Rogers and her students started a project to save an endangered species—the California freshwater shrimp. Say,

Ms. Rogers and her class were on a mission to save the California freshwater shrimp, but they accomplished more than that and learned a great deal about the environment and themselves. As you read the article, pause to think about how you would feel if you were one of Ms. Rogers students.

Read the article aloud or have students read independently. If students enjoy reading with a partner, suggest they trade off (for example, one reads the question, the other reads the response and then switch).

For reading strategies to support English learners and below-grade-level readers, see the Science-Centered Language Development chapter.

13. Discuss the reading

Have students make two columns in their notebook. Title the left-hand column "What the article is about" and the other "What the article makes me think about." Tell students to start with the right column first. They should record how the article makes them feel, what it reminds them of, and any personal connections or questions they have. Give students a few minutes to share their thoughts with a partner.

Then, ask students to record a summary of what the article was about. Tell students to keep the summary brief, pick out the most important ideas, and to say it in their own words.

Conduct a whole class discussion using these questions as a guide.

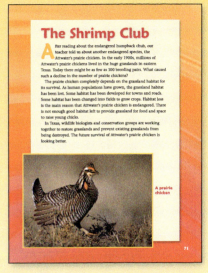

SCIENCE AND ENGINEERING PRACTICES

Obtaining, evaluating, and communicating information

ELA CONNECTION

These suggested reading strategies address the Common Core State Standards for ELA.

RI 2: Determine the main idea of a text and explain how it is supported by key details; summarize the text.

RI 5: Describe the overall structure of information in a text.

RI 7: Interpret information presented visually, and explain how the information contributes to an understanding of the text.

W 8: Gather relevant information from experiences and print, and categorize the information.

Environments Module—FOSS Next Generation

INVESTIGATION 3 — Brine Shrimp Hatching

➤ *What is a watershed?* [The area of land that drains into a particular body of water. The watershed is usually named for the body of water into which the water flows. The watershed starts at the highest point of land, and all the water that flows into a single creek defines a watershed.]

➤ *What is a watershed restoration project?* [A process of returning the land to a more natural environment. This often involves planting native plants to provide habitats for native animals.]

➤ *What activities did the class do as part of the Shrimp Club restoration project?* [Besides restoring the natural environment, they raised money to purchase materials, developed a newsletter to share information with the school community, made public presentations, met with ranchers to get their participation, wrote to government officials, testified at local governmental hearings, visited with their congressional representatives, and worked with professional restoration scientists.]

➤ *Why are watershed restoration projects important?* [To protect native plants and animals and sustain native ecosystems.]

➤ *Is there a restoration project going on in your community?*

➤ *How can you get involved in a Citizen Science Project?*

ELA CONNECTION

These suggested reading strategies address the Common Core State Standards for ELA.

RI 9: Integrate information from two texts on the same topic.

SL 1: Engage in collaborative discussions.

14. **Read more about human impact**
    This is a good opportunity for students to learn more about another environmental project. See FOSSweb for a list of recommended books. One suggestion is *Once a Wolf: How Wildlife Biologists Fought to Bring Back the Gray Wolf* by Jim Brandenburg.

15. **View online activity: "Virtual Investigation: Trout Range of Tolerance"**
    In small groups or as individuals, have students engage with the online activity focusing on trout survival when freshwater conditions change due to the construction of a dam. The link to this activity for teachers is in the Resources by Investigation on FOSSweb, and for students, in the Online Activities.

262

Full Option Science System

Part 3: Determining Viability

WRAP-UP/WARM-UP

16. Share notebook entries

Conclude Part 3 or start Part 4 by having students share notebook entries. Ask students to open their science notebooks to the most recent entry. Read the focus question together. Ask students to pair up with a partner to discuss what they found out about the viability of brine shrimp eggs.

To help students use data to evaluate cause-and-effect claims, try the multiple corners strategy. Post three or four different claims about the brine shrimp investigation without supporting data or evidence, each on a separate piece of chart paper. For example:

1. Brine shrimp will hatch if the water is salty enough.
2. Brine shrimp will hatch in water if there is some salt in it.
3. Brine shrimp will hatch if you put them in water.
4. Brine shrimp need water and salt to hatch.

Place each claim poster in a different corner of the room. Divide the class so there is an even number of students at each claim. In the first round, have students write two points of data under the claim on the poster that either supports or refutes the claim. They should not identify whether the data supports or refutes the claim, just provide the data. The class then rotates to a different claim.

In the second round, students verify that the data posted accurately represent the class findings and identifies whether the data supports or refutes the claim by writing "supports" or "refutes." The class rotates one more time.

At the last claim poster, students edit the claim for accuracy or revise the claim to best represent available information.

Allow time for students to revise their own responses.

TEACHING NOTE

Go to FOSSweb for Teacher Resources and look for the Science and Engineering Practices—Grade 4 chapter for details on how to engage students with the practice of engaging in argument from evidence.

SCIENCE AND ENGINEERING PRACTICES

Engaging in argument from evidence

CROSSCUTTING CONCEPTS

Cause and effect

ELA CONNECTION

These suggested reading strategies address the Common Core State Standards for ELA.

SL 1: Engage in collaborative discussions.

W 5: Strengthen writing by revising.

Environments Module—FOSS Next Generation

INVESTIGATION 3 – Brine Shrimp Hatching

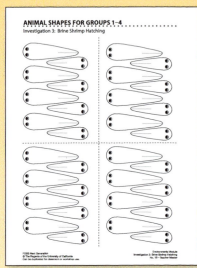

No. 16—Teacher Master

MATERIALS for
Part 4: Variation in a Population

For each student

- 1 *FOSS Science Resources: Environments*
  - "Variation and Selection"
- 1 Zip bag, 1 L

For the class

- 16 Boundary marker flags
- 2 Basins, 8 L
- • Marking pens ★
- • Chart paper ★
- 2 Animal shape sheets, gray
- 2 Animal shape sheets, green
- 2 Animal shape sheets, brown
- 2 Animal shape sheets, red
- 1 Camera (optional) ★
- ❏ 4 Teacher master 16, *Animal Shapes for Groups 1–4* (optional)

For benchmark assessment

- ❏ • *Investigation 3 I-Check*
- ❏ • *Assessment Record*

★ Supplied by the teacher.   ❏ Use the duplication master to make copies.

Part 4: Variation in a Population

GETTING READY for
Part 4: *Variation in a Population*

1. Schedule the investigation
This part will take one outdoor session. Plan another session for the reading and two sessions for review and benchmark assessment.

2. Preview Part 4
Students go to the schoolyard in two teams, to place a population of imaginary animals in a suitable habitat based on a description of the population's natural history. Through a predator-prey simulation, students find out how variations in color and size within a population affect survival of the population. The focus question is **What are some benefits of having variation within a population?**

3. Select your outdoor site
Select an area in the schoolyard with two different habitats that are each about 10 meters (m) × 10 m (33' × 33'). That's about half the size of a tennis court. Suitable areas may contain a lawn at the edge of the asphalt, shrubs, packed dirt, bark mulch, rocks, and so forth. The two areas should be separated by approximately 10 m.

For example, your schoolyard might have packed soil with a few grass spots for one habitat and the weedy edge along a fence for another. The shape of your selected sites may be determined by your schoolyard—a long thin rectangle may work best in some schoolyards and a square area might suit in others.

4. Plan for animal shapes
For a class of 32 students, you will need two full sheets of each color of animal shapes: gray, green, red, and brown, for a total of eight sheets. These full sheets are provided in the kit. The organisms differ in color and size.

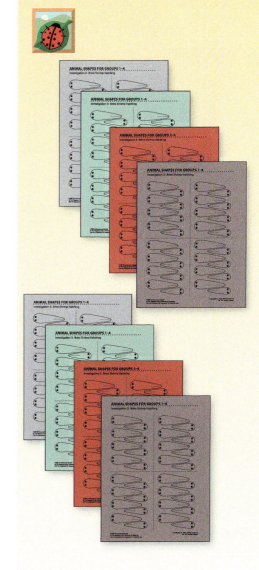

The class will be divided into two teams of 16 students each. Each team will use four full sheets, one of each color, to prepare their imaginary animal population.

Prepare the animal shapes for one team (16 students). Cut each of four sheets (one of each color) into quarters so that you end up with 16 quarter sheets, four of each color. Each student on the team will receive one of the quarter sheets and cut out the eight shapes on the quarter sheet. All of the animal shapes for one team are placed in a basin. Once outdoors, team members will randomly select 8 animals so that each student ends up with a few animals of each color to place in the team's habitat.

Environments Module—FOSS Next Generation

INVESTIGATION 3 – Brine Shrimp Hatching

If you need to prepare your own sheets of animal shapes, use teacher master 16, *Animal Shapes*, and colored paper to prepare the sheets for the teams. You can make the copies on regular paper or on card stock. These sheets can also be purchased as replacement part items from Delta Education.

5. Plan for materials
Students will cut out their organisms indoors from the sheets you provide. Prepare to instruct students how to cut a little slit (about 2 cm long) along the dotted line at the tail end of the organism. Students can slide the slit onto a blade of grass or a leaf to help keep the organisms from blowing away in the wind or to put them in more creative places. Practice this so you can demonstrate it to students.

6. Prepare population table on chart paper
During the simulation, you will record and display the population results on a piece of chart paper or portable whiteboard. If necessary, you could use a large, flat cardboard box. Design the table indoors and take it outdoors for the activity.

7. Check the site
The morning before taking students to your selected site, do a quick search for potentially distracting or unsightly items.

8. Enlist additional adults
If possible, seek out an additional adult to join you outdoors. Remind the adult that students will need time to struggle with the challenges and that his or her job is to lend support but not to solve the problems for students.

9. Bring a camera
If possible, bring a camera outdoors so students can take photos of the camouflaged animals in the team's habitat. Students could print copies of the photos and place them in their notebooks.

10. Plan benchmark assessment: I-Check
Plan to give *Investigation 3 I-Check* at the conclusion of this part. Print or copy the assessment masters or schedule students on FOSSmap. Students should complete their responses to the items independently.

The Assessment Coding Guides are on FOSSweb. After you review and code students' work, plan which items you will bring back for additional discussion and self-assessment. See the Assessment chapter for guidance and information about online tools.

Variation in a Population Activity Results

	Habitat 1		Habitat 2	
	Starting number	Ending number	Starting number	Ending number
Brown				
Gray				
Green				
Red				

Part 4: Variation in a Population

GUIDING *the Investigation*
Part 4: *Variation in a Population*

1. **Review** *population*
   Ask students to name some of the animal populations they have learned about. Remind students that a population is a group of animals or plants of one kind that lives in the same area. Ask,

   ➤ *Do our goldfish look exactly the same? What differences might we see within this population?*

   ➤ *Do our guppies look exactly the same? What differences might we see within this population?*

   ➤ *Do our isopods look exactly the same? What differences might we see within this population?*

   ➤ *Do our brine shrimp look exactly the same? What differences might we see within this population?* [Any observable differences in appearance will be really hard to see because they are so small. But we do know that not all of the eggs hatched at the same time so there is some variation in the brine shrimp.]

   Explain that the differences between individuals of the same species are called **variations**. Consider a few other animals and name some of the variations (e.g., cows vary in color, spot locations, and size).

2. **Introduce the outdoor experience**
   Explain to students that they are going to introduce a new population of animals into the schoolyard. Hold up a large sample of the paper organism students will work with. Pull out a smaller shape of the same color, and ask the class how the organisms differ. Make sure students know that size is often a variation within a population. Hold up several shapes of different colors and ask how the organisms vary. Say,

   Today you will work with a population of animals that vary in color and size. We will introduce this population into our schoolyard. We have green, red, gray, and brown organisms.

3. **Read populations' natural history**
   Read the natural history of the organisms in the population based on the two outdoor areas you selected.

   These organisms are members of a population called _____. They live in schoolyards. They live in several habitats in those schoolyards including _____ and _____. They eat _____. Their predators include several kinds of birds, such as jays and _____. They protect themselves by trying to blend in with their environment.

FOCUS QUESTION

What are some benefits of having variation within a population?

Materials for Step 2
- *Animal shapes of four colors*

EL NOTE

If necessary, show sample pictures of cows, dogs, cats, etc. Write the word variation *on the word wall.*

TEACHING NOTE

You can assure students this is a simulation activity. No real animal populations will be introduced into the schoolyard.

Environments Module—FOSS Next Generation

267

INVESTIGATION 3 – Brine Shrimp Hatching

> **NOTE**
> Each team will observe as the members of the opposite team become predators and attempt to find the hidden organisms in the habitat.

Materials for Step 4
- *Animal shapes for groups*
- *Scissors*

Explain that the class will be divided into two teams and that each team will have its own habitat. Each team will place equal numbers of these organisms (and equal number of each color) in the appropriate habitats so their eyes are showing. Then the teams will switch habitats and act as predators trying to find the organisms placed by the other team.

4. **Focus question: What are some benefits of having variation within a population?**
Write or project the focus question on the board as you say it aloud.

   ➤ *What are some benefits of having variation within a population?*

   Have students write the focus question in their notebooks.

5. ## Prepare cutout shapes
   Tell students that before they go outdoors to place their populations, they need to cut out the shapes they will use. Divide the class into groups. Make scissors available. Distribute the appropriate animal shapes to each group, making sure that you distribute an equal number to both teams and the same number of each color. Each student in a group should take one of the quarter sheets and carefully cut out the eight shapes.

   Instruct students to make the cut on the dotted line to create a slit that can be used to hold an organism onto a blade of grass or onto the branches of a plant.

   When everyone has cut out about eight shapes, have all the members from one team place their organisms in a basin and mix up the color. Then each student should randomly select about eight animals to take outdoors.

6. ## Introduce the results table
   Display the table you prepared on chart paper and make sure students understand what should be recorded. Insert the numbers for the starting population for each color of organisms. The starting numbers for each color should be the same.

7. ## Review outdoor safety
   Remind students of the rules and expectations for the outdoor work. Refer to the *FOSS Outdoor Safety* poster.

8. ## Go outdoors
   Make sure students are dressed appropriately and have their organisms in their hand. Move outdoors to a gathering place somewhere between the two selected habitats.

Part 4: Variation in a Population

9. **Circle up and review**

   Gather students together in a circle. Review the details of the natural history of the organisms (refer to Step 3).

   Explain that when the organisms are placed in the appropriate habitats, they cannot be hidden but their bodies must be visible with their eyes showing. Introduce the two habitats and mark the boundaries with marker flags as the students watch.

10. **Place animal populations**

    Students should spread out within their defined habitat for the population of animals and begin placing their organisms in "safe" locations. Remind students that part of the body must be fully visible. Give students several minutes to place their organisms.

11. **Release the predators**

    When populations are in their habitats, bring one team to the other team's site. Have the team who placed the organisms in that site sit down on the edge of the boundaries to watch the predators hunt. Give each of the predators a plastic bag (1 L). The plastic bags will represent the predator's stomach. Give predators about 45 seconds to find as many organisms as possible. You might want to stop the search before all the organisms have been collected.

    Count and record on the chart paper table the number of survivors of each color. Then move to the second site and repeat the hunt. Make sure to give both teams the same amount of hunting time (about 45 seconds).

12. **Have a sense-making discussion**

    After you record the results for the second team, ask students to return to the gathering spot and sit so they can see the table on the chart paper. Ask,

    ➤ *What variations did we have in the starting population?* [Color and size.]

    ➤ *Which colors and which sizes of organism were easiest for predators to find?*

    ➤ *What do the data tell us?*

    ➤ *What variation in the population was most successful?*

    ➤ *If that variation was passed on from parents to offspring, what might that mean for the survival of the next generation?*

    ➤ *How do you think the results might change during another season? Why?*

    ➤ *What if the environment changed? How would that impact the survival of each variation?*

SCIENCE AND ENGINEERING PRACTICES

Planning and carrying out investigations

Analyzing and interpreting data

Constructing explanations

Variation in a Population Activity Results

	Habitat 1		Habitat 2	
	Starting number	Ending number	Starting number	Ending number
Brown	32	23	32	27
Gray	32	8	32	18
Green	32	16	32	8
Red	32	1	32	14

Data from one class. Results will vary depending on the nature of the two habitats.

CROSSCUTTING CONCEPTS

Systems and system models

Environments Module—FOSS Next Generation

INVESTIGATION 3 – Brine Shrimp Hatching

SCIENCE AND ENGINEERING PRACTICES

Constructing explanations

inherited trait
reproduce
survive
thrive
variation

13. Return to class

Have students return to the habitats and gather all remaining organisms. Gather the equipment and return to the classroom.

14. Discuss variation

Ask,

➤ *What are some benefits of having variations within a population?* [Individuals of the same kind differ slightly in their characteristics (variations), and sometimes the variations give individuals an advantage in surviving and reproducing. Adults that have the advantage might **reproduce** and the offspring will have characteristics that allow them to **survive** and **thrive** in the changed environment. The offspring might inherit the variation from the parents. That is called an **inherited trait**. Variations within the individuals in a population make it possible for a kind of organism (species) to survive when the environment changes.]

Bring the concept of variation back to the brine shrimp hatching investigation. Tell students that not all shrimp in a population have exactly the same tolerance to salinity. That is true for the shrimp eggs and for other stages in the life cycle of the brine shrimp.

➤ *How might variation in salt tolerance affect survival of a population of brine shrimp?* [If salt concentration changes, some eggs will hatch, grow, and reproduce, keeping the population going.]

15. Review vocabulary

Review the vocabulary dealing with variation in individuals in populations. Add new words to the word wall.

16. Answer the focus question

Have students answer the focus question in their notebooks. You might suggest that they consider the populations they have studied in the classroom to answer the question.

➤ *What are some benefits of having variation within a population?*

Review a sample of student notebooks after completing this part.

Part 4: Variation in a Population

READING *in Science Resources*

17. Read "Variation and Selection"

Read the article "Variation and Selection," using the strategy that is most effective for your class. The article introduces the concept of natural selection, based on variation of individuals in a population.

Tell students that this article will give them more information about the concept of variation and selection. Have students preview the text by looking at and discussing the photographs and the illustration.

Next, tell students to set up their notebooks in order to record questions and notes for each section. They could use one page for each section or divide a page in halves or fourths. Describe the procedure:

1. *For each section write the subtitle.*
2. *Scan each section reading only the captions and bolded words.*
3. *Write at least one question for each section.*

Read the article aloud or have students read independently. Pause after each section for students to write a brief summary and record any information that might help answer the question for that section.

18. Discuss the reading

Give students a few minutes to compare their summaries and to share their questions and answers. To check for understanding, write each of the questions below one at a time on the board. Give students time to talk over the answers with a partner before sharing out with the whole group.

▶ *Why don't all dogs look exactly alike?* [All dogs are thought to have come from a common ancestor, a wolf. Over 2,500 years, dogs have been selectively bred for certain characteristics. Once you have a male and female dog that both have the desirable characteristics, you breed those dogs and the resulting pups will most likely also have the same desirable characteristics. Over many, many generations of dogs, different characteristics were selected for, and the result is that today, we have many different kinds of dog breeds. Selective breeding depends on having variation in the population.]

▶ *In nature, what is the mechanism that selects the individuals that will breed to produce offspring in a population?* [The living or nonliving environment—natural selection, as compared to artificial selection.]

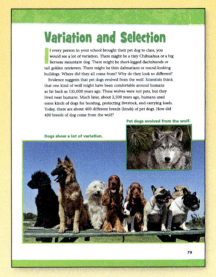

SCIENCE AND ENGINEERING PRACTICES

Obtaining, evaluating, and communicating information

ELA CONNECTION

These suggested reading strategies address the Common Core State Standards for ELA.

RI 2: Determine the main idea of a text and explain how it is supported by key details; summarize the text.

RI 7: Interpret information presented visually, and explain how the information contributes to an understanding of the text.

SL 4: Report on a text in an organized manner, using appropriate facts and relevant, descriptive details to support main ideas or themes; speak clearly at an understandable pace.

Environments Module—FOSS Next Generation

INVESTIGATION 3 – Brine Shrimp Hatching

➤ *Describe what is meant by "a change in the environment might apply a new pressure on a population."* [Individuals in a population have to meet their needs in order to survive—air, water, food, space including shelter—and they also need to find a suitable mate and successfully produce offspring in order for the population to remain viable. When the living or nonliving environment changes, individuals in the population might have to find new food sources or a new habitat. This is the pressure placed on the population. Those individuals that are adapted to withstand the pressure will reproduce. The traits of the survivors will be passed to the next generation as inherited traits.]

➤ *What information has come from the studies of Darwin's finches on the Galápagos Islands?* [The environmental pressure of changing food sources over thousands of years has resulted in 13 different species of finches, each with a slightly different size or shape of beak that can eat different kinds of food. This is an example of natural selection.]

After the discussion, have students draw a line of learning under their last entry and write a reflection on something new they learned from the article. If necessary, give students a prompt such as,

"I used to think _____ , but now I think _____ ."

"I still have a question about _____ ."

"Some interesting things I learned today were _____ ."

"I was surprised by _____ ."

ELA CONNECTION

This suggested reading strategy addresses the Common Core State Standards for ELA.

W 9: Draw evidence from informational texts to support reflection.

Part 4: Variation in a Population

WRAP-UP

19. Review Investigation 3

Distribute one or two self-stick notes to each student. Ask students to cut each note into three pieces so each piece has a sticky end.

Ask students to take a few minutes to look back through their notebook entries to find the most important things they learned in Investigation 3. Students should include at least one science and engineering practice, one disciplinary core idea, and one crosscutting concept. They should tag those pages with self-stick notes. They might use a highlighter or colored pencils to call out the key points.

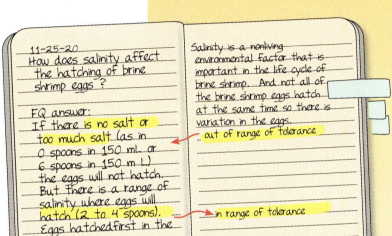

Lead a short class discussion to create a list of three-dimensional statements that summarize what students have learned in this investigation. Here are examples of the big ideas that should come forward in this discussion.

- We planned and carried out an investigation to determine that brine shrimp have ranges of tolerance for salt concentration in their aquatic environment. (Planning and carrying out investigations; using mathematics and computational thinking.)
- We used data to support the claim that brine shrimp eggs can hatch in a range of salt concentrations, but more hatch in environments with optimum salt concentration. (Analyzing and interpreting data; systems and system models.)
- We obtained information through readings that individuals of the same kind differ in their characteristics, and sometimes the differences give individuals an advantage in surviving and reproducing. (Obtaining, evaluating, and communicating information.)
- We obtained information through readings that human activities can impact ecosystems, but that communities working together can help protect Earth's environments. (Obtaining, evaluating, and communicating information.)

DISCIPLINARY CORE IDEAS

LS1.A: Structure and function

LS2.C: Ecosystem dynamics, functioning, and resilience

LS4.B: Natural selection

LS4.D: Biodiversity and humans

ESS3.C: Human impacts on Earth systems

Environments Module—FOSS Next Generation

INVESTIGATION 3 – *Brine Shrimp Hatching*

20. Discuss investigation guiding question

Students should discuss the investigation guiding question with a partner before responding to it in their notebooks.

➤ *How is optimum environment related to organism and population survival?*

B R E A K P O I N T

21. Assess progress: I-Check

Give the I-Check assessment at least one day after the wrap-up review. Distribute a copy of *Investigation 3 I-Check* to each student. You can read the items aloud, but students should respond to the items independently in writing. Alternatively, you can schedule the I-Check on FOSSmap for students to take the test online.

If students take the assessment on paper, collect the I-Checks. Code the items using the coding guides on FOSSweb, and plan for a self-assessment session, identifying disciplinary core ideas, science and engineering practices, and crosscutting concepts students may need additional help with. Refer to the Assessment chapter for more information.

> **TEACHING NOTE**
>
> *During or after these next steps with the I-Check, you might ask students to make choices for possible derivative products based on their notebooks for inclusion in a summative portfolio. See the Assessment chapter for more information about creating and evaluating portfolios.*

Interdisciplinary Extensions

INTERDISCIPLINARY EXTENSIONS

Math Extensions

- ### Problem of the week
  A boy wants to set up an experiment to find out the best salt concentration to hatch brine shrimp. He has six containers that hold 0.75 liter of water. He will use a spoonful of brine shrimp eggs per container. He starts with 1/8 spoon of salt for the first container:

  Amount of Salt Used

Container 1	1/8 spoon of salt
Container 2	Twice as much salt as he put in container 1
Container 3	Twice as much salt as he put in container 2
Container 4	Twice as much salt as he put in container 3
Container 5	Twice as much salt as he put in container 4
Container 6	Twice as much salt as he put in container 5

  How much salt does the boy need for all six containers? Show all your work. Use drawings, tables, and/or charts to help you.

  A girl set up a series of six experiments. She wanted to find out the best salt concentration for hatching brine shrimp. She put 1 liter of water and 1 little spoon of brine shrimp eggs in six containers. Then she added a different amount of salt to each container.

  Amount of Salt Used

Container 1	8 spoons of salt
Container 2	Half as much salt as she put in container 1
Container 3	Half as much salt as she put in container 2
Container 4	Half as much salt as she put in container 3
Container 5	Half as much salt as she put in container 4
Container 6	Half as much salt as she put in container 5

  How much salt did the girl need for all six containers? Show all your work. Use drawings, tables, and/or charts to help you.

No. 17—Teacher Master

Environments Module—FOSS Next Generation

275

INVESTIGATION 3 – Brine Shrimp Hatching

Notes on the problem. The first pattern is a series of doubling, or multiplying by two. Once the products are calculated for the amount of salt in each container, the total amount of salt needed is the sum of those amounts of salt.

This problem provides opportunities for students to work with fractions with like denominators, and reinforces the concept of a unit (8/8 = 1 full spoon).

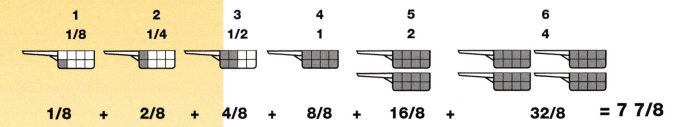

Container #	Amount of Salt
1	1/8
2	1/8 + 1/8 = 1/8 × 2 = 2/8
3	2/8 + 2/8 = 2/8 × 2 = 4/8
4	4/8 + 4/8 = 4/8 × 2 = 8/8 = 1
5	1 + 1 = 1 × 2 = 2
6	2 + 2 = 2 × 2 = 4

Sum of amounts of salt: 1/8 + 2/8 + 4/8 + 1 + 2 + 4 = 7 7/8

The second pattern is a series of divisions by two. Once the divisions are calculated for the amount of salt in each container, the total amount of salt needed is the sum of those amounts of salt.

To solve the problem, students need to understand the meaning of fractions. In this case, the spoon is the unit, and they will have to determine the fractional parts of a spoon beyond the one-half quantity. Area models (from the math problem in Investigation 1) are helpful to understand "half of a half" of a fraction. Students will also need to know how to add fractions with unlike denominators. In this case, they are using common fractions whose equivalents can be drawn easily.

Interdisciplinary Extensions

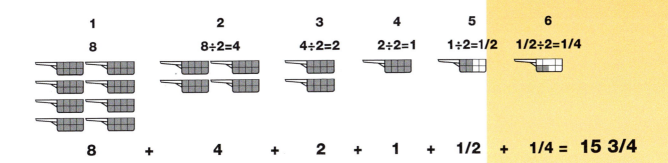

Container #	Amount of Salt		
1	8		
2	8 ÷ 2 =		4
3	4 ÷ 2 =		2
4	2 ÷ 2 =		1
5	1 ÷ 2 =		1/2
6	1/2 ÷ 2 = 1/4		

Sum of amounts of salt: 8 + 4 + 2 + 1 + 1/2 + 1/4 = 15 3/4 spoons

- **Estimate brine shrimp larvae**

  This activity provides a context for students to learn about using rounded numbers to create fractional parts of a whole and to practice multiplication of multidigit numbers.

  Select a large, clear, flat-bottomed container that will be good for viewing brine shrimp larvae. Have students trace the bottom of the container and set their outline aside.

  Pour the brine shrimp into the container. Ask students to estimate the number of larvae in the container. Have them share their counts and how they determined them.

  Ask students how they could divide their outlines of the container shape into 16 equal parts. Allow them to agree on a way to divide it. Have them divide their outlines into 16 equal parts and lightly shade one of the 16 parts.

  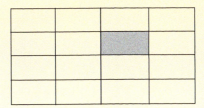

  Have students situate the container with brine shrimp on top of the outline. Count the number of brine shrimp larvae in the shaded area. Ask students how they can get a fairly accurate count of larvae in the entire container by using that number. [Multiply the number in that part by 16.]

  Have students multiply to determine the number of brine shrimp larvae. In this case, a rounded solution provides a close enough estimate to the total number in the container.

Environments Module—FOSS Next Generation

INVESTIGATION 3 – Brine Shrimp Hatching

> **TEACHING NOTE**
>
> Encourage students to use the Science and Engineering Careers Database on FOSSweb.

Science Extensions

- ### Observe the life cycle of brine shrimp
  Grow brine shrimp as long as possible in one of the 6-liter basins. Refer to teacher master 18, *Home/School Connection*, for information about the optimum salt recipe, temperature, and feeding requirements. With luck, you should be able to maintain a self-sustaining colony for a long time, observing swimming, mating life cycle, and light response. Have students observe and describe the life cycle of brine shrimp and compare the cycle to that of other organisms they have studied.

- ### Observe adult brine shrimp
  Purchase live adult brine shrimp at a pet store (they are food for tropical fish). Have students use a microscope to observe the structures and behaviors of the shrimp. Students should also set up an appropriate environment for maintaining the brine shrimp in the classroom.

  You might ask students what they think will happen if they introduce brine shrimp to the goldfish aquarium. Using the fish net, scoop out a few adult brine shrimp, rinse them off in aged or treated tap water thoroughly, and add them to the aquarium. Students should observe the goldfish eat the brine shrimp as they ate the *Gammarus*. Ask students why they think goldfish will eat brine shrimp when they aren't freshwater organisms.

- ### Evaluate accuracy of advertising
  Have students use their knowledge of the brine shrimp life cycle and environmental factors to draw inferences and evaluate the accuracy of product claims about sea monkeys. Advertisements are available on the Internet as well as on products purchased at local stores.

  Here's an example of an excerpt from an advertisement available on the Internet:

  Sea-Monkeys® are real Time-Travelers asleep in biological time capsules for their strange journey into the future!

- ### Investigate brine shrimp and light
  Have students design an experiment to find out how brine shrimp respond to the environmental factor of light.

Full Option Science System

Interdisciplinary Extensions

Environmental Literacy Extensions

- **Investigate other environmental factors**
  Have students use other chemicals such as detergent or vinegar in a range of very dilute concentrations to observe their effect on brine shrimp hatching. Make sure students use the same optimum salt concentration in each container.

- **Show *The Mono Lake Story***
  The Mono Lake Story is a 27-minute video that showcases the Mono Basin's natural beauty through all four seasons and tells the story of the Mono Lake Committee's work to preserve and restore this special place for future generations. If you did not show the video in Part 2, show it now. There are four extra features in addition to the main video (*Science and Restoration at Mono Lake, Education in the Mono Basin, The Public Trust,* and *Mono Lake through the Seasons*). You will find the video on FOSSweb in the streaming video section.

Home/School Connection

Students can raise brine shrimp at home with their families. Send little paper packets of eggs home with students to get started. This may exhaust your supply of eggs, which is fine. It is good to start each year with a fresh supply of eggs.

Print or make copies of teacher master 18, *Home/School Connection* for Investigation 3, and send it home with students after Part 3.

No. 18—Teacher Master

Environments Module—FOSS Next Generation

279

INVESTIGATION 3 – Brine Shrimp Hatching

Investigation 4: Range of Tolerance

INVESTIGATION 4 – *Range of Tolerance*

PURPOSE

Students return to the phenomenon of different plants surviving in desert and rain forest environments. They then conduct experiments with seed plants to answer the guiding question.

Content

- Organisms have ranges of tolerance for environmental factors. Within a range of tolerance, there are optimum conditions that produce maximum growth.
- Organisms have specific requirements for successful growth, development, and reproduction. A relationship exists between environmental factors and how well organisms grow.
- Adaptations are structures and behaviors of an organism that help it survive and reproduce.

Practices

- Conduct controlled experiments with four kinds of plants to discover their range of tolerance for water and their range of salt tolerance.
- Graph and interpret data from multiple trials from plant experiments.

Part 1
Water or Salt Tolerance and Plants............................ 294

Part 2
Plant Patterns 314

Part 3
Plant Adaptations 325

Guiding questions for phenomenon:
What environmental conditions result in the best growth and survival of different plants?
How do the structures of plants function to support the survival of the organisms in a particular environment?

Science and Engineering Practices
- Planning and carrying out investigations
- Analyzing and interpreting data
- Constructing explanations
- Engaging in argument from evidence
- Obtaining, evaluating, and communicating information

Disciplinary Core Ideas
LS1: How do organisms live, grow, respond to their environment, and reproduce?
LS1.A: Structure and function
LS2: How and why do organisms interact with their environment and what are the effects of these interactions?
LS2.C: Ecosystem dynamics, functioning, and resilience
LS4: How can there be so many similarities among organisms yet so many different kinds of plants, animals, and microorganisms?
LS4.D: Biodiversity and humans

Crosscutting Concepts
- Patterns
- Cause and effect
- Structure and function

Full Option Science System

INVESTIGATION 4 – Range of Tolerance

Investigation Summary	Time	Focus Question for Phenomenon, Practices
PART 1 — **Water or Salt Tolerance and Plants** Half the class sets up a controlled experiment to determine the range of water tolerance for the early growth of four different plants (barley, corn, pea, and radish). Students make observations after 5, 8, and 13 days of growth. They disassemble their planters and compare the growth of each plant in the different environments. The other half of the class sets up a controlled experiment to test the effect of salinity on the same four plants. They water each container with a different concentration of salt water. Students monitor growth of their plants at 5, 8, and 13 days after planting. They disassemble their planters and compare the growth of each plant in the different environments to determine the salt tolerance of the four plants.	**Active Inv.** 4–5 Sessions * **Reading** 2 Sessions	**How much water is needed for early growth of different kinds of plants?** **What is the salt tolerance of several common farm crops?** **Practices** Planning and carrying out investigations Analyzing and interpreting data Constructing explanations Engaging in argument from evidence Obtaining, evaluating, and communicating information
PART 2 — **Plant Patterns** Students observe and map plant-distribution patterns in the schoolyard. They discuss the environmental factors that might be responsible for these patterns.	**Active Inv.** 2 Sessions **Reading** 1 Session	**How does mapping the plants in the schoolyard help us to investigate environmental factors?** **Practices** Planning and carrying out investigations Analyzing and interpreting data Constructing explanations Obtaining, evaluating, and communicating information
PART 3 — **Plant Adaptations** Students review environmental factors that influence plant growth (water, light, nutrients). They are introduced to different adaptations of plants that allow some to thrive in dry environments and others to thrive in wet environments.	**Active Inv.** 1 Session **Reading** 1 Session **Assessment** 2 Sessions	**What are some examples of plant adaptations?** **Practices** Obtaining, evaluating, and communicating information

* A class session is 45–50 minutes.

At a Glance

Content Related to DCIs	Writing/Reading	Assessment
• Every organism has a range of tolerance for each factor in its environment. • Organisms have specific requirements for successful growth, development, and reproduction. A relationship exists between environmental factors and how well organisms grow. • Optimum conditions are those most favorable to an organism.	**Science Notebook Entry** Plant Experiment—Water Tolerance Plant Experiment—Salt Tolerance Plant Experiment Setup Plant Observations A and B Plant Profile (optional) **Science Resources Book** "Environmental Scientists" "Range of Tolerance" "How Organisms Depend on One Another" **Online Activity** "Tutorial: Analyzing Environmental Experiments"	**Embedded Assessment** Performance assessment
• Organisms have specific requirements for successful growth, development, and reproduction. A relationship exists between environmental factors and how well organisms grow.	**Science Notebook Entry** Answer the focus question **Science Resources Book** "Animals from the Past" (optional)	**Embedded Assessment** Response sheet
• Adaptations are structures and behaviors of an organism that help it survive and reproduce. • A relationship exists between environmental factors and how well organisms grow.	**Science Notebook Entry** Answer the focus question **Video** All about Plant Adaptations	**Benchmark Assessment** Posttest **NGSS Performance Expectations addressed in this investigation** 4-LS1-1 3-LS4-4

Environments Module—FOSS Next Generation

283

INVESTIGATION 4 – Range of Tolerance

BACKGROUND for the Teacher

Organisms depend on a supportive environment for life itself. If even one environmental factor is beyond the limits that an organism can withstand, the organism will perish. Within the limits, the organism can survive. Those limits are called an organism's range of tolerance for the environmental factor. Spruce trees have a very large range of tolerance for temperature. They are able to survive cold well below 0 degrees Celsius (°C) and heat up to 50°C. Tropical orchids, on the other hand, have a much narrower range of tolerance for temperature. They will fail and die long before the temperature falls to 0°C.

While an organism might survive a wide range of values for any one factor, it will thrive within a smaller range. The conditions that are most favorable to an organism's survival, growth, and reproduction are optimum conditions.

Environmental scientists spend a great deal of time observing organisms in natural environments. This, in part, is how they learn about requirements for survival and the relationships among organisms. When field observation does not provide enough information, scientists design an experiment and bring nature into the laboratory.

"The conditions that are most favorable to an organism's survival, growth, and reproduction are optimum conditions."

How Much Water Is Needed for Early Growth of Different Kinds of Plants?

One such experiment might investigate how much water is needed to germinate the seeds of the cottonwood tree. The scientist would conduct an experiment in the laboratory, where all the factors can be controlled. A set of identical containers could have identical samples of substrate and the same number of seeds from the same seed source. In a laboratory where the temperature, light, and air can all be controlled, the only factor that varies for the seeds is water quantity.

Based on such an experiment, it is possible to determine the range of water quantity that supports seed sprouting. Such controlled experiments, where one factor is manipulated incrementally, lead to an understanding of the organism's range of tolerance, and perhaps determine the optimum conditions for germinating cottonwood seeds.

Seed germination is the first important step in the life of a plant. Water enters the seed, the seed swells, and the seed coat splits open. The main parts of the seed (two halves in

Radish   Clover   Pea   Corn   Barley

Background for the Teacher

some seeds, as in beans) are called cotyledons, and they supply the food energy for the young plant. The plant develops from the single embryo that lies between the two cotyledons. Other plants (such as corn) have a single cotyledon, but growth proceeds in a similar manner.

During the first week of growth most seedlings derive their nourishment from the cotyledons. As it uses up that food source, the plant begins to make its own food, using carbon dioxide from the air, water, nutrients from the soil, and light.

What Is the Salt Tolerance of Several Common Farm Crops?

Water is essential for seed germination and plant growth. But will any kind of water do the job? Will plants grow in a saltwater environment? We know that many nonagricultural plants thrive in saline environments (e.g., pickleweed, cordgrass, and eel grass), but can we use salt water to **irrigate** agricultural crops during a **drought** when fresh water is limited?

Much of the world's land is marginally suitable for agriculture, because the supply of fresh water is low, or because a high concentration of salt is in the soil or in potential irrigation water. These conditions exist in many coastal areas where other conditions, such as light intensity, temperature, long growing season, and soil fertility, make the area desirable for farming. Scientists are experimenting with many agricultural crops to find **salt-tolerant** varieties that can be irrigated with saline water or grown in saline soils.

For growth, plants require a number of different chemicals or nutrients in varying amounts, including chlorine, sodium, magnesium, iron, and manganese. If one of the chemicals is missing or below a particular plant's range of tolerance, the plant will fail. On the other hand, if an excessive amount of a particular chemical is in a plant's environment, it might also have an adverse effect on the plant. Different kinds of plants have different ranges of tolerance for chemicals in their environment.

The concentration of salt (NaCl) in the most concentrated solution (4 spoons per liter) used in Part 1 of this investigation is about two-thirds the concentration of seawater. The concentration of all dissolved salts in seawater averages about 35 parts per 1000. About 27 parts per 1000 are NaCl, or about 27 grams (g) in 1000 milliliters (mL) of water. In this activity, 4 spoons of NaCl have a mass of about 16 g and are added to 1000 mL of water.

"The concentration of salt in the most concentrated solution used in Part 1 of this investigation is about two-thirds the concentration of seawater."

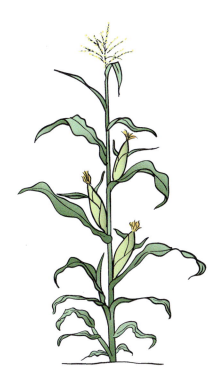

Environments Module—FOSS Next Generation

285

INVESTIGATION 4 – *Range of Tolerance*

Salt is a nuisance in agriculture. When most living cells are exposed to an environment in which the salt concentration is greater outside the cell than inside, osmosis draws water through the cell membrane from inside to outside. If salt water is used to irrigate a field of corn, the plant tissues will lose water because of osmosis, and the plant will die from dehydration. Many plants are **salt-sensitive**.

If mildly salty water is used to irrigate a field, the plants might indeed survive, but such a practice is a giant step toward trouble. NaCl does not leach out of soil easily, so when the water evaporates from the field, it leaves its load of salt behind. Next time the field is irrigated, the residual salt will dissolve, making the concentration of salt higher than before. Each subsequent watering adds to the salt load until the salt concentration exceeds the crop's salt tolerance and the salt becomes toxic to the plants.

In both the water-tolerance and salt-tolerance experiments that students conduct, they start with seeds. An interesting follow-up question would be to find the water and salt tolerance for mature plants. It might be different from the tolerance of germinating seeds for those particular conditions. Understanding the needs of organisms at different stages in their life cycle is an important concept that could be developed.

How Does Mapping the Plants in the Schoolyard Help Us to Investigate Environmental Factors?

Plant distribution is the arrangement of plants in an area. Plants grow in certain places because environmental factors are suitable for the germination of seeds and continued growth of developing plants. Environmental factors include temperature, light, moisture, soil type, available minerals, wind, and other plants competing for the same resources. Animals, including humans, also affect distribution patterns.

Specific plants clearly show the effects of these factors. For example, the wind widely disperses cattail seeds, but only those seeds that land in or around fresh water will germinate and grow. Willow trees grow in moist ravines but cannot survive on the drier hilltops where only drought-resistant plants grow. Palm trees thrive in temperate climates but are not able to survive where temperatures fall below freezing for extended periods of time.

On a smaller scale, these same environmental factors influence plant distribution. In a backyard, schoolyard, or park, slight variations in light, moisture, and temperature will favor the success of different plants. One kind of plant may cover more area and is considered to be a **dominant plant**. To a trained observer, the presence of certain plant types is an indicator of the prevailing environmental conditions in a small area.

Background for the Teacher

What Are Some Examples of Plant Adaptations?

It has been about 400 million years since the first plants ventured out of the sea and colonized the land. Look around at the result. Plants cover the planet except for a few regions of extreme drought, such as the Sahara desert, and the polar ice fields. Plants live everywhere, including hot arid deserts, windy high plains, fresh water, seacoasts, frozen tundra, and tropical islands.

The diversity of **adaptations** in the plant kingdom is astounding. Duckweed (barely more than a millimeter across—the smallest flowering plant in the world), saguaro cactus, corn, pansy, and sequoia redwood (the largest organism in the world), are all examples of the vast array of plants. At first glance, these plants might appear to be very different from one another; however, they have many things in common. Just about all plants have stems, leaves, and roots.

Duckweed

Stems may dominate the plant, like the grapevine, or they may be reduced almost into nonexistence, as in the dandelion. They may be green and flexible, or rigid and covered by rough bark. Stems establish the basic shape of the plant and provide a structure for presenting leaves. Leaves come in many shapes, sizes, and colors, mostly shades of green. Leaves are responsible for gas exchange and most of the photosynthesis that the plant carries out.

Stems support and position leaves so that they receive just the right exposure to sunlight. Trees in the rain forest that require direct exposure to sunlight need very long stems (trunks) to compete with other sun-loving plants. Such trees may be more than 60 meters (m) tall. Extreme height, however, is also a liability because tropical soil is very shallow. Such a tall, top-heavy tree could easily be toppled in a storm. A number of tropical trees have large bracing structures on their trunks, called buttresses, to broaden the base for additional stability. Smaller plants in a dense forest have other ways of gaining access to sunlight.

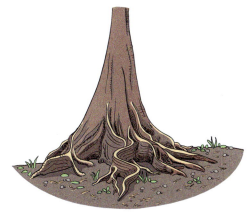

Tree buttress

Vines have specialized structures called tendrils that allow them to attach to other structures. When a tendril touches an object, it twines tightly around it. As a result, the vine can direct more energy to growing long and less energy to building a rigid structure. It simply takes advantage of the structure of its support and climbs up to position its leaves in the sunlight.

Other plants have features that allow them to assume a lifestyle free of soil as an anchorage. They grow right on the high branches of an established tree. No need for long or massive stems. Such plants rely on rain to provide the water they need to survive. This strategy works in the rain forest. Many orchids live this way. Some branch dwellers that live in

Environments Module—FOSS Next Generation 287

INVESTIGATION 4 – Range of Tolerance

areas where water is not so readily available have specialized roots that go into the tissue of the host tree's branches and steal the water they need. Mistletoe is a parasitic plant that does this.

Plants like the philodendron have another alternative to long, massive stems. They invest their energy in huge leaves. In this way, they can use the filtered light that makes its way down to the forest floor.

Plants in the desert face a different problem. Because leaves lose so much water in a dry, hot environment, many desert plants, such as cacti, do not have green leaves. Their leaves are reduced to defensive spines. Others, such as the paloverde tree, sprout leaves for only a short time in response to rain. How do these plants carry out photosynthesis and produce the food they need without leaves? The stem of the cactus and the trunk and branches of the paloverde are green. The cells in the outermost layer of the stems contain chloroplasts, which carry out photosynthesis for the plant.

Barrel cactus

The stems of some plants have thorns, bristles, or hairlike coverings that protect them from being eaten. The redwood tree has very thick bark that protects the inside of the tree from getting too hot during a forest fire. The bark also contains fire-retardant chemicals, so that it is less likely to burn. The stems of some plants, including trees such as cedar, pine, creosote, and redwood, contain chemicals that are poisonous to many insects and fungi, protecting the stem from attack.

All the cells in the plant need a supply of sugar produced by photosynthesis to keep growing and building new tissue. For photosynthesis, plants need a lot of carbon dioxide coming in through the stomata (openings in the leaves) and water coming up from the roots. However, water can quickly evaporate from the cells in the leaves. The process of water vapor leaving the cells through the stomata is called transpiration.

If the cells in the leaf lose too much water, the cells will shrink (a condition we call wilting), and if water is not soon restored, the cells will die. Even so, a little bit of wilting can actually benefit a plant when the water supply is low. When the leaf is drooping, it is not in position to capture the maximum amount of sunlight. Reduced sunlight slows the rate of water loss in the leaves. When the weather is hot, the Sun is shining, a breeze is blowing, and the humidity is low, water evaporates very quickly. In these conditions, plants can rapidly lose valuable water. Plants are able to respond to dry conditions to reduce the amount of water loss from their leaves by changing the shape of the stomata, so that they close and reduce water loss significantly.

Background for the Teacher

Some plants have pits in the surfaces of their leaves. The pits are filled with numerous hairlike structures. Stomata are clustered in these pits. Because the breeze can't blow directly across the stomata, less water transpires from stomata in pits. Most plants have leaves with a cuticle, a layer of waxy material on the surface of the leaf. This cuticle reduces the amount of water that evaporates out of the cells and into the air. In dry climates, the cuticle can be very thick.

Some plants have very thick leaves that can hold a lot of water. Because the leaves are so thick, most of the water in the leaf is farther from the surface and the stomata. Many desert plants have small leaves, resulting in less surface for transpiration.

Succulent plant leaves

Some plants actually collect the water they need. For example, the redwoods on the West Coast obtain about half their water from the fog that comes off the ocean. The tiny droplets of water collect on the short, thick needles of the redwood trees and drip off. During one night of heavy fog, as much water can drip off a redwood as during a drenching rain. This keeps the trees and the plants under them alive during the summer months when there is little rain, but plenty of fog.

Other plants growing in dry climates have fuzzy leaves that collect moisture from dew. The fuzz increases the surface area of the leaf, creating more area for vapor to condense into liquid water. Dew collected by hairy-leaved plants keeps the soil moister than around plants with smooth leaves.

Plants all over the world have leaf adaptations that help them survive and reproduce. Plants growing in windy areas often have slender, flexible leaves to yield to the force of the wind. Trees living in rainy regions often have leaves with points along the margin to speed the flow of water off the leaves. Some plants that grow in cold regions have hairy leaves that fold up around the plant at night, acting like a blanket to hold warmth.

Redwood needles

Next time you are in a garden, park, woods, or field, look closely at the leaves on the plants. Compare the size, shape, surface, flexibility, thickness, color, and pattern of the leaves. They all serve the same functions—photosynthesis and water management—but each different leaf performs those functions in a way that contributes to the success of the plant on which it grows.

Maple leaf

Environments Module—FOSS Next Generation

INVESTIGATION 4 – Range of Tolerance

TEACHING CHILDREN about Range of Tolerance

> **NGSS Foundation Box for DCI**
>
> **LS1.A: Structure and function**
> - Plants and animals have both internal and external structures that serve various functions in growth, survival, behavior, and reproduction. (4-LS1-1)
>
> **LS2.C: Ecosystem dynamics, functioning, and resilience**
> - When the environment changes in ways that affect a place's physical characteristics, temperature, or availability of resources, some organisms survive and reproduce, others move to new locations, yet others move into the transformed environment, and some die. (3–LS4-4, extended from grade 3)
>
> **LS4.D: Biodiversity and humans**
> - Populations live in a variety of habitats, and change in those habitats affects the organisms living there. (3–LS4-4, extended from grade 3)

Developing Disciplinary Core Ideas (DCI)

In this investigation, students experiment with terrestrial plants to determine their tolerance for water and salt. They discover that plants cannot survive if they are exposed to too little or too much water. Within specific limits, plants will survive. Those limits describe the plant's range of tolerance. Nested in the range is the plant's optimum for a condition.

This is an important concept for students to understand. When environmental conditions fall within an organism's range of tolerance, it is able to survive. Within that range, it may thrive, depending on how close to optimum the conditions are. And outside the range, the organism will not survive.

In the classroom, students will be eager to plant seeds and observe one of the miracles of the world. They will probably become very attached to their plants after caring for them for many days. In each investigation, it is necessary to sacrifice the plants to look at roots or weigh the products. Be prepared. Offer students the opportunity to replant and take the plants home. Or have them set up extras at the start—some to keep and some to study.

This module does not take up issues or promote positions on environmental problems. It does try, however, to prepare students to recognize and understand local and global environmental issues, and to think productively about them. The responsibility for making thoughtful, informed decisions about the environment will fall to them in just a few years. FOSS wants scientific knowledge and flexible, probing thinking to be among the tools and experiences tomorrow's citizens bring to bear on important issues affecting the future of the planet.

As you draw this module to a close, ask questions from time to time that extend beyond the plants in the cups and the fish in the aquariums. Ask students to think about the plants in the woods and on the deserts, and the fish in the sea. How are their lives and environments similar to those of your study organisms? How are they different? Who is managing the global environmental variables, and what is our role in the whole picture? And listen to the children's emerging ideas about a larger world.

The experiences students have in this investigation contribute to the disciplinary core ideas **LS1.A: Structure and function; LS2.C: Ecosystem dynamics, functioning, and resilience;** and **LS4.D: Biodiversity and humans.**

Teaching Children about Range of Tolerance

Engaging in Science and Engineering Practices (SEP)

In this investigation, students engage in these practices.

- **Planning and carrying out investigations** collaboratively as a class to determine the range of tolerance for water and salt for germination and growth of a variety of seeds; evaluating experimental design methods and ways of collecting data; making firsthand observations to produce data on plant growth under incremental changes in environmental conditions of a specific factor to serve as the basis for evidence for an explanation; making predictions about what would happen if an environmental factor changed.

- **Analyzing and interpreting data** by representing data in tables and visual displays to reveal growth patterns and making sense of data using logical reasoning.

- **Constructing explanations** based on their data when monitoring plant growth under different environmental conditions; constructing explanations of observed relationships between environmental factors and the distribution of kinds of plants in the schoolyard; using evidence from plant experiments to support explanations, and identifying the evidence that support specific points in the explanation.

- **Engaging in argument from evidence** by using data to evaluate claims about cause (amount of moisture or amount of salt in water used for watering) and growth patterns; constructing an argument about plant growth using data; use data from other students to evaluate claims about cause and effect.

- **Obtaining, evaluating, and communicating information** from books and media, and integrating that with their firsthand experiences to construct explanations about plant growth and survival patterns in terms of environmental factors.

NGSS Foundation Box for SEP

- **Plan and conduct an investigation** collaboratively to produce data to serve as the basis for evidence.
- **Evaluate appropriate methods** and/or tools for collecting data.
- **Make observations and/or measurements to produce data** to serve as the basis for evidence for an explanation of a phenomenon or test a design solution.
- **Make predictions** about what would happen if a variable changes.
- **Represent data in tables and/or various graphical displays** to reveal patterns that indicate relationships.
- **Analyze and interpret data** to make sense of phenomena using logical reasoning.
- **Organize simple data sets** to reveal patterns that suggest relationships.
- **Construct an explanation** of observed relationships.
- **Use evidence** (e.g., measurements, observations, patterns) to support an explanation.
- **Identify the evidence** that supports particular points in an explanation.
- **Construct an argument** with evidence, data, and/or a model.
- **Use data to evaluate claims** about cause and effect.
- **Read and comprehend grade-appropriate complex texts** and/or other reliable media to summarize and obtain scientific and technical ideas and describe how they are supported by evidence.
- **Obtain and combine information** from books and other reliable media to explain phenomena.
- **Communicate scientific and/or technical information** orally and/or in written formats, including various forms of media and may include tables, diagrams, and charts.

Environments Module—FOSS Next Generation

INVESTIGATION 4 – Range of Tolerance

> **NGSS Foundation Box for CC**
> - **Patterns:** Similarities and differences in patterns can be used to sort, classify, communicate, and analyze simple rates of change for natural phenomena. Patterns can be used as evidence to support an explanation.
> - **Cause and effect:** Cause-and-effect relationships are routinely identified and used to explain change.
> - **Structure and function:** Substructures have shapes and parts that serve functions.

Exposing Crosscutting Concepts (CC)

In this investigation, the focus is on these crosscutting concepts.

- **Patterns.** Organisms have ranges of tolerance to environmental factors and these patterns can be used as evidence to support explanations and determine solutions to environmental problems.
- **Cause and effect.** Organisms respond to environmental conditions and these responses are predictable and can be used to explain change.
- **Structure and function.** The form of an organism's structures determines the function.

Connections to the Nature of Science

- **Scientific investigations use a variety of methods.** Scientific methods are determined by questions. Scientific investigations use a variety of methods, tools, and techniques.
- **Scientific knowledge is based on empirical evidence.** Science findings are based on recognizing patterns. Scientists use tools and technologies to make accurate measurements and observations.
- **Science is a way of knowing.** Science is both a body of knowledge and processes that add new knowledge. Science is a way of knowing that is used by many people.
- **Science is a human endeavor.** Men and women from all cultures and backgrounds choose careers as scientists and engineers. Most scientists and engineers work in teams. Science affects everyday life. Creativity and imagination are important to science.

Adaptation
Dominant plant
Drought
Irrigate
Plant distribution
Salt-sensitive
Salt-tolerant

Teaching Children about Range of Tolerance

Conceptual Flow

Students return to the desert and rain forest ecosystems to observe the phenomenon of different plants surviving in different environments. The guiding questions are what environmental conditions result in the best growth and survival of different plants? and how do the structures of plants function to support the survival of the organisms in a particular environment?

The **conceptual flow** for this investigation starts in Part 1 when students are challenged to plant several kinds of seeds in identical planter cups, varying the environments only with regard to the amount of water the seeds receive or the amount of salt in the water (two separate experiments, each conducted by half the groups in the class).

Students analyze the results of the experiments (water tolerance and salt tolerance). The seeds do not grow in the absence of water, nor do they grow in the presence of excess water. All plants have a **range of tolerance** for the factors in their environment. Within the range is a **condition that is optimum** for the growth and development of plants. Students compare the number of **leaves**, **length of stems**, and **development of roots** to determine the optimum amount of water for each kind of plant. In the second experiment, students compare the plants to determine a **range of tolerance and optimum conditions for salt content** (a toxin) in the water used to irrigate the plants.

In Part 2, students continue their investigation of range of tolerance for plants. Students go outdoors to inventory plants and map their distribution in the study site. Students then attempt to determine the **environmental factors that affect the distribution of the plants** in the site. Distribution patterns are determined by the optimum conditions preferred by each kind of plant: **shade tolerance**, **moisture tolerance, temperature tolerance**, and **physical impact tolerance**.

In Part 3, students view a video about plants living in a diverse range of environments (e.g., desert, rain forest). Students are introduced to the concept of **adaptation, structures and behaviors that affect an organism's chances of survival in their environment**. Adaptations are natural features of organisms that make it possible for them to live in their environments.

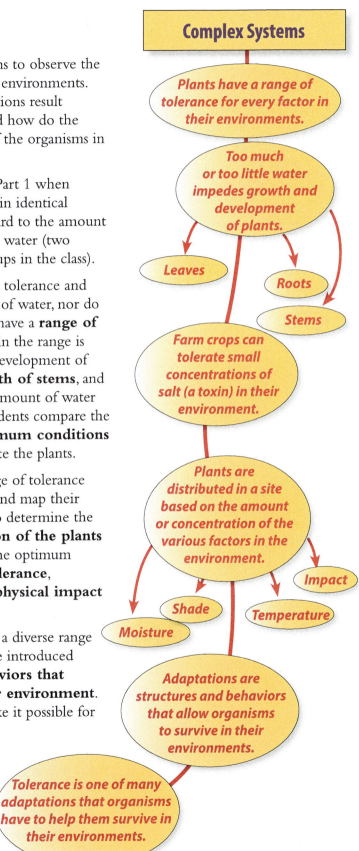

Environments Module—FOSS Next Generation

293

INVESTIGATION 4 – Range of Tolerance

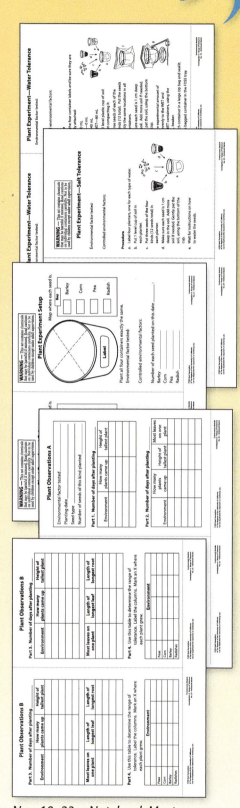

Nos. 19–23—Notebook Master

MATERIALS for
Part 1: Water or Salt Tolerance and Plants

For each student

- [ ] 1 Notebook sheet 19, *Plant Experiment—Water Tolerance* (Water Tolerance groups)
- [ ] 1 Notebook sheet 20, *Plant Experiment—Salt Tolerance* (Salt Tolerance groups)
- [ ] 1 Notebook sheet 21, *Plant Experiment Setup*
- [ ] 1 Notebook sheet 22, *Plant Observations A*
- [ ] 1 Notebook sheet 23, *Plant Observations B*
- [ ] 1 Notebook sheet 24, *Plant Profile* (optional)
  - 1 *Plant Profile* (See Steps 36–37 of Guiding the Investigation.) ★
  - 1 *FOSS Science Resources: Environments*
    - "Environmental Scientists"
    - "Range of Tolerance"
    - "How Organisms Depend on One Another"

For each group

- 4 Containers, 1/2 L
- 4 Self-stick notes or prepared labels
- 2 Plastic cups
- 12 Seeds of each kind: barley, corn, pea, radish
- 1 FOSS tray
- 1 Metric ruler, 15 cm
- 1 Beaker, 100 mL
- 4 Zip bags, 4 L (Water Tolerance groups only)

For the class

- 16 Tray columns, plastic
- 3 Basins, 8 L
- 8 Plastic cups
- 1 Pitcher
- • Moist potting soil, 21 L ★
- 8 Self-stick notes
- • Transparent tape
- • Newspaper ★
- • Water ★

294 Full Option Science System

Part 1: Water or Salt Tolerance and Plants

- ❏ • Teacher master 15, *Container Labels* (optional)
- ❏ • Teacher master 19, *Plant Profile*
- • Computers with Internet access ★

For the class for Water Tolerance groups
- Dry potting soil, 3 L (See Step 4 of Getting Ready.) ★
- 3 Seeds of each kind: barley, corn, pea, radish
- 4 Zip bags, 4 L
- 1 Container, 1/2 L

For the class for Salt Tolerance groups
- 1 Beaker, 1 L
- 4 Containers with lids, 1 L
- 1 Spoon, 5 mL
- Kosher salt
- Safety goggles ★

For embedded assessment
- *Performance Assessment Checklist*

★ Supplied by the teacher.   ❏ Use the duplication master to make copies.

No. 15—Teacher Master

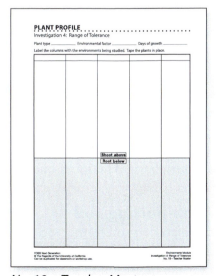

No. 19—Teacher Master

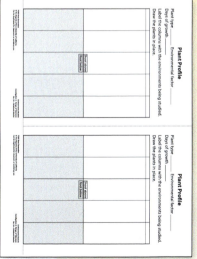

No. 24—Notebook Master

Environments Module—FOSS Next Generation

INVESTIGATION 4 – Range of Tolerance

TEACHING NOTE

The best way of scheduling the investigation is to have half the groups investigate water tolerance in plants and the other half of the class investigate salt tolerance in plants. Through a jigsaw activity, groups share their investigations and findings. The suggested schedule has two days for the experimental setup, but it could be done on one day. Consider what would work best in your class.

▶ **NOTE**
To prepare for this investigation, view the teacher preparation video on FOSSweb.

GETTING READY for
Part 1: Water or Salt Tolerance and Plants

1. **Schedule the investigation**
   This part will take 4–5 sessions of active investigation over a 2-week period—one or two sessions to set up the two different experiments and three observations at about day 5, 8, and 13. Half the groups will set up a water tolerance experiment; the second group will set up the salt tolerance experiment. You can have the groups set up the experiments at the same time or you can introduce the experiments to the whole class and then have half the class set up the experiments while the other half completes the first reading. Plan on two reading periods during the 2 weeks of plant growth, but before the final observation session.

   Don't wait to complete Part 1 before going on to Parts 2 and 3. Refer to the suggested teaching schedule in the Overview chapter for more details on scheduling.

2. **Preview Part 1**
   Half the class sets up a controlled experiment to determine the range of water tolerance for the early growth of four different plants (barley, corn, pea, and radish). Students make observations after 5, 8, and 13 days of growth. They disassemble their planters and compare the growth of each plant in the different environments. The focus question is **How much water is needed for early growth of different kinds of plants?**

   The other half of the class sets up a controlled experiment to test the effect of salinity on the same four plants. They water each container with a different concentration of salt water. Students monitor growth of their plants at 5, 8, and 13 days after planting. They disassemble their planters and compare the growth of each plant in the different environments to determine the salt tolerance of the four plants. The focus question is **What is the salt tolerance of several common farm crops?**

3. **Acquire potting soil**
   Purchase potting soil from a local store. Three 8-liter (L) bags (or the equivalent) will be needed. Potting soil right from the bag is used for the moist soil in the experiment. Keep the bags sealed to retain the moisture.

4. **Prepare dry soil for water tolerance**
   Start 3 to 4 days before the investigation to dry 3 L of soil. Each group will need one plastic cup full of completely dry soil. Spread the soil on newspaper to dry; once dry, store it in a labeled bag.

▶ **NOTE**
Be sure to dry the soil 3 to 4 days before starting the investigation.

Part 1: Water or Salt Tolerance and Plants

5. Consider the type of salt for salt tolerance
The salt provided in the kit is kosher salt—pure sodium chloride (NaCl) with no iodine or other additives. This is the same kosher salt that was used in the brine shrimp investigation.

Table salt will work for this activity, but because it is finer grained, 1 spoon of table salt in a liter of water will make a more concentrated saltwater solution than will 1 spoon of kosher salt.

6. Make four saltwater solutions for salt tolerance
Students use four saltwater solutions to irrigate the plants. You or a few students can make up the solutions ahead of time.

- 0 spoons of salt in 1 L water
- 1 spoon of salt in 1 L water
- 2 spoons of salt in 1 L water
- 4 spoons of salt in 1 L water

Make salt solutions as follows:

a. Fill the 1 L beaker with water to the 1000 mL line.

b. Use the 5 mL spoon to add the indicated number of level spoons of salt to the water. Stir.

c. Pour the salt solution into a liter container so it is filled to a safe level to avoid spillage. There might be a small amount of solution remaining in the beaker that you can toss.

d. Put the lid on, and label the container.

7. Plan soil distribution
Transfer the moist soil from the potting soil bag to two basins just before using it. Place the dry soil in a separate basin for the water tolerance experiment. Return unused soil to the bag, and seal it tightly to preserve the moisture.

8. Provide newspaper
For easy cleanup, have students cover their desks with newspaper when they are working with soil.

9. Plan to distribute seeds
Set up two seed-distribution locations. Place each kind of seed in two labeled cups. Getters can count the 12 seeds of each kind into another plastic cup at one of the two materials stations.

10. Decide on labels for planters
Students can use self-stick notes with a bit of tape to make labels for their planters, or they can use prepared labels copied from teacher master 15 (one set of labels is for water tolerance and a second set of labels is for salt tolerance). If students use the prepared labels, they will need to tape them on the planters.

▶ **SAFETY NOTE**
Check to see if any student has an allergy to seeds used in this investigation—corn, radish, barley, and pea (legume).

Environments Module—FOSS Next Generation

INVESTIGATION 4 – Range of Tolerance

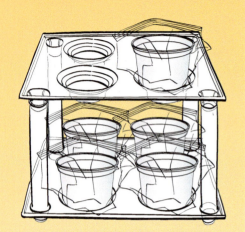

11. Plan for placement of plant experiments

Plan where to place the FOSS planter trays while the seeds grow. Place them by a window if possible. However, because the plants will grow for only 12 to 14 days, however, they will do just fine with regular room light.

The FOSS planter trays are designed to be stacked two high to conserve space. Use four plastic tubes as support columns between two trays.

12. Plan cleanup

The soil in the salt tolerance experiments will be contaminated with salt and therefore *cannot* be used again. At the end of the experiment, have students throw the soil in a green bin to be professionally composted or in the trash.

The soil in the water tolerance experiments can be used again for planters.

Remove the labels from all the planters. All containers should be rinsed with water (no soap) to remove soil (and salt).

13. Plan to read *Science Resources*: "Environmental Scientists," "Range of Tolerance," and "How Organisms Depend on One Another"

Plan to have students read "Environmental Scientists" while setting up the plant experiments (one half of the class reads while the other half sets up one of the two experiments).

Plan to read "Range of Tolerance" and "How Organisms Depend on One Another" after observation sessions on days 5 and 8.

14. Plan for online activity

Students can gain more experience with the design of experiments that test for preferred environment, range of tolerance, and optimum conditions with the "Tutorial: Analyzing Environmental Experiments" online activity. The link is on FOSSweb.

15. Plan assessment: performance assessment

Students conduct investigations in this part to continue their exploration of environmental factors. This will provide a good opportunity for you to observe students' scientific practices. Carry the *Performance Assessment Checklist* with you as you visit groups while they are working. (See Step 12 of Guiding the Investigation for what to look for.)

Part 1: Water or Salt Tolerance and Plants

GUIDING the Investigation
Part 1: Water or Salt Tolerance and Plants

1. Introduce the phenomenon of plant diversity
Ask students to review how the plants of the rain forest and desert environments compare. You might have them turn to the first reading in the *FOSS Science Resources* book, "Two Terrestrial Environments," and have them look at the first and last page of the article for side-by-side photos. You can also project the interactive eBook so the class can view and discuss these photos. Explain that they will be investigating plant diversity and gathering data to answer the guiding questions what environmental conditions result in the best growth and survival of different plants? and how do the structures of plants function to support the survival of the organisms in a particular environment?

2. Access prior knowledge
Ask students to think about a time when they grew plants from seeds. Ask,

➤ *What do you think would have happened if you did not water the seeds? What if you gave them a lot of water?*

➤ *How much water do different seeds need to start growing?*

3. Focus question: How much water is needed for early growth of different kinds of plants?
Write or project the focus question on the board as you say it aloud. Have students write it in their notebooks.

➤ *How much water is needed for early growth of different kinds of plants?*

4. Discuss water experiment
Ask students to suggest an experimental design the class can use to find out how much water is the right amount of water to start plant growth. Have them discuss possible experimental designs in their groups.

Remind them that a controlled experiment is a set of compared investigations in which one factor is changed by steps while all other factors are kept the same, or controlled. Ask each group to share some ideas they have for the design. After the class discussion, build on students' ideas to summarize and present this procedure.

- Set up four planters.
- Put the same amount of soil and three seeds of each kind in each planter.
- Prepare the planters exactly the same except for the amount of water.

FOCUS QUESTIONS

How much water is needed for early growth of different kinds of plants?

What is the salt tolerance of several common farm crops?

TEACHING NOTE

Students can revisit previous investigations, such as the brine shrimp investigation, for ideas on how to set up a controlled experiment changing only one environmental factor.

SCIENCE AND ENGINEERING PRACTICES

Planning and carrying out investigations

TEACHING NOTE

Go to FOSSweb for Teacher Resources and look for the Science and Engineering Practices—Grade 4 chapter for details on how to engage students with the practice of planning and carrying out investigations.

TEACHING NOTE

Make sure that students understand that the four planters are set up exactly the same.

Environments Module—FOSS Next Generation

INVESTIGATION 4 – Range of Tolerance

5. Show the materials
Show students the two supplies of soil, one dry and one moist. Hold up one of the 1/2 L containers students will use for planters and the 100 mL beaker they will use to measure water. Point out the cups containing the four kinds of seeds, barley, corn, pea, and radish.

6. Discuss the water environments
Each of the group's four identical planters will have a different moisture environment. One will be dry, created with the dry soil and no water added. The moist environment will use moist soil, with no water added. The wet and very wet environments will use moist soil, and they will have water added. To summarize, draw a table on the board showing the environment, soil, and amount of water to add to each planter.

Environment	Soil	Water
Dry	dry	none
Moist	moist	none
Wet	moist	40 mL
Very wet	moist	80 mL

7. Describe the swamp environment
Tell students that you will maintain one additional environment called "swamp." It will have moist soil and 120 mL of water. This is the soaked, flooded, or swamp environment. You or a team that finishes its planters quickly can set up this fifth environment.

8. Describe becoming a plant expert
Explain that each student in a group will focus on one of the four kinds of plants (barley, corn, peas, or radishes). For now, the plant expert will plant his or her kind of seed in each of the four containers, being sure that the planting locations are exactly the same in each of the containers. Later, the expert will record observations for his or her type of plant.

300

Full Option Science System

Part 1: Water or Salt Tolerance and Plants

9. **Review the procedure**
   Distribute notebook sheet 19, *Plant Experiment—Water Tolerance*, and review the design.

   a. Apply the four container labels and be sure they are securely attached:

   DRY—0 mL
   MOIST—0 mL
   WET—40 mL
   VERY WET—80 mL

   b. Put in 1 level plastic cup of soil without compacting it.

   c. Add three seeds of each of the four kinds (12 total). Put the seeds in exactly the same locations in all four containers.

   d. Make sure each seed is 1 cm deep in the soil. Add additional soil on top if necessary. Gently pat the soil, using the bottom of the cup.

   e. Add the experimental amount of water evenly to the WET and VERY WET containers, using the 100 mL beaker. (Point out that the beaker is calibrated in 10 mL increments, with numbers at 20, 40, 60, 80, and 100 mL.)

   f. Put each container in a large zip bag and seal it.

   g. Put the bagged container in the FOSS tray.

10. **Discuss mapping seed locations**
    Tell students that they will also need to make a map to show where they planted the seeds. Keeping accurate records is important to scientists and will be especially important when they make observations and need to identify the seeds. Distribute a copy of notebook sheet 21, *Plant Experiment Setup*, to each student. Review the information they should record.

    - Environmental factor tested
    - Controlled factors
    - Planting date
    - Number of seeds
    - Location of planted seeds

    The label on the front of each container can be used for orientation.

11. **Set up the water experiment**
    Allow 20 to 25 minutes for the groups to set up the experiment.

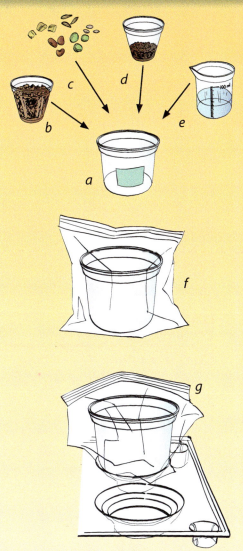

EL NOTE

If necessary, model the procedure and/or project the notebook sheet with the illustration. If necessary, model how to fill out the sheet on the board or by using a projection system.

Materials for Step 11
- Containers
- Cups of seeds
- Beakers
- Plastic cups
- FOSS trays
- Labels
- Zip bags
- Soil
- Water
- Newspaper
- Tape

Environments Module—FOSS Next Generation

INVESTIGATION 4 – Range of Tolerance

SCIENCE AND ENGINEERING PRACTICES

Planning and carrying out investigations

Analyzing and interpreting data

Constructing explanations

Obtaining, evaluating, and communicating information

DISCIPLINARY CORE IDEAS

LS2.C: Ecosystem dynamics, functioning, and resilience

CROSSCUTTING CONCEPTS

Cause and effect

Materials for Step 14
- *Tray columns*

12. **Assess progress: performance assessment**
    Visit students as they work and note how well they are following conventional science practices.

    What to Look For
    - Students plan and set up a well-reasoned investigation. (Planning and carrying out investigations.)
    - Students accurately map the seeds they've planted. (Obtaining, evaluating, and communicating information.)
    - Students analyze and interpret data to explain effects of water and salt tolerance. (Analyzing and interpreting data; constructing explanations; LS2.C: Ecosystem dynamics, functioning, and resilience; cause and effect.)

13. **Make predictions**
    Ask students to predict what will happen in the planters, based on their experience with the seeds in the terrarium.

    ➤ *Which environment do you think will produce the tallest plants?*

    ➤ *Which environment will produce the plants with the most leaves?*

    This is a good opportunity for a science talk where students can discuss their predictions based on their prior experience.

14. **Clean up**
    Use the clear plastic tubes to stack the trays. Set them in a safe place. Have students clean up the materials and toss out or recycle the newspapers.

POSSIBLE BREAKPOINT

302   Full Option Science System

Part 1: Water or Salt Tolerance and Plants

15. Set the scene for the salt experiment
Introduce the investigation by telling a story.

*A farmer lives near the ocean. He usually **irrigates** (waters) his crops with river water, but there is a **drought** (dry spell) and there's not enough river water for his radishes, barley, corn, and peas. But there is plenty of seawater. His question is, "Can I use a mix of seawater and fresh water to grow these plants?"*

Students will conduct an experiment to help this farmer decide what to do about water.

16. Discuss salinity of seawater
Ask what is in seawater that might affect the growth of plants. Tell students that there are several different salts in seawater, but the most common is sodium chloride, better known as table salt. Ask students how they think the salinity (saltiness) of the water might affect radishes, barley, corn, or peas.

17. Focus question: What is the salt tolerance of several common farm crops?
Write or project the focus question on the board as you say it aloud.

➤ *What is the salt tolerance of several common farm crops?*

18. Discuss the experimental design
Have students discuss in their groups how they could set up an experiment to test the effects of sodium chloride on these four crops. Remind them that in a controlled experiment, all conditions must be the same except for the one factor being tested (salinity of the water used to irrigate the plants). Discuss the groups' ideas.

19. Introduce the four salt solutions
Bring out the four labeled containers of salt solution and explain how they were made. They represent four mixtures of fresh water and seawater, from pure fresh water (0 spoons of salt) to pure salt water (4 spoons of salt).

20. Omit the bags covering the plants
Suggest that in this experiment the plants may be open to the air; no plastic bags are needed to isolate the plants. In this case, students will be watering the plants more often than just at the beginning of the investigation.

EL NOTE
If necessary, use illustrations along with the story. Write the new vocabulary words on the word wall.

SCIENCE AND ENGINEERING PRACTICES
Planning and carrying out investigations

TEACHING NOTE
Go to FOSSweb for Teacher Resources and look for the Science and Engineering Practices—Grade 4 chapter for details on how to engage students with the practice of planning and carrying out investigations.

TEACHING NOTE
Students can revisit previous investigations, such as the brine shrimp investigation, for ideas on how to set up a controlled experiment changing only one environmental factor.

Environments Module—FOSS Next Generation

INVESTIGATION 4 – Range of Tolerance

> **TEACHING NOTE**
> The amount of water must be agreed to and used by everyone. A good amount to use is 60 mL.

SCIENCE AND ENGINEERING PRACTICES

Planning and carrying out investigations

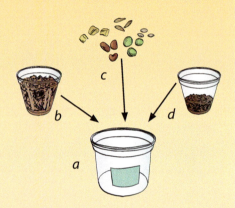

> **EL NOTE**
> If necessary, model how to fill out the sheet on the board or using a projection system.

21. Set the amount of water to use

Students need to add the same amount of water each time they water the containers. Tell students they should use 60 mL of water. Remind them that the planters will be open to the air.

The plants should be watered as needed, but should *not* be overwatered. All plants must be watered on the same day, and all planters must be watered with the same amount of the appropriate solution.

22. Discuss a planting procedure

Let students decide how the planters should be prepared. Discuss how the water-tolerance investigation was set up. Students should suggest that they must set up all four planters exactly the same and make a map of seed locations. If they have trouble coming up with a procedure, guide them to do the following:

a. Label four planters, one for each type of water.

b. Put 1 level cup of soil in each planter.

c. Plant three seeds of the four kinds (12 seeds total) in each cup.

d. Push the seeds into the surface of the soil. Add additional soil to cover the seeds if necessary. Gently pat the soil, using the bottom of the cup.

Use notebook sheet 20, *Plant Experiment—Salt Tolerance*, to scaffold the design.

23. Discuss record sheet

Distribute a copy of notebook sheet 21, *Plant Experiment Setup*, to each student. Review the information students should record now. Once again, explain that each student in a group will become an expert on one of the four kinds of plants (peas, corn, barley, or radishes). The plant expert will plant that kind of seed *in each of the four planters*.

Remind students that scientists conduct multiple trials to improve the accuracy of results. Ask them how the class is providing for multiple trials.

Part 1: Water or Salt Tolerance and Plants

24. Distribute the planting materials
Let each group divide up these tasks:
- Spreading newspaper on the worktable.
- Labeling planters.
- Putting 1 level cup of soil in each planter.
- Getting 12 of each kind of seed.

Students plant their own kind of seed in the same location in each container.

25. Plant the seeds
Allow 20 minutes for students to plant.

26. Water the seeds
Caution students to make sure the correct salt solution is used to water each labeled planter. Describe the procedure.
a. One student takes the 100 mL beaker to the materials station, gets the standard amount of the saltless water, and pours the water evenly on the planter with that label.
b. The next student takes the 100 mL beaker to the station and gets 1-spoon-of-salt water and pours the standard amount on the proper planter.
c. Repeat with the 2-spoons-of-salt water and the 4-spoons-of-salt water.

27. Clean up
Each group should put its four plant containers on a FOSS tray. Put the trays in the selected spot in the classroom. Trays can be stacked two high using the clear plastic tubes.

Have Getters return other materials to the materials station.

28. Review watering schedule
The plants should be watered during the growing period. They will probably need water after about 7 days and again after 11 or 12 days. Remind the class that all the containers should be watered with the same amount of the appropriate solution; even if a planter cup seems to still be moist, add the measure of water. Adding water is an all-or-none proposition. Students should record on the *Plant Experiment Setup* sheet when the plants are watered.

Materials for Steps 24–25
- *Containers of salt solutions*
- *Labels*
- *Planting containers*
- *Cups of seeds*
- *Beakers*
- *Soil*
- *Newspaper*
- *Tape*
- *Safety goggles*

> **TEACHING NOTE**
> By progressing from the least concentrated to the most concentrated salt solution, students do not have to wash the beaker between solutions.

Materials for Step 28
- *FOSS trays*
- *Tray columns*

BREAKPOINT

Environments Module—FOSS Next Generation

INVESTIGATION 4 – Range of Tolerance

TEACHING NOTE

Remind students that they will each be experts for **only one kind of plant**. Each student will record data for his or her kind of plant in each of the different conditions for the environmental factor they are investigating.

Materials for Step 30
- *Planters*
- *Metric rulers*

▶ **NOTE**

Have one group investigate the 120 mL water (swamp) planter. Be aware that the seeds usually mold and the decay can make the environment smell bad. Be aware of students who have seed or mold allergies.

SCIENCE AND ENGINEERING PRACTICES

Planning and carrying out investigations

29. **Observe the plant growth on day 5**

    Distribute to students a copy of notebook sheet 22, *Plant Observations A*, and have students complete the setup information.

    For the water tolerance groups, have the students find the "Environment" column in Part 1 on the sheet. They should record "dry" in the first box, followed by "moist," "wet," and "very wet."

    For the salt tolerance groups, have the students find the "Environment" column in Part 1 on the sheet. They should write "0 spoons of salt" in the first box, followed by the other concentrations of salt.

30. **Record observations on day 5**

    Have students record in Part 1 (for day 5). They can take the containers out of the bags briefly in order to make observations and to measure plant height with a metric ruler.

31. **Discuss observations on day 5**

    ➤ *In which environment did the most plants come up?*

    ➤ *Which environment has the tallest plant? What is its height?*

    ➤ *In which environment did plants have the most leaves?*

B R E A K P O I N T

306　　　　　　　　　　　　　　　　　　　　　　　　**Full Option Science System**

Part 1: Water or Salt Tolerance and Plants

32. **Observe plant growth on day 8**
    Have the groups observe plant growth after 8 days. Remind students to discuss the purpose of their experiment: to find out how the one factor affects plant growth—water or salt. Ask,

    ➤ *What similarities and differences do you observe in the planters?*

    ➤ *What parts of the plants (leaves, roots, stems) should we watch closely as we continue to observe our experiment?*

33. **Record observations on day 8**
    Have students record in Part 2 (for day 8) of the *Plant Observations A* sheet. They can take the containers out of the bags briefly in order to make observations.

34. **Discuss observations on day 8**

    ➤ *In which environment did the most plants come up?*

    ➤ *Which environment has the tallest plant? What is its height?*

    ➤ *In which environment did plants have the most leaves?*

BREAKPOINT

SCIENCE AND ENGINEERING PRACTICES
Planning and carrying out investigations

EL NOTE
If necessary, model how to record using the board or by using a projection system.

Environments Module—FOSS Next Generation

INVESTIGATION 4 – Range of Tolerance

Materials for Steps 36–37
- Planters
- Newspaper
- Metric rulers
- **Plant Profile** sheets (optional)
- Tape (optional)

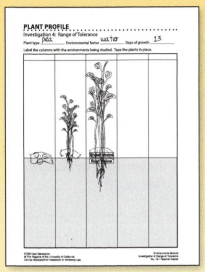

No. 19—Teacher master, completed

> **TEACHING NOTE**
>
> There is also a notebook sheet called Plant Profile (sheet 24). Students can draw what their plants look like on this sheet (instead of taping them on the teacher sheet).

35. Observe plant growth on day 13
After 13 days or so, it is time to end the experiment and observe the roots. Each group member will take one planter at a time and carefully turn it on its side over newspaper to remove the plants. The plants should be teased apart and each type given to the appropriate expert.

36. Introduce the *Plant Profile* sheet (optional)
Distribute a copy of teacher master 19, *Plant Profile*, to each student (full sheet). Point out the five columns with a box at the top of each. Students will write the environments in the boxes.

37. Profile plants (optional)
Each plant expert will tape *all* of his or her plants on one profile sheet in the appropriate columns. For example, if three pea plants grew in the moist environment, all three pea plants should be taped to the pea expert's profile sheet in the "moist" column. The line in the middle of the profile is the soil level.

If only two pea plants grew in the very wet environment, both will be taped in the "very wet" column. Dry pea seeds can be taped in the "dry" column. Have Getters get tape, and let students make their plant profiles.

38. Fill in *Plant Observations B* sheet
Once plant profiles are completed, have Getters get metric rulers from the materials station. Each plant expert can then count, measure, and record observations in Parts 3 and 4 on notebook sheet 23, *Plant Observations B*.

39. Clean up
The salty soil and any plants not taped to *Plant Profile* sheets should be wrapped in newspaper and composted. Have students return the rest of the soil to two basins. Store it in a plastic bag. Remove the labels from the planting containers, and rinse the containers thoroughly. Leave the bags open to dry.

Part 1: Water or Salt Tolerance and Plants

40. **Have a sense-making discussion**
    Ask students to compare the *Plant Profiles* in their groups. For the water experiment groups ask,

    ➤ *What is the least amount of water needed to grow these plants? Is the least amount the same for all kinds of plants?*

    ➤ *Can you give plants too much water? What happens to the plants?*

    ➤ *What is the range of water in which your kind of plant survived?*

    ➤ *Which water environment was best for each kind of plant?*

    For the salt experiment groups ask,

    ➤ *Did the seeds germinate the same in all salt concentrations?*

    ➤ *Did plants grow to the same height in all salt concentrations?*

    ➤ *Did the same number of plants grow in all salt concentrations?*

    ➤ *How does salt affect each kind of plant?*

    Students should find that all the plants are affected by salt. It takes the plants longer to germinate, fewer seeds germinate, and the plants grow more slowly, if at all. There will be variations in the responses, depending on the kind of plant. Plants that can survive in salty conditions are **salt-tolerant**; plants that fail in slightly salty conditions are **salt-sensitive**. Salt is harmful to these plants.

41. **Review range of tolerance**
    Explain that for both conditions, there is a boundary between the lower limit and the higher limit for that condition in which a seed can grow. Those limits define the range of tolerance.

42. **Review the concept of optimum**
    Explain that within the range of tolerance, the condition or conditions where a plant grows best is called the optimum. *Optimum* means "best." Ask the plant experts if they can determine the optimum condition for the plants.

43. **Review vocabulary**
    Review the term *controlled experiment*—a set of compared investigations in which only one factor is changed by steps.

44. **Answer the focus questions**
    Ask students to record an answer to their specific focus questions using a claim and providing evidence for their claim.

    ➤ *How much water is needed for early growth of different kinds of plants?*

    ➤ *What is the salt tolerance of several common farm crops?*

TEACHING NOTE

Students in each group will need to observe and discuss each other's data to compare the growth among all four plants.

TEACHING NOTE

Refer to the Sense-Making Discussions for Three-Dimensional Learning chapter in Teacher Resources *on FOSSweb for more information about how to facilitate this with students.*

CROSSCUTTING CONCEPTS

Cause and effect

TEACHING NOTE

Ask students to compare the experimental results to previous investigations with mealworms, isopods, and brine shrimp.

EL NOTE

For students who need scaffolding to answer the focus question, provide a sentence frame such as, We observed _____. Based on these observations, I think _____.

 drought
irrigate
salt-sensitive
salt-tolerant

SCIENCE AND ENGINEERING PRACTICES

Constructing explanations

Environments Module—FOSS Next Generation

INVESTIGATION 4 – Range of Tolerance

READING in Science Resources

45. Read "Environmental Scientists"

Read the article "Environmental Scientists" at any time during this investigation. If you have the class set up the two plant tolerance experiments sequentially, this reading will engage the salt-tolerance experimental groups while they are waiting their turn to set up their experiment. The reading consists of four short biographical snapshots of environmental scientists past and present—Edward O. Wilson, Rachel Carson, Tyrone B. Hayes, and Wangari Muta Maathai.

One strategy is to have students jigsaw the reading. Each student in a group of four reads one of the snapshots and then presents the information to the others in the group. The presentation should answer the discussion questions below.

For reading strategies to support English learners and below-grade-level readers, see the Science-Centered Language Development chapter.

46. Discuss the reading

Use these questions to guide the discussion about each environmental scientist.

➤ *What was the focus of this person's work?*

➤ *What contribution did this person make to our understanding of the natural world?*

➤ *What more would you like to know about this person?*

ELA CONNECTION

These suggested reading strategies address the Common Core State Standards for ELA.

SL 4: Report on a text in an organized manner, using appropriate facts and relevant, descriptive details to support main ideas or themes.

BREAKPOINT

47. View online activity

Have students engage with the online activity "Tutorial: Analyzing Environmental Experiments" to help them clarify the difference between testing for preferred environment, range of tolerance, and optimum conditions. Students can work in small groups or as individuals.

Full Option Science System

Part 1: Water or Salt Tolerance and Plants

48. Read "Range of Tolerance"

Read the article "Range of Tolerance," using the strategy that is most effective for your class. This reading can be done after students make their plant observations on day 5.

Tell students that this article will give them more information and examples of the concept, "range of tolerance." Have students take turns defining and telling a partner an example of the range of tolerance of an organism [brine shrimp] for an environmental factor [water salinity]. If necessary, give them a prompt such as,

Range of tolerance means _____ .

For example, _____ .

Have students preview the text by looking at and discussing the photographs and reading the captions with a partner. Ask students what they know about the chaparral ecosystem from past experience and/or looking at the pictures. Ask,

➤ *What do you think are the nonliving environmental factors that might affect the organisms that live there?*

Tell students to read the article to find out more about how organisms survive in the chaparral ecosystem. To support comprehension, give students self-stick notes to jot down symbols and phrases about the text. Here are some examples.

* * interesting
* ? question
* L learning something new
* W wondering
* S surprising

49. Discuss the reading

Give students a few minutes to share their self-stick note responses and to review the questions at the end of the article with their table group. Call on Reporters to share their group's answer to a question, along with details and examples from the text. Allow other group Reporters to respond by either agreeing, disagreeing and offering alternative explanations or evidence, or adding more information. After the discussion, let students revise or add to their original answers and record them in their notebooks.

➤ *All plants need water. What does optimum water mean for a plant? What does range of tolerance mean for a plant?* [Optimum water conditions produce the best growth results for a healthy plant. The range of tolerance for water is the condition that allows the plant to survive.]

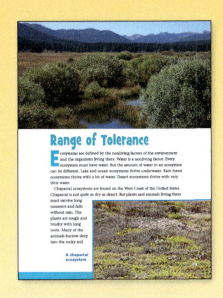

ELA CONNECTION

These suggested reading strategies address the Common Core State Standards for ELA.

RI 1: Refer to details and examples in a text when explaining what the text says explicitly and when drawing inferences from text.

RI 7: Interpret information presented visually, and explain how the information contributes to an understanding of the text.

W 8: Gather relevant information from experiences and print, and categorize the information.

SL 3: Identify the reasons and evidence a speaker provides to support particular points.

L 5: Demonstrate understanding of word relationships.

SCIENCE AND ENGINEERING PRACTICES

Obtaining, evaluating, and communicating information

Environments Module—FOSS Next Generation

INVESTIGATION 4 – Range of Tolerance

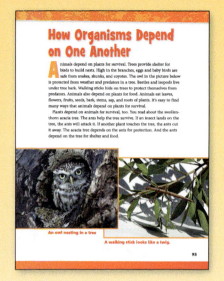

ELA CONNECTION

These suggested reading strategies address the Common Core State Standards for ELA.

RI 2: Determine the main idea of a text and explain how it is supported by key details; summarize the text.

RI 4: Determine the meaning of general academic domain-specific words or phrases.

RI 8: Explain how an author uses reasons and evidence to support particular points in a text.

SL 1: Engage in collaborative discussions.

SCIENCE AND ENGINEERING PRACTICES

Engaging in argument from evidence

Obtaining, evaluating, and communicating information

50. Read "How Organisms Depend on One Another"

"How Organisms Depend on One Another" revisits some of the observations in the Amazon rain forest article and gives more examples of how plants and animals interact. Students can read this article after they make their plant observations on day 8.

Start by having students discuss the question,

▶ *Do you think plants and animals need each other to survive and reproduce? If so, what are some examples?*

Read the first sentence of the article and tell students that this is the main point of the article. Ask,

▶ *How do you think the author will support this idea in the article?* [Explain and give examples of how plants and animals depend on one another.]

Have them preview the text by looking at the photographs and reading the captions. Remind students of the acacia tree and the leaf cutter ants they read about in the "Amazon Rain Forest Journal" and ask why the author would use this as an example. Read the article aloud or have students read independently. Pause after each section for students to take turns summarizing the text in their own words with a partner.

51. Discuss the reading

Review any unknown words or phrases and what strategies students used to determine their meaning. Give students a few minutes to write the answers to the questions at the end of the article in their notebooks. Have students share their answers with their group. For the last question, ask students to come to consensus and for the Reporter of each group to be prepared to share the group's claim and the evidence and reasoning that supports the claim. Encourage students to use details and examples from the text. If necessary, provide argumentation language structures such as:

We think _____ . Our evidence is _____ .

Some people in our group thought _____ .

We agree/disagree with _____ because _____ .

We wonder if _____ .

Why do you think _____ ?

▶ *Describe three examples of how animals depend on plants for survival.* [Trees provide shelter for birds to nest. Holes in trees provide protection for birds and eggs. Animals hide in plants to escape from predators, such as a beetle under bark or a walking stick on a branch. Plants provide food for animals.]

Part 1: Water or Salt Tolerance and Plants

➤ *Describe three examples of how plants depend on animals for survival.* [Bees pollinating a flower, seeds traveling on the fur of an animal, and seeds traveling in the stomach of an animal.]

➤ *Do you think animals pollinate flowers and disperse seeds on purpose or by accident? Explain why you think so.* [The animals are getting food (pollen or nectar) from the flower and get pollen on their legs, bodies, or head. By accident, the pollen sticks to them and travels to another flower. They take seeds to eat and drop them by accident.]

Ask this additional question about humans.

➤ *How do people depend on plants and animals for survival?*

WRAP-UP/WARM-UP

2. Share notebook entries

Conclude Part 1 by having students share notebook entries. Ask students to open their science notebooks to the most recent entry. Conduct a jigsaw by having students pair up with a partner who did a different plant experiment. Ask them to

- discuss the design of the tolerance experiment;
- share their answers to the focus question.

To help students compare and refine their arguments by evaluating the relevance of the data, try this argumentation strategy.

Present two positions to students on the board, both containing evidence from students' collected data on their plant growth. One position should contain sufficient evidence to support the claim. The other position has a credible claim, but insufficient evidence or irrelevant data. For example,

Salt water is not good for germinating seeds because only a few seeds germinate.

Barley is more salt-tolerant than peas because more barley seeds germinated in the salt water than did peas.

Ask students to identify the evidence in both positions. Have students discuss which position has better evidence supporting the claim being made. Ask how the evidence supports the claim. Then, have students work with a partner to refine the other position to incorporate stronger evidence. Allow students to revise their own responses.

SCIENCE AND ENGINEERING PRACTICES

Engaging in argument from evidence

TEACHING NOTE

See the **Home/School Connection** for Investigation 4 at the end of the Interdisciplinary Extensions section. This is a good time to send it home with students.

Environments Module—FOSS Next Generation

INVESTIGATION 4 – Range of Tolerance

MATERIALS for
Part 2: Plant Patterns

For each student
- 1 *FOSS Science Resources: Environments*
  - "Animals from the Past" (optional)

For each group
- 8 Colored pens or crayons (1 black, 1 brown, 1 red, 1 green, 1 blue, 1 yellow, 1 orange, 1 purple) ★
- 2 Clipboards ★
- 2 Pieces of white paper ★
- 2 Zip bags, 4 L

For the class
- 2 Maps of schoolyard in sections (See Step 4 of Getting Ready.) ★
- 1 Roll of transparent tape
- 25 Boundary marker flags (See Step 4 of Getting Ready.)
- 1 Camera (optional) ★
- 2 Sheets of cardboard or poster board (about 76 cm × 60 cm) ★
- 2 Sheets of chart paper ★
- • Scissors ★
- • Carry bag ★

For embedded assessment
- ❏ • Notebook sheet 25, *Response Sheet—Investigation 4*
- ❏ • *Embedded Assessment Notes*

★ Supplied by the teacher.   ❏ Use the duplication master to make copies.

No. 25—Notebook Master

Part 2: Plant Patterns

GETTING READY for
Part 2: Plant Patterns

1. **Schedule the investigation**
   This part will take three sessions. Plan on one outdoor session followed by an active session indoors for the response sheet. Ideally, do the outdoor session on a day when the ground is dry and all plants have their leaves. Plan one session for the reading.

2. **Preview Part 2**
   Students observe and map plant-distribution patterns in the schoolyard. They discuss the environmental factors that might be responsible for these patterns. The focus question is **How does mapping the plants in the schoolyard help us to investigate environmental factors?**

3. **Select your outdoor site**
   Look closely at all sides of your schoolyard to select one or two study areas. It would be good to have one study area that is heavily managed by humans (shrubs, gardens and lawn) and a more natural study area that will show the influence of the natural environmental factors on plant distribution. Four groups will work in one study area and four groups will work in a second study area. Each group of four students will restrict their observations to a small area (about 5 m × 5 m) in one of these two outdoor areas. The total size of each study area will be determined by the area you have available, and could be larger or smaller. If necessary, get permission to collect plant leaf samples.

4. **Make a map and set out boundaries**
   Once you have selected the best study areas in your schoolyard for mapping plant patterns, draw the map of each area. Each map will be divided into sections for students to take with them during the activity. The maps should be placed on firm board like poster board, flattened cardboard, or the firm backing of a pad of chart paper.

   For each map, you can use one large piece of chart paper and cut it into smaller group sections or you can use individual sheets of 8.5" × 11" paper and place the sheets next to each other on a large piece of cardboard or poster board in order to draw each map. Use tiny pieces of tape to hold the sheets in place as you draw the schoolyard map.

▶ **NOTE**
If you have 24 or fewer students, map only one study area. If you have more than 24 students, consider mapping two areas.

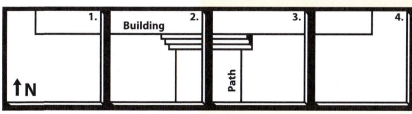

Sample map in one study area for four groups.

Environments Module—FOSS Next Generation 315

INVESTIGATION 4 – Range of Tolerance

Take the boundary flags to your study areas and stake out the four corners of each section of the map. Some of the flags will be marking the boundaries of two sections.

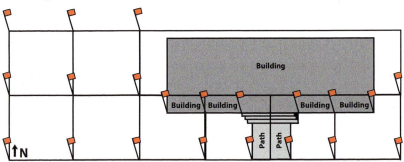

Using a pencil, draw in the largest physical features of the area first. Include roads, buildings, boulders, dumpsters, sidewalks, stairs, ramps, fences, and so forth. If possible, use existing maps of the school grounds as a guide. Once you are confident that the features are in the correct location, go over the pencil lines on the map with a black marker.

Once the map has been created, divide it equally into sections that correspond with the number of groups you will have.

5. Prepare grid on chart paper
In Step 6 of Guiding the Investigation, students sort the plants they found. Draw a 4 × 6 grid on a piece of chart paper for this sorting activity.

6. Organize supplies for outdoor work
There are two parts to the outdoor work. For the plant hunt, students will need a bag to collect samples from plants in their assigned area. For the mapping part, prepare bags of materials ahead of time for easy distribution outdoors. Each bag will contain a set of eight distinctly different colored marking pens or crayons. Each group should have the identical set of eight colors of pens or crayons. Each group will also need a clipboard and two pieces of paper for making the key. Organize all the materials in a bag or crate to be carried to the outdoor location.

7. Check the site
The morning before taking students to your selected site, do a quick search for potentially distracting or unsightly items. This is a good time to put the boundary marker flags in place.

Full Option Science System

Part 2: Plant Patterns

8. **Plan to review outdoor safety rules**
   Before going outdoors, remind students of the rules and expectations for the outdoor work. Remind students to not put anything from the ground or from plants in their mouths. If the unmanaged areas have poison ivy, poison oak, poison sumac, or stinging nettles, be prepared to point them out to students and warn them to avoid these plants.

9. **Enlist additional adults**
   If possible, seek out an additional adult to join you outdoors. Remind the adult that students will need time to struggle with the challenges and that his or her job is to lend support but not to solve the problems for students.

10. **Bring a camera (optional)**
    If possible, bring a camera to have students take photos of the plants in each area and the corresponding map sections when they are completed.

11. **Plan to read *Science Resources*: "Animals from the Past"**
    Plan to read "Animals from the Past" during a reading period after the active investigation.

12. **Plan assessment: response sheet**
    Use notebook sheet 25, *Response Sheet—Investigation 4*, for a closer look at students' understanding of how to design an experiment and how to draw conclusions from the results. Plan to spend time discussing the sheets with students after you have reviewed them. For more information about next steps and self-assessment strategies, see the Assessment chapter.

Environments Module—FOSS Next Generation

INVESTIGATION 4 – Range of Tolerance

FOCUS QUESTION

How does mapping the plants in the schoolyard help us to investigate environmental factors?

Materials for Step 4
- *Map of schoolyard*
- *Transparent tape*
- *Selected drawing utensils*
- *Clipboards*
- *Chart paper*
- *Zip bags*

GUIDING *the Investigation*
Part 2: *Plant Patterns*

1. **Introduce** *plant distribution*
   Ask students,

   ➤ *What environmental factors influenced our indoor plant experiments?* [Water and salt concentration.]

   ➤ *What determined the location where our indoor plants would grow?* [We did!]

   Tell them,

   We've been studying plants and how various environmental factors affect them. Today we will go to the schoolyard to observe and map **plant distribution**, *that is, how plants are spread out or arranged in an area. And we will investigate the environmental factors that might be responsible for the distribution.*

   Add the new term to the word wall.

2. **Focus question: How does mapping the plants in the schoolyard help us to investigate environmental factors?**

   Write or project the focus question on the board as you say it aloud.

   ➤ *How does mapping the plants in the schoolyard help us to investigate environmental factors?*

   Have students write the focus question in their notebooks.

3. **Review outdoor safety**
   Remind students of the rules and expectations for the outdoor work. Refer to the *FOSS Outdoor Safety* poster.

4. **Go outdoors**
   Gather the sectioned map you created, chart paper, the bags of materials, and clipboards (two per group). Give each group one empty bag for the first part. Make sure students are dressed appropriately, and head out to your home base.

5. **Introduce the plant search**
   Once outdoors, form a sharing circle near a few plants. Model how to pinch off a leaf sample. Explain that each group will collect a leaf or small representative sample from the eight most important plants in their area. Most of the time one leaf is enough, but when the plant is too small, they need to collect a leaf cluster. Explain where the boundaries are for the two study areas and make it clear which areas are off limits (e.g., the bed of tulips). Before sending

Part 2: Plant Patterns

students off, briefly discuss what makes a plant important (its size, abundance, economic significance, beauty, aroma, etc.). Allow 5 minutes for exploring the plants and collecting representative samples.

6. Sort leaf samples by kind

Display the 4 × 6 grid you prepared. Facilitate sorting the leaf samples by kind on the chart-paper grid. Have students in one group place their leaf samples in eight of the squares. Ask the other groups to do the same. If subsequent groups have leaves from the same species as one already placed in the grid, students should place the samples they collected in the same grid square with the other leaves. The result will be a sorting of leaves by kind.

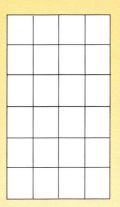

7. Identify the eight most important plants

Briefly discuss why students selected certain plants. Have the class decide which eight plants are the most important plants in the area. (The squares with the most leaves will most likely identify those plants.)

8. Make a plant-identification key

On two clean pieces of chart paper, tape the eight identified important plants in one column. Across from each sample, draw a colored circle with a marker or crayon. Use a different color for each plant. Distribute the clipboards, and have groups make their own replica of the plant-identification key. Students can use a small leaf sample from the grid or draw the leaf sample to create the group plant-identification key. Groups will need to share tape.

9. Introduce the mapping activity

Divide the class into half (four groups in one half, four groups in the other half). Lay the maps of each study site in front of the appropriate half of the class, properly oriented with the actual area. Explain that each group will receive a section of their map and use their set of eight colored pens or crayons. Outline the procedure.

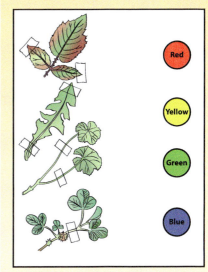

Half of the plant-identification key displaying four of the eight important plants.

a. Find your group's section of the study area map.

b. Locate all the places where the eight most important plant species are growing in your section.

c. Map out where the plants are located by using the appropriate colored pen to mark Xs at the appropriate places on the map. Each plant receives one X.

Distribute the map sections, and point out that each area has been marked off at each corner with a boundary marker. Mention that it is easier if each student concentrates on mapping one or two plants in the area.

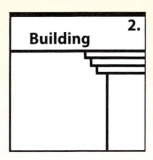

Environments Module—FOSS Next Generation

319

INVESTIGATION 4 – Range of Tolerance

SCIENCE AND ENGINEERING PRACTICES

Analyzing and interpreting data

CROSSCUTTING CONCEPTS

Patterns

▶ **NOTE**
Go to FOSSweb for *Teacher Resources* and look for the Crosscutting Concepts—Grade 4 chapter for details on how to engage students with the concept of patterns.

SCIENCE AND ENGINEERING PRACTICES

Constructing explanations

dominant plant
plant distribution

10. **Map plant species**
    Give groups about 12 minutes to map the plants in their areas. Wander among the groups, making sure students have oriented their maps correctly, are working on mapping the correct plants, and are making progress.

11. **Look for plant patterns**
    When everyone has finished mapping, call the class back together. Reassemble the sections of each map. Each group should briefly describe the plant patterns revealed on their section of the map.

    Ask students to observe the entire map and look for patterns. Let several students share observations.

    Ask,

    ▶ *What is the most common color on each plant-distribution map, and which plant does it represent?*

    Explain that the most abundant plants in an area are called dominants. **Dominant plants** cover more space or are larger than others and usually have a significant influence on other organisms in the area. Ask,

    ▶ *Do you see any patterns of colors?*

    Review the idea of plant distribution as the arrangement of plants in an area. Ask,

    ▶ *How can you explain the distribution of plants in the schoolyard?* [e.g., humans planted them and care for them.]

    ▶ *Which environmental factors (for example, light, wind, rain, soil, human or other animal intervention) might be affecting the distribution of the plants that were found?* [Buildings offer shade and protection from wind, and this might help certain plants that are shade tolerant. Other plants need lots of sunshine so they will thrive in open areas. Some plants grow where there is lots of moisture all of the time; other plants are water-tolerant and will grow where there is little moisture.]

    ▶ *Were there any places that did not have plants growing? Why do you think this is so?*

12. **Return to class**
    Have students return all the materials to the bags, gather up all the equipment, and head back to the classroom to discuss the results.

13. **Review vocabulary**
    Review the vocabulary dealing with plant distribution and patterns and add the new words to the word wall.

Part 2: Plant Patterns

14. Answer the focus question
Have students answer the focus question in their notebooks.

➤ *How does mapping the plants in the schoolyard help us to investigate environmental factors?*

15. Assess progress: response sheet
Distribute a copy of notebook sheet 25, *Response Sheet— Investigation 4*, to each student. Have students glue the sheet on a left-hand page and write their responses on the right-hand page.

Collect the science notebooks after class and review students' responses. Plan to spend time discussing the sheets with students after you review them.

What to Look For

- Students recognize that observations should be thorough—it is important to include observations from all conditions (moist, wet, super wet). For example, if there were no sprouts in the wet and super wet cups on May 12, the student should have included that information.

- Students write that observations should include quantitative information such as the height of the plants, the length of the leaves, and so forth.

- Students suggest improving the data by including the planting date to indicate the elapsed time since planting.

SCIENCE AND ENGINEERING PRACTICES
Planning and carrying out investigations

EL NOTE

For students who need scaffolding on the focus question response, provide a sentence frame such as,
We found out _____ .
I think _____ , because _____ .

Environments Module—FOSS Next Generation

INVESTIGATION 4 – Range of Tolerance

READING *in Science Resources*

16. Read "Animals from the Past" (optional)

Read the article "Animals from the Past," using the strategy that is most effective for your class. The article introduces the concept of extinction, using examples from recent geological history in the United States.

Give students a few minutes to preview the text by looking at the photographs and illustrations. Tell students to choose three pictures they find particularly interesting.

On your signal, have students get up and find a partner from another group. Have each student in the pair take turns discussing their first picture using these prompts:

➤ What do you notice?

➤ What do you think it represents?

➤ What does it remind you of?

Give students another signal and have them find new partners to discuss their second pictures. Repeat the process for the third picture.

Tell students to read the article looking for examples of animals that have become extinct and the causes for their extinction.

Read the article aloud or have students read independently. Pause after each section for students to take turns summarizing and sharing questions they have with a reading partner.

17. Use the Frayer model

Use the Frayer model to help students develop their understanding of extinction. Start by coming up with a definition for the word "extinct" and writing it in the definition box [when a species no longer exists].

Record information in the Frayer model either as a class on chart paper or in small groups. Students should make their own in their notebooks. Give students a few minutes to review the questions at the end of the article using the information from their Frayer model in their small groups.

ELA CONNECTION

These suggested reading strategies address the Common Core State Standards for ELA.

RI 1: Refer to details and examples in a text when explaining what the text says explicitly and when drawing inferences from text.

RI 2: Determine the main idea of a text and explain how it is supported by key details; summarize the text.

SL 4: Report on a text in an organized manner, using appropriate facts and relevant, descriptive details to support main ideas or themes; speak clearly at an understandable pace.

L 4: Determine or clarify the meaning of unknown and multiple-meaning words and phrases.

Part 2: Plant Patterns

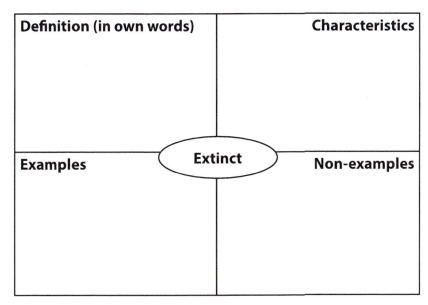

Call on Reporters to share their group's answer to a question, along with details and examples from the text. Allow other group Reporters to respond by either agreeing, disagreeing and offering alternative explanations or evidence, or adding on more information. After the discussion, let students revise or add to the original answers and record them in their notebooks.

➤ *What does* extinct *mean?* [No longer living on Earth.]

➤ *What are some animals that once lived in the United States, but are now extinct?* [Ground sloth, saber-toothed cat, mastodon, mammoth.]

➤ *What are some animals that are similar to animals that are now extinct?* [Ground sloths of rain forest of South America, mountain lion of North America, elephant of Africa and Asia.]

➤ *What can cause animals to become extinct?* [Changes in the environment. Organisms that are not adapted to survive in the changed environment become extinct.]

18. Read more about animal extinction

This is a good opportunity for students to learn more about animal extinction. See FOSSweb for a list of recommended books. One suggestion is *Last Chance to See* by Douglas Adams and Mark Carwardine.

SCIENCE AND ENGINEERING PRACTICES

Obtaining, evaluating, and communicating information

ELA CONNECTION

These suggested reading strategies address the Common Core State Standards for ELA.

SL 1: Engage in collaborative discussions.

RI 9: Integrate information from two texts on the same topic.

Environments Module—FOSS Next Generation

INVESTIGATION 4 – *Range of Tolerance*

WRAP-UP/WARM-UP

19. Share notebook entries

Conclude Part 2 or start Part 3 by having students share notebook entries. Ask students to open their science notebooks to the most recent entry. Have them pair up with a partner to share what they learned about plant distribution in the schoolyard.

Encourage students to give each other constructive feedback on their entries. If necessary, provide prompts such as:

Why do you think so?

Is there another reason for _____ ?

Have you thought about _____ ?

ELA CONNECTION

This suggested reading strategy addressed the Common Core State Standards for ELA.

SL 1: Engage in collaborative discussions.

Part 3: Plant Adaptations

MATERIALS *for*
Part 3: *Plant Adaptations*

For the class
- 1 Computer with Internet access ★
- Self-stick notes and scissors (for review session) ★

For benchmark assessment
- ❏ • *Posttest*
- ❏ • *Assessment Record*

★ Supplied by the teacher.   ❏ Use the duplication master to make copies.

Environments Module—FOSS Next Generation

INVESTIGATION 4 – Range of Tolerance

GETTING READY for
Part 3: Plant Adaptations

1. **Schedule the investigation**
   Plan on four sessions for this part—one session for the video and review discussion, one for a final review session, and one for the *Posttest*. Before the final review session, complete the final observations of the plant experiments from Part 1 (day 13 observations). Refer to the suggested teaching schedule in the Overview chapter for details on scheduling.

   Also plan one session to read about and revisit the mealworms that are now darkling beetles (from Investigation 1).

2. **Preview Part 3**
   Students review environmental factors that influence plant growth (water, light, nutrients). They are introduced to different adaptations of plants that allow some to thrive in dry environments and others to thrive in wet environments. The focus question is
   What are some examples of plant adaptations?

3. **Preview the video**
   Preview the video *All about Plant Adaptations* (duration 22 minutes), and plan how best to have students view it. This video is best shown at one sitting, so that students can follow Dusty's quest to investigate the environmental factors affecting his plant. The link to this video for teachers is in Resources by Investigation on FOSSweb.

4. **Plan benchmark assessment:** *Posttest*
   To help students prepare for the *Posttest*, plan a day for groups to review the important points from students' science notebooks. You might want to spend a little extra time on Investigation 4, given that students will not take an I-Check for this investigation. Later in the week, have students take the *Posttest*, working independently to respond to the items. Print or copy the assessment masters or schedule students on FOSSmap to take the *Posttest*. Review students' responses, using the coding guides found on FOSSweb.

▶ **NOTE**
Students also revisit the mealworms and discuss the life cycle of beetles. This brings Investigation 1 to a close.

Part 3: Plant Adaptations

GUIDING *the Investigation*
Part 3: *Plant Adaptations*

1. **Review the reading "Variation and Selection"**
   Ask students to think back to the reading "Variation and Selection." Ask,

   ➤ *What is natural selection?* [A process in nature where organisms that have traits that allow them to thrive when facing specific environmental pressures (conditions) will tend to survive and reproduce in greater numbers than other individuals of the same kind. Over many generations, those traits that allow the organisms to survive will be seen in the whole population.]

   ➤ *What is it about populations that allow natural selection to occur?* [The natural variation among individuals in the population.]

   ➤ *What is an **adaptation**?* [Any structure or behavior that improves an organism's chances of survival in an environment.]

   ➤ *What happens to a population of organisms when the environment changes?* [Some individuals in the population might have adaptations that allow them to survive in the changed environment. They will reproduce, and over the generations, those traits will be observed in the population. Other individuals that can't survive will no longer pass their traits on to future generations.]

2. **Introduce *All about Plant Adaptations* video**
   Tell students that they are going to view a video to learn more about plant adaptations. Some of the information in the video will be a review of desert and rain forest environments that they read about earlier. But some of the information about specific plant adaptations might be new.

3. **Focus question: What are some examples of plant adaptations?**
   Write or project the focus question on the board as you say it aloud.

   ➤ *What are some examples of plant adaptations?*

   Have students write the focus question in their notebooks. They should take notes so they can answer the focus question at the end of the video and discussion.

FOCUS QUESTION
What are some examples of plant adaptations?

TEACHING NOTE
Note that the video's definition of adaptation *is slightly different from the one we have used throughout the module. The video states that an adaptation is a change in a plant's structure occurring over long periods of time that make the plant more fit for living in its environment ("changing to suit the environment"). Both definitions are correct—both the process of change and the structure or behavior that results from the change.*

Environments Module—FOSS Next Generation

327

INVESTIGATION 4 – Range of Tolerance

SCIENCE AND ENGINEERING PRACTICES
Obtaining, evaluating, and communicating information

CROSSCUTTING CONCEPTS
Structure and function

TEACHING NOTE
Some examples from the video include going dormant during harsh conditions of temperature or moisture; deep, sturdy roots as anchors; large leaves or vines to capture sunlight; color or odor to attract pollinators; thorns or oil to keep predators away.

4. **View the video**
   Have students watch the 22-minute video.

   In *All about Plant Adaptations*, young members of a plant club try to find out why an indoor plant (Venus flytrap) is not thriving with tap water and light. As they try to get information, the kids discover that some plants live in cold, arctic environments and others live in dry, arid regions. These plants survive in these environments by remaining inactive or dormant for long periods of time. The kids discover that some plants have adaptations to survive under the lush, thick canopy of the rain forest (large leaves or vines that climb to reach the light), and that epiphytes, like some orchids, are able to obtain their water and nutrients out of the air. The kids learn about other adaptations plants exhibit either to protect themselves from predators or to help them reproduce to ensure their survival. In the end, the kids determine the environmental factors that provide the proper conditions for the Venus flytrap (in the range of tolerance).

5. **Discuss the video**
   Ask,

   ➤ *What did Dusty do to find out why his Venus plant was not thriving?* [He gathered information from experts. He conducted investigations by changing one environmental factor at a time and recorded observations over time, continuing to find out more information until he was able to isolate the factor that was responsible for the plant's problem.]

   ➤ *What kind of plant did Dusty have and in what environment is it found in nature?* [Venus flytrap, a carnivorous plant found in moist bogs of the southeastern United States. This plant is adapted to get nutrients from insects that it catches in its leaves.]

   ➤ *What environmental factor did Dusty change to improve the plant's growth?* [Changed from mineral water to distilled water.]

   ➤ *How could Dusty do a controlled experiment to find out more about the range of tolerance for the Venus plant?* [Dusty had just two plants in his experiment, one watered with tap water and one with mineral water. He could get more plants and vary the concentration of the minerals in the water that each plant received. He could have multiple trials to improve the results.]

   ➤ *What questions do you have about plants and their adaptations?*

6. **Answer the focus question**
   Have students answer the focus question in their notebooks.

   ➤ *What are some examples of plant adaptations?*

328 Full Option Science System

Part 3: Plant Adaptations

WRAP-UP

7. Review Investigation 4

Distribute one or two self-stick notes to each student. Ask students to cut each note into three pieces, making sure that each piece has a sticky end.

Ask students to take a few minutes to look back through their notebook entries to find the most important things they learned in Investigation 4. Students should include at least one science and engineering practice, one disciplinary core idea, and one crosscutting concept. They should tag those pages with self-stick notes. They might use a highlighter or colored pencils to call out the key points.

Lead a short class discussion to create a list of three-dimensional statements that summarize what students have learned in this investigation. Here are examples of the big ideas that should come forward in this discussion.

- We conducted investigations to add evidence to the idea that every organism has a range of tolerance for each factor in its environment. Optimum conditions are those most favorable to the success of an organism. (Planning and carrying out investigations; analyzing and interpreting data.)

- We built explanations based on investigations throughout the module to understand that organisms have specific requirements for successful growth, development, and reproduction. A relationship exists between environmental factors and how well organisms grow. (Constructing explanations; systems and system models; energy and matter.)

- We learned that plants have structures (adaptations) that help them survive and reproduce in their environments. (Obtaining, evaluating, and communicating information; structure and function.)

Complete the review by discussing other concepts from other parts of the module that students indicate they need clarified before taking the *Posttest*.

8. Discuss guiding questions for Investigation 4

Students should use the information from the firsthand investigations, readings, and videos about plants in different environments to discuss the guiding questions in their groups.

➤ What environmental conditions result in the best growth and survival of different plants?

➤ How do the structures of plants function to support the survival of the organisms in a particular environment?

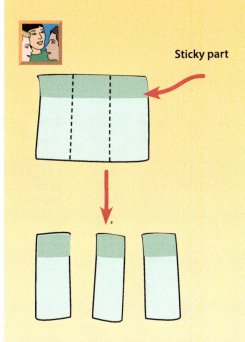

Sticky part

DISCIPLINARY CORE IDEAS

LS1.A: Structure and function

LS2.C: Ecosystem dynamics, functioning, and resilience

LS4.D: Biodiversity and humans

Environments Module—FOSS Next Generation

INVESTIGATION 4 – Range of Tolerance

9. **Discuss module driving question**
   Students should discuss the module driving question with a partner before responding to it in their notebooks.

   ➤ *How do the structures of an organism allow it to survive in its environment?*

BREAKPOINT

10. **Assess progress:** *Posttest*
    Give the *Posttest* at least one day after the wrap-up review. Distribute a copy of the *Posttest* to each student. You can read the items aloud, but students should respond to the items independently. Alternatively, you can schedule the *Posttest* on FOSSmap for students to take the test and then you can access data through helpful reports to look carefully at students' progress.

    If students take the test on paper, collect the *Posttests*. Code the items using the coding guides on FOSSweb.

Interdisciplinary Extensions

INTERDISCIPLINARY EXTENSIONS

Language and Environmental Literacy Extensions

- **Research drought areas**
  Have students use the Internet to research drought impact around the world, the process of land conversion to desert, and the resulting loss of food production.

- **Research salt on roads**
  In many areas of the country, salt is spread on icy roads to speed up melting. Have students call or write local agencies to find out what kind of salt is used and what effect it has on roadside plants.

- **Keep an environmental-news bulletin board**
  Have students read the newspaper and search the Internet to find articles that deal with environmental factors or environmental issues and their influence on organisms. Make a bulletin board of these articles. Ask students to report on some of the issues. Ask a few students to take the lead in organizing this bulletin board. Invite all students to contribute news articles.

Math Extension

- **Problem of the week**
  A student needs water for his cabbage garden. On a hillside above the garden there is a spring with water flowing underground. The student built a spring box to collect the underground water for his cabbage. He put a pipe from the spring box to his garden.

  The spring box needs to fill to the top with water before it will flow into the garden.

  On the first day, the water level came up 5 centimeters (cm) in the spring box during the night, and then went down 3 cm during the day. The second day, the water level went up another 5 cm at night and down 3 cm during the day. This continues every night and day. The spring box measures 20 cm from the bottom to the top of the box where the pipe is attached.

  On what day or night will water first flow into the student's cabbage garden?

  Show all your work. You can use drawings, a chart, or a number line to help you.

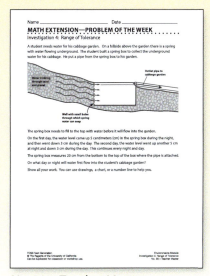

No. 20—Teacher Master

Environments Module—FOSS Next Generation

331

INVESTIGATION 4 – Range of Tolerance

Notes on the problem. Introduce the context to this problem before students work on it independently. Describe a spring box. You might also physically model the rise and fall of the water level in the spring box. This will be especially helpful to those students who do not have strong language skills.

There are at least four strategies students can use to find a solution. Plan to have students share their solutions with the class. If students do not suggest different solutions, use the time to teach these new strategies.

Strategy 1: Drawing a Picture

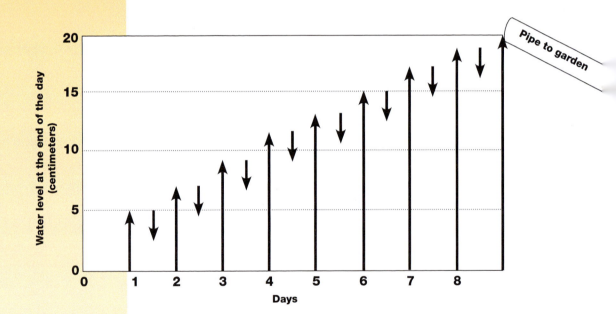

Interdisciplinary Extensions

Strategy 2: Using a Table

Water Level in Centimeters			
Day	a.m. (-3)	p.m. (+5)	End of day
1		5	5
2	5 – 3	5	7
3	7 – 3	5	9
4	9 – 3	5	11
5	11 – 3	5	13
6	13 – 3	5	15
7	15 – 3	5	17
8	17 – 3	5	19

Strategy 3: Using a Number Line

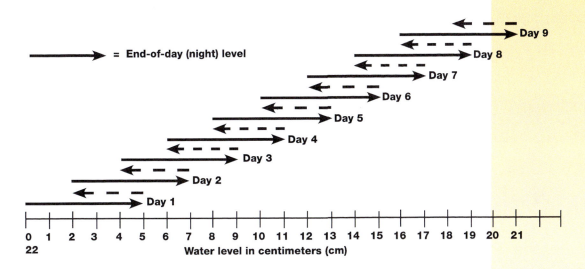

Strategy 4: Computing
During one night and day, the spring box gains 5 cm of water and loses 3 cm of water. The net gain is 2 cm in a 24-hour period. If the water level must be at 20 cm for flow to begin, students are likely to simply divide 20 cm by the 2-cm gain each day and conclude that it will take 10 days. However, the flow will actually begin during the ninth night.

On the ninth day, the water level falls to 16 cm during the day. That night, it only has to rise 4 cm to reach the top—at 20 cm—to begin flowing.

INVESTIGATION 4 – Range of Tolerance

TEACHING NOTE

Encourage students to use the Science and Engineering Careers Database on FOSSweb.

Science and Environmental Literacy Extensions

- **Visit a nursery**
  Ask students to visit a nursery to see what kinds of devices are available to monitor soil moisture in houseplants. Some are simple and inexpensive. Bring one to class for students to use with a plant in the classroom.

  Have students inquire at the local nursery about cold- and heat-tolerant, drought-tolerant, or sun- and shade-tolerant plants.

- **Simulate acid rain**
  Have students design and conduct an experiment to investigate the effect of acid rain on plants. They can use different concentrations of a common household acid, vinegar.

- **Investigate effects of gray water**
  Have students find out what happens to plants if kitchen-sink gray water is the water source. Have them use different concentrations of the gray water.

- **Plant a school garden**
  Work with the school administration and custodian to find a small patch of soil that students can use as a flower or vegetable garden. Have students decide what kind of plants to grow. They could start plants from seed indoors and transplant them outdoors. They could also move their terrarium plants to the garden when they are finished with the classroom study.

- **Make terrariums from around the world**
  Find some containers that are appropriate for terrariums, and set them up as permanent resources in the classroom. Containers should be transparent and colorless. Two-liter soft-drink bottles are good for small terrariums, and large spring-water containers are good for large terrariums.

  Some students may be from countries where the prevailing environmental conditions are different from those in North America. Encourage them to design a terrarium that models that environment. Help them locate plants that are representative of that environment. Label the terrariums for the countries they represent. Some of the terrariums may be arid and include cacti and succulent plants, and others might be moist and humid with ferns and mosses.

Interdisciplinary Extensions

Home/School Connection

Planting a garden is a good home activity for students. Here are some math problems to help in the planning. Print or make copies of teacher master 21, *Home/School Connection* for Investigation 4, and send it home with students after Part 1.

Four friends want to plant a garden. One of their fathers said they could use an area in the backyard that was a rectangle 8 meters by 4 meters.

Fair plots. The friends want to divide the area into four equal parts. Each friend has a different way to divide the garden plot fairly. Show at least four different ways the garden could be divided. Which one would you pick for the garden? Why?

How much area to plant? The friends decided on a garden plan. They wanted to know how much area each person has for planting. One friend calculated and said that each would get 12 square meters to plant. Another said they would each get 8 square meters. A third said they would get 8 meters each. Only one of the friends had the correct answer.

What is the area of each person's piece of the garden?

A row of marigolds. The friends decided to plant a single row of marigolds along one of the 8-meter sides. They will start the row at one corner of the garden plot and plant to the opposite corner. Marigolds need to be planted 10 centimeters apart. How many marigolds will they need for the row?

A border for the garden. The friends really like the row of marigolds along one side of the garden. They want to plant marigolds around the whole garden. The border of marigolds will be planted along the outer edges, or the perimeter, of the garden. How many marigolds will they need in all to make a border for the entire garden?

No. 21—Teacher Master

Environments Module—FOSS Next Generation

335

INVESTIGATION 4 – Range of Tolerance

Here are a number of simple garden plots students might design. Many other ways are possible.

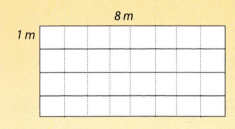

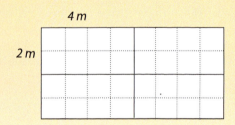

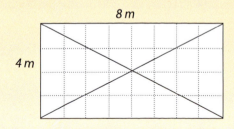

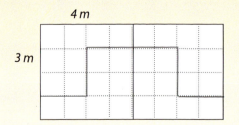

Notes on the problem.

Fair plots. Each gardener will have 8m^2.

This activity provides an opportunity to look at an area model to represent fractions. The key is that each of the parts must have an equal area. When you pose this problem initially, we suggest that you have your students draw the rectangles freehand and divide them into fractional parts on the data sheet (or unlined paper). This will give you an informal assessment of their understanding of fractions as part of a whole. The divisions that are commonly done are shown in the sidebar.

Students often divide a shape into congruent pieces (same shape, same size) when creating fractional parts.

In this problem, the way to prove that the fractional parts are equivalent is to determine the area of each part. You may want to provide grid paper so that students can redraw four rectangles to scale, divide them into four parts, and finally determine the area of each piece in square units. This will deepen their understanding of fractions as equal parts of a whole and reinforce a square unit as a measurement of area.

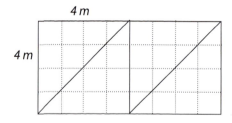

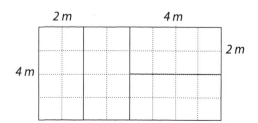

336 Full Option Science System

Interdisciplinary Extensions

How much area to plant? Students deepen their understanding of area and perimeter as they learn that area is measured in square units, while perimeter is a linear unit. They also use and apply the formulas for area (A = l × w) and perimeter.

Depending upon what time of the year you teach geometry and measurement in relation to this module, you might have to review the meaning of area and perimeter before students approach these problems. It might be helpful to model how to cover a defined rectangular area with unit tiles to define the unit of measure and how to determine the area.

A row of marigolds. To determine how many marigolds are needed for a single row, students will need to convert the two given measurements into a common unit—either meters or centimeters.

In addition, students will need to take into account planting from corner to corner. Given the length of the side as 800 cm and plants will be situated 10 cm apart, if they only plant 80 marigolds, the corner where they end will not have a marigold! An additional marigold is needed to have a plant at the opposite corner (where the planting ends).

There are many ways to solve the problem. Below are common incomplete solutions you might see:

Using the conversion:

100 centimeters = 1 meter

Row = 8 m = 800 cm

Marigold planted 10 cm apart from others.

800 cm row ÷ 10 cm per plant = 80 marigolds needed

Using the conversion:

10 centimeters = 1/10 of meter = distance between marigolds

Row = 8 m = 80 × 1/10

There are 80 "1/10" units in 8 m.

Therefore 80 plants are needed.

In each case, the extra plant needed to go at the end of the row was not added in. Since at one corner there is a plant, as you continue to put plants in, the 80th plant will be at the 70-cm point. There is a final plant needed for the 80-cm point, making the total 81 plants.

Environments Module—FOSS Next Generation

INVESTIGATION 4 – Range of Tolerance

Students will benefit from drawing a model of what is happening as the marigolds are planted. Simulate the planting using a smaller number. Have students draw a line segment 10 units long. Draw a dot at one end of the line segment. Continue making dots at one-unit intervals. How many dots were needed to go from one end to the other? (11 dots) Though the length of the line is 10 units, the number of points that represents the plants is one more than 10.

• = marigold

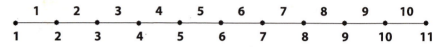

A border for the garden. This time instead of just planting a single row, students need to look at the perimeter or the total distance around the garden. Depending upon your students' backgrounds and experiences, you might want to review the meaning of perimeter concretely as well as the formulas for perimeter:

$P = l + w + l + w$

$P = 2l + 2w$

$P = 2(l + w)$

Students will again need to convert to a common unit of measure to solve the problem. They can do that before or after they determine the perimeter.

In planting a row of marigolds, they needed an extra marigold. In this problem, the marigolds are planted continuously around the perimeter and the corners of the rectangle are not to be counted twice! To determine the number of marigolds needed, divide the perimeter by 10.

Calculating the perimeter in meters first:

$P = 2(8\text{ m}) + 2(4\text{ m}) = 16\text{ m} + 8\text{ m} = 24\text{ m}$

Then converting to cm, 24 cm × 100 = 2400 cm

Converting to cm first and then substituting:

$P = 2(800\text{ cm}) = 2(400\text{ cm}) = 1600\text{ cm} + 800\text{ cm} = 2400\text{ cm}$

In both cases, divide 2400 cm by 10 cm to determine 240 marigolds plants are needed for the border.

Assessment

ENVIRONMENTS – *Assessment*

THE FOSS ASSESSMENT SYSTEM *for Grades 3–5*

"Assessment is like science. …To assess our students, we plan and conduct investigations about student learning and then analyze and interpret data to develop models of what students are thinking. These models allow us to predict the effect of additional teaching addressing the patterns we notice in student understanding and misunderstanding. Assessment allows us to improve our teaching practice over time, spiraling upward" (*2016 Science Framework for California Public Schools, Kindergarten through Grade 12*, chapter 9, page 3).

An important rule of thumb in educational assessment is that assessments should be designed to meet specific purposes. One size does not fit all. The FOSS assessment system provides ample opportunities for both formative and summative assessment. Formative assessments provide short-term information about learning by making students' thinking visible in order to guide instructional decisions. Summative assessments provide valid, reliable, and fair measures of students' progress over a longer period of time, at the end of a module, or the end of the year. The purpose for the assessment determines the choice of instruments that you will use.

The FOSS assessment system is designed to assess students in cycles: short, medium, and long. The assessment tasks allow students to demonstrate their facility with three-dimensional understanding of science.

Short cycle. Embedded assessment opportunities are incorporated into each part of every investigation. These assessments use student-generated artifacts, including science notebook entries, answers to focus questions, response sheets, and **performance assessments**. Embedded assessments provide daily monitoring of students' learning and practices to help you make decisions about instructional next steps. Embedded assessments using science notebooks provide evidence of students' conceptual development. Performance assessments focus on science and engineering practices and crosscutting concepts, as well as disciplinary core ideas.

Contents

The FOSS Assessment System for Grades 3–5	339
Assessment for the NGSS	341
Embedded Assessment	346
Benchmark Assessment	350
Next-Step Strategies	356
FOSSmap and Online Assessment	360
Sample Assessment Items	362

> **NOTE**
> For coding guides and student work samples, go to the Assessment section in *Teacher Resources* on FOSSweb.

Full Option Science System

ENVIRONMENTS — Assessment

I-Check opportunities occur every 1–2 weeks. These assessments are actually hybrid tools that provide summative information about students' achievement, and we have found they are even more powerful when used for formative assessment. Daily embedded assessments provide a quick snapshot of students' immediate learning, and I-Checks challenge students to put this learning into action in a broader context. Now students must think about the science and engineering practices, disciplinary core ideas, and crosscutting concepts they have been learning, and know when, where, and how to use them. I-Checks (short for "I check my own understanding") also provide opportunities for guided self-assessment, an important skill for future learning and development of a growth mindset. Properly executed feedback can help a student focus attention on areas that need strengthening. When a student responds to feedback, you can develop an even more precise understanding of the student's learning. A feedback/response dialogue can develop into a highly differentiated path of instruction tailored to the learning requirements of individual students.

Medium and long cycle. *Survey/Posttest*, **interim benchmark assessments, and portfolios** are tools provided for medium- and long-cycle assessment. Students take the *Survey* before instruction begins. This entry-level assessment provides you with information about students' prior knowledge and developing practices. It indicates modifications that might be needed for specific students or groups of students.

The *Posttest* is given at the end of a module. It provides summative information about students' three-dimensional learning. It also lets students compare their *Survey* responses to those on the *Posttest* and see how their understanding has grown. You can also use the *Posttest* for formative instructional evaluation by making notes about things you might want to focus on or do differently next time you teach the module.

Interim benchmark assessments can be given during a module when you want specific information about how students are progressing toward a specific performance expectation, at the end of the module, or at the end of the school year. These assessments specifically address NGSS performance expectations, and require students to apply what they have learned to new contexts. The interim assessment tasks can also inform instruction when used formatively to plan for next year's instruction.

Students can also collect work samples in a portfolio as they work through the module. At the end of each investigation, they can create derivative products to document their three-dimensional learning.

Assessment for the NGSS

ASSESSMENT *for the* NGSS

A Framework for K–12 Science Education (National Research Council, 2012), *Next Generation Science Standards* (National Academies Press, 2013), *Developing Assessment for the Next Generation Science Standards* (National Research Council, 2014), and many state frameworks provide a new vision for science education. These documents emphasize the idea that science education should resemble the way that scientists work. Students plan and conduct investigations, gather data to construct explanations, and engage in argumentation to build their understanding of the natural world. They apply that knowledge to engineering problems and design solutions. Students are expected to construct and discuss explanations and model systems in more and more sophisticated ways as they move through the grades. Assessment plays an important role in this new vision of science education—assessment is the bridge between teaching and learning.

Several key points in these foundational documents provide guidance for a well-designed assessment system. FOSS has followed these guidelines to ensure a robust assessment system that provides valuable diagnostic information about students' learning.

Assessment tasks should consist of multiple components in order to measure all three dimensions of science and engineering learning. The FOSS assessment system provides multiple tools and strategies to assess the three dimensions: (1) science and engineering practices, (2) disciplinary core ideas, and (3) crosscutting concepts. These tools and strategies provide evidence about what students can do, their developing conceptual understandings, and the connections that they are making among disciplines. Entry-level assessments, given before instruction begins, show what students can do and what they know before they begin a new module.

Assessment systems should include formative and summative tasks. Formative assessment tasks are embedded in the curriculum at key stages in instruction. These tasks are designed to support teachers in collecting and analyzing data about students' conceptual understanding and growing practice. Notebook entries, answers to focus questions, response sheets, performance assessments, and oral presentations/interviews provide the information you need to decide what students need to do next to move toward a learning goal. FOSS suggestions for next-step strategies help you address students' developing conceptions and provide information for differentiated instruction as needed.

"Assessment plays an important role in this new vision of science education—assessment is the bridge between teaching and learning."

Environments Module—FOSS Next Generation

ENVIRONMENTS — *Assessment*

Summative assessments are designed to provide valid, reliable, and fair measures of students' progress. The FOSS system includes three types of summative assessments: *Posttests*, interim assessments, and portfolios. These assessments include multicomponent tasks including open-ended constructed-response problems as well as some multiple-choice and short-answer items. *Posttests* are given at the end of a module and interim assessments can be given twice during a module or as an end-of-year assessment. Students can also collect work products at the end of each investigation for inclusion in a portfolio. Coding guides found on FOSSweb provide teachers with guidance when evaluating these assessments. All written assessments are consistent with grade-level writing and mathematics in the Common Core State Standards for ELA/Literacy and for Mathematics.

Assessment systems should support classroom instruction. The main purpose of the FOSS assessment system is to support classroom instruction—to provide the bridge between teaching and learning. Teachers need information daily about what students have learned or may be confused about. FOSS has developed a technique in which teachers spend only 10 minutes after a lesson, using a reflective-assessment practice (explained in detail later in this chapter), to gather data to determine instructional next steps. Are students ready to move on to the next lesson, or do they need some additional clarification? Our research has shown that the reflective-assessment practice provides evidence-based information that is crucial for differentiating instruction for all students—this practice can make a significant difference in students' overall achievement.

Assessment developers need to take a rigorous approach to the process of designing and validating assessments. The FOSS assessment system is based on a construct-modeling approach for assessment design. That means that we have done the research needed to describe a conceptual framework and learning performances that provide evidence of students' progressive learning (see the Framework and NGSS chapter), and we have done the technical work needed to ensure that assessment tasks provide valid, reliable, and fair evidence of students' learning. (See the Benchmark Assessment section for a more detailed description of the design behind the FOSS assessment system.)

> *"Our research has shown that the reflective-assessment practice provides evidence-based information that is crucial for differentiating instruction for all students—this practice can make a significant difference in students' overall achievement."*

Assessment for the NGSS

Assessment systems should include an interpretive system and locate students along a sequence of progressively more complex understanding. FOSS provides extensive support for interpreting assessment information. For each embedded assessment, specific information in the Getting Ready and Guiding the Investigation sections describes what students do and what to look for in student responses to assess progress. Coding guides, found on FOSSweb, are provided for each item on benchmark assessments (I-Checks and the *Survey/Posttest*, as well as intermin assessments). Samples of student work, especially for open-response questions, are also available on FOSSweb. Resources in this chapter and on FOSSweb provide you with information about how to use the coding guides, as well as what to do for next steps when students need to spend more time on a practice or core idea, or to look at it from a different perspective to see more connections.

You can use FOSSmap to have students take assessments online and generate a number of diagnostic and summary reports (delivered as PDFs). Some of these reports provide information about class progress; others provide individual students and parents with information about what students know and what they still need to work on. (See the FOSSmap and Online Assessment section in this chapter.)

The FOSS assessment system was developed over a period of 5 years with data from more than 500 teachers and their students. We know that teachers can employ this assessment system for the benefit of their students, and we know that students achieve more. Perhaps even more important is the change in classroom culture that occurs when assessment is thoughtfully employed as the bridge between teaching and learning. Assessment is no longer a stress factor for students or teachers. It encourages all to adopt a growth mindset—if I know where my strengths and weaknesses are and I continue to be thoughtful and work hard, I can make progress. It models what scientists do. Scientists use the information they have to argue for the best explanation, but they keep an open mind, so when new evidence emerges, they can incorporate that into their thinking, too. That's also what good curriculum and assessment are all about.

> *"Assessment is no longer a stress factor for students or teachers. It encourages all to adopt a growth mindset—if I know where my strengths and weaknesses are and I continue to be thoughtful and work hard, I can make progress."*

ENVIRONMENTS — Assessment

NGSS Performance Expectations

"The NGSS are standards or goals, that reflect what a student should know and be able to do; they do not dictate the manner or methods by which the standards are taught.... Curriculum and assessment must be developed in a way that builds students' knowledge and ability toward the PEs [performance expectations]" (*Next Generation Science Standards*, 2013, page xiv). The FOSS assessment system includes embedded, performance, and benchmark assessments. The chart displayed on this and the next page provides an overview of these assessments across the three fourth-grade modules. These assessments help students build knowledge and ability in concert with active investigations and readings to meet the goals of the NGSS.

Grade 4 NGSS Performance Expectations	FOSS Module Embedded Assessment	Benchmark Assessment
4-PS3-1. Use evidence to construct an explanation relating the speed of an object to the energy of that object.	Energy • Inv 2, Part 2: notebook entry	Energy • *Investigation 4 I-Check* • *Survey/Posttest*
4-PS3-2. Make observations to provide evidence that energy can be transferred from place to place by sound, light, heat, and electric currents.	Energy • Inv 1, Part 1: notebook entry • Inv 4, Part 1: notebook entry	Energy • *Investigation 1 I-Check* • *Investigation 2 I-Check* • *Investigation 3 I-Check* • *Investigation 4 I-Check* • *Survey/Posttest*
4-PS3-3. Ask questions and predict outcomes about the changes in energy that occur when objects collide.	Energy • Inv 4, Part 3: response sheet	Energy • *Investigation 4 I-Check* • *Survey/Posttest*
4-PS3-4. Apply scientific ideas to design, test, and refine a device that converts energy from one form to another.	Energy • Inv 5, Part 3: performance assessment	Energy • *Investigation 1 I-Check* • *Investigation 3 I-Check* • *Survey/Posttest*
4-PS4-1. Develop a model of waves to describe patterns in terms of amplitude and wavelength and that waves can cause objects to move.	Energy • Inv 5, Part 1: response sheet	Energy • *Survey/Posttest*
4-PS4-2. Develop a model to describe that light reflecting from objects and entering the eye allows objects to be seen.	Energy • Inv 5, Part 2: notebook entry	Energy • *Survey/Posttest*
4-PS4-3. Generate and compare multiple solutions that use patterns to transfer information.	Energy • Inv 3, Part 3: notebook entry	Energy • *Investigation 3 I-Check*

Full Option Science System

Assessment for the NGSS

Grade 4 NGSS Performance Expectations	FOSS Module Embedded Assessment	FOSS Module Benchmark Assessment
4-LS1-1. Construct an argument that plants and animals have internal and external structures that function to support survival, growth, behavior, and reproduction.	**Environments** • Inv 1, Part 1: notebook entry • Inv 4, Part 1: performance assessment	**Environments** • *Investigation 1 I-Check* • *Survey/Posttest*
4-LS1-2. Use a model to describe that animals receive different types of information through their senses, process the information in their brain, and respond to the information in different ways.	**Environments** • Inv 2, Part 4: response sheet • Inv 3, Part 1: performance assessment	**Environments** • *Investigation 2 I-Check* • *Survey/Posttest*
4-ESS1-1. Identify evidence from patterns in rock formations and fossils in rock layers to support an explanation for changes in a landscape over time.	**Soils, Rocks, and Landforms** • Inv 2, Part 4: notebook entry	**Soils, Rocks, and Landforms** • *Investigation 2 I-Check* • *Survey/Posttest*
4-ESS2-1. Make observations and/or measurements to provide evidence of the effects of weathering or the rate of erosion by water, ice, wind, or vegetation.	**Soils, Rocks, and Landforms** • Inv 1, Part 2: response sheet • Inv 1, Part 3: performance assessment • Inv 2, Part 1: notebook entry • Inv 2, Part 2: performance assessment • Inv 2, Part 3: response sheet	**Soils, Rocks, and Landforms** • *Investigation 1 I-Check* • *Investigation 2 I-Check* • *Survey/Posttest*
4-ESS2-2. Analyze and interpret data from maps to describe patterns of Earth's features.	**Soils, Rocks, and Landforms** • Inv 3, Part 1: notebook entry • Inv 3, Part 2: response sheet	**Soils, Rocks, and Landforms** • *Investigation 3 I-Check* • *Survey/Posttest*
4-ESS3-1. Obtain and combine information to describe that energy and fuels are derived from natural resources and their uses affect the environment.	**Energy** • Inv 5, Part 3: notebook entry	**Energy** • *Survey/Posttest* **Soils, Rocks, and Landforms** • *Survey/Posttest*
4-ESS3-2. Generate and compare multiple solutions to reduce the impacts of natural Earth processes on humans.	**Soils, Rocks, and Landforms** • Inv 3, Part 4: notebook entry	**Soils, Rocks, and Landforms** • *Investigation 3 I-Check* • *Survey/Posttest*
3-5-ETS1-1. Define a simple design problem reflecting a need or a want that includes specified criteria for success and constraints on materials, time, or cost.	**Energy** • Inv 1, Part 4: performance assessment	**Energy** • *Investigation 1 I-Check*
3-5-ETS1-2. Generate and compare multiple possible solutions to a problem based on how well each is likely to meet the criteria and constraints of the problem.	**Energy** • Investigation 3 **Soils, Rocks, and Landforms** • Inv 3, Part 3: performance assessment	**Energy** • *Investigation 2 I-Check* • *Investigation 3 I-Check*
3-5-ETS1-3. Plan and carry out fair tests in which variables are controlled and failure points are considered to identify aspects of a model or prototype that can be improved.	**Environments** • Inv 1, Part 2: response sheet • Inv 3, Part 3: response sheet **Soils, Rocks, and Landforms** • Inv 1, Part 3: performance assessment • Inv 2, Part 2: performance assessment	**Energy** • *Investigation 3 I-Check* • *Investigation 4 I-Check*

Environments Module—FOSS Next Generation

ENVIRONMENTS — *Assessment*

EMBEDDED *Assessment*

Assessment is the bridge that connects teaching and learning. Assessing students on a regular basis gives you valuable information that guides instruction and keeps families and other interested members of the educational community informed about students' progress.

Embedded assessments are suggested for most investigation parts. You will find a description of what and when to assess in the **Getting Ready** section of each part of each investigation. For example, here is the Getting Ready assessment step for Part 1 of the first investigation.

> **20. Plan assessment: notebook entry**
>
> At the first breakpoint, collect students' notebooks after class to check their answer to the questions on notebook sheet 1, *Mealworm Observations* (see Step 16 of Guiding the Investigation). In Step 25 of Guiding the Investigation, review students' answers to the focus question to see if they understand how structures and behaviors function to help mealworms survive.

As you progress through the lesson and it is time for students to create a work product to be assessed, you will find a step in **Guiding the Investigation**. It provides a bulleted list of what to look for when you review the students' work for disciplinary core ideas, science and engineering practices, and crosscutting concepts.

> **16. Assess progress: notebook entry**
>
> Have students turn in their notebooks after class, open to the page you will be reviewing. Check students' answers to the questions on notebook sheet 1, *Mealworm Observations*.
>
> **What to Look For**
>
> - *Structures of a mealworm are drawn and labeled.*
> - *Functions of the mealworm structures are described.*
> - *Observed mealworm behaviors and how those help the organism survive are discussed.*
> - *Students have asked one question of their own about mealworms.*
> - *Students describe what they need to know about the mealworms to keep them alive and healthy.*

Embedded assessment is enacted through two strategies: science notebook entries and teacher observation of science practices.

Embedded Assessment

Performance Assessments

Assessing the three dimensions envisioned in the NRC *Framework* and the NGSS performance expectations challenges students to be engaged in science and engineering practices in order to build disciplinary core ideas bridged by crosscutting concepts. This is an everyday occurence in the FOSS curriculum. One part in each investigation has been designated as a performance assessment for you to formatively check students' progress for all three dimensions at the same time. You peek over students' shoulders while they are in the act of doing science or engineering. You take note of what they are doing and discussing and sometimes conduct short interviews. Observing the rich conversation among students and the actions they are taking to investigate phenomena or design solutions to problems provides important information about student progress. At times, you might step in with a 30-second interview to ask a few carefully crafted questions to learn more about students' deeper conceptual understanding and practices. The What to Look For bullets in each performance assessment step will help you focus on pertinent science and engineering practices, disciplinary core ideas, and crosscutting concepts. You can record student progress on the *Performance Assessment Checklist* for each part.

Science Notebook Entries

Making good observations and using them to develop explanations about the natural world is the essence of science. This process calls for critical thinking and honed communication skills. Science notebook entries are designed specifically to help you understand the practices, crosscutting concepts, and scientific explanations and knowledge that students are developing.

Four kinds of notebook entries serve as assessments for learning. Each part of each investigation is driven by a **focus question**. Each part usually concludes with students writing an answer to the focus question in their notebooks. Their answers reveal how well they have made sense of the investigation and whether they have focused on the relevant actions and discussions. Prepared **notebook sheets** or **free-form notebook entries** provide information about how students make and organize observations and how they think about analyzing and interpreting data. Finally, **response sheets** provide more formal embedded-assessment data once in each investigation. These assessments focus on specific scientific knowledge, practices, or crosscutting concepts that students often struggle with, giving you an additional opportunity to help students untangle concepts that they might be overgeneralizing or have difficulty differentiating.

> **NOTE**
> You only need 10 minutes after a lesson to review student work and gather evidence of learning. See the reflective-assessment practice on the next page.

Environments Module—FOSS Next Generation

347

ENVIRONMENTS — Assessment

Using the Reflective-Assessment Practice

Successful teachers incorporate a system of continuous formative assessment into their standard teaching practices. One of the keys to formative assessment is frequency. The more often you can gather evidence about students' progress, the more able you will be to guide each student's path to understanding. The **reflective-assessment practice** provides a proven method for gathering that information, in a way that takes little time, but has a big impact on students' learning.

Reflective-Assessment Practice

1. Anticipate — Use the Investigations Guide to plan for each part and determine embedded assessment.

2. Teach — Use Guiding the Investigation to teach the lesson. Collect student notebooks.

3. Review — Review students work (10 minutes). Use "What to Look For" in Guiding the Investigation.

4. Reflect — Note trends and patterns you see in student understanding.

5. Adjust — Plan next instructional steps based on assessment reflection. Make notes for next year.

AFTER EACH PART

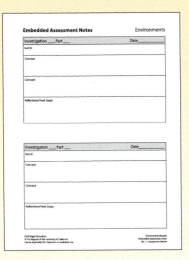

You can record notes for two lessons on each page.

Print or make copies of **Embedded Assessment Notes** to record students' progress with the embedded assessments in each part of an investigation (see the What to Look For bullets). If you don't think you can review every student's work in 10 minutes, choose a random sample. Many teachers are pleasantly surprised to discover how many students they can review in that short amount of time. The important thing is that you are looking at student work on a daily basis as often as possible and looking for patterns that reveal strengths and areas that need additional support.

Embedded Assessment

1. **Anticipate.** Check the Getting Ready section for the suggested embedded assessment for the part you are planning. Before class, fill in the investigation and part number along with the date on *Embedded Assessment Notes*. Check the assessment step in Guiding the Investigation, and fill in the things you are looking for in student thinking. Limit your assessment to one or two important ideas.

2. **Teach.** Follow the steps in Guiding the Investigation.

3. **Review.** Collect the notebooks at the end of class. Have students turn in their notebooks *open to the page you will be reviewing.* (This may sound trivial, but it will save you a lot of time.) If you didn't write in your "What to look fors" before teaching the lesson, do that now by checking the assessment step in Guiding the Investigation. Use *Embedded Assessment Notes* to record what you observe. Make a tally mark for each student who "got it"; write in names and notes for students who need help. Spend no more than 10 minutes on this review.

> **TEACHING NOTE**
>
> We encourage you to stop after 10 minutes even though you may be tempted to spend more time. If you keep this process quick and easy you are more likely to use it frequently. Frequency is key to the success of this practice.

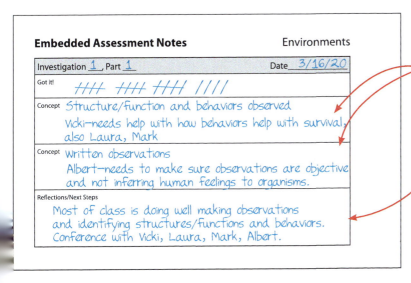

Add what to look for and make notes as you review student work.

Write reflections and next steps here after 10 minutes of review.

4. **Reflect.** Take 5 minutes to summarize the trends and patterns (highlights and challenges) you saw, and record notes in the "Reflections/Next Steps" section.

5. **Adjust.** In the same section of the sheet, describe the next steps you will take to clarify any problems, or note highlights you saw in students' progress. Some suggestions appear later in this chapter and accompany the coding guides. This is the defining factor in formative assessment. You must take some action to help students improve. If you do this process frequently, the next steps required should take only a few minutes of class time when the next part begins.

Environments Module—FOSS Next Generation

ENVIRONMENTS – *Assessment*

BENCHMARK *Assessment*

The FOSS benchmark assessments are carefully designed to help you look at students' progress across the module. Each of these assessments includes items that incorporate the three dimensions described in the *NRC Framework*: science and engineering practices, disciplinary core ideas, and crosscutting concepts. Benchmark assessments provide a broader focus on students' learning than do embedded assessments. Benchmark assessments require students to determine when and how pieces of knowledge, and larger conceptual relationships need to be recalled or applied.

Survey

The *Survey* is administered a few days before instruction begins. Students are often uneasy about having to take a "test" when they haven't yet had the instruction they need in order to do well. It is important for them to know that the *Survey* will help you determine what they already know and what they need to learn, so that you can plan extra support when needed—students will not be graded on this assessment. Help students view the *Survey*, as well as all of the assessments, as learning tools. At the end of the module, have students compare their answers from the *Survey* to their answers on the *Posttest*. Students have few opportunities to see how their knowledge changes.

I-Checks

At the end of each investigation, students take an I-Check benchmark assessment. I-Checks (short for "I check my own understanding") provide students with an opportunity to demonstrate their learning. When you return the I-Checks, students have another opportunity to reflect on their understanding through self-assessment activities.

If students complete an I-Check using paper and pencil, determine their progress by reviewing each item using the coding guides provided on FOSSweb. When you code the open-response items, we recommend that you code them one item at a time; for example, code item 8 for all students, then move on to item 9, then 10, and so on. Even though you have to shuffle papers more, you will find that it takes less time overall to code the assessments. Proceeding one item at a time, rather than one student at a time, allows you to establish a mind-set for each item, think about the whole class's performance on that item, and plan next steps you might take. Another benefit to following this procedure is that you will code more consistently.

> **NOTE**
> Alternatively, students can take the *Survey*, I-Checks, and *Posttest* online. Most of the items will be automatically coded for you. You can print students' answers from the *Survey* for comparison with the *Posttest*. See the FOSSmap and Online Assessment section in this chapter.

Benchmark Assessment

I-Checks are most valuable when you use them as formative assessments. Studies have shown that students learn more when they take part in evaluating their own responses. To do this, code students' tests, but do not put any marks on them. Thinking about the content of the questions comes to a halt when students see marks on their papers (especially grades) made by their teachers. Using the self-assessment strategies described later in this chapter is an important step in developing a class culture focused on growth mindset that values assessment as a learning process.

Posttest

Students take the *Posttest* after all the investigations are completed. It can be administered in any of the ways described for the other benchmark assessments. After coding the *Posttests*, return them and the *Surveys* to students. Have students compare their *Survey* and *Posttest* responses. Discuss the changes that have occurred.

Use the *Posttests* for formative evaluation of the module. Make notes about things you might want to focus on or do differently the next time you teach the module.

When Grades Are Required

"If we are always focusing on formative assessment, how do we give grades?" is a frequent question. Our recommendation is that you use derivative products for giving grades. That way students always know when they will be graded, and they always have a chance to improve their work before turning something in for a grade. For example, if you need to grade a notebook entry, have students rewrite that entry on a separate sheet of paper to be turned in for the grade. In this process, they can work with other students or use a class discussion to improve their entry before turning it in.

When grading I-Checks, rather than figuring a percentage, choose three or four items that students reflect on through self-assessment strategies, then have the opportunity to rewrite before turning in for a grade. When you follow this process, students are always clear about when they will be graded, and know that they always have the opportunity to make their work better, which helps develop a growth mindset.

If you must base grades on a percentage, first subtract 1 from each code (except zero, no negative numbers) to transform codes to points. Then you can add points and determine a percentage. We suggest 80% should be an A, rather than the traditional 90%, because these assessments are designed for diagnostic purposes, not minimum mastery.

Environments Module—FOSS Next Generation 351

ENVIRONMENTS — Assessment

Portfolios

Students can choose work at the end of each investigation to include in a summative portfolio for end-of-course presentations or grades. Give students a *Portfolio Checklist* and a folder to hold the checklist and their work samples. At the end of each investigation, suggest an item or two that they might add to the portfolio to demonstrate their three-dimensional learning. Students can photograph and print notebook pages, or create a derivative product from the notebook page to be included in the portfolio. As always, notebooks should not be graded—they should be a risk-free environment for students to record their thinking. If students are going to include something from their notebook, they should have a chance to improve it before adding it to the portfolio. A sample *Portfolio Checklist* can be found with the other FOSS assessment charts, or you could make your own with students.

Interim Assessments

These assessments directly target specific NGSS performance expectations. They are not meant to be diagnostic for daily instruction using FOSS curriculum, rather they are generic tasks that students should be able to answer given any curriculum used to teach the NGSS. They are a good tool for students to use to practice for state or district tests. Interim assessments can be given after students complete an investigation focusing on a particular performance expectation, or you can give them at the end of the year for program evaluation, especially when a state or district test is not given at that grade level. (These tasks will evolve and be updated as we learn more about large-scale assessments.)

These tasks begin with a scenario that students read to set the context for the practices and items that will be involved in the task. Teachers can opt in many cases to include a hands-on experience, and in a few cases a computer-simulation. Each task then consists of a number of related items that students answer. Most are constructed-response items but a few may be multiple-choice, multiple-answer, or short answer.

Benchmark Assessment

Benchmark Assessment Design

The foundation of the FOSS assessment system rests on three pillars: a conceptual framework, a student-progress model, and the NGSS performance expectations. It is important to remember that the performance expectations are a sampling of expected student proficiencies and that a full curriculum will always necessarily include more. The **conceptual framework** for this module can be found in the Framework and NGSS chapter. It is a segment from a larger learning progression that FOSS has carefully engineered for grades K–8. The concepts you see in the framework span grades K–8, and the bullets designate learning that is important for fourth graders. The **student-progress model** is displayed in the table below. It describes in general terms how FOSS uses four levels to categorize student learning. The **NGSS performance expectations** also guide assessment design, providing a sampling of end goals and always reminding us that learning science is a three-dimensional enterprise that includes science and engineering practices, disciplinary core ideas, and crosscutting concepts.

Level	Description
Strategic (4)	Performance on an item or assessment shows exceptional understanding of three-dimensional learning. Students are able to apply their knowledge of practices, disciplinary core ideas, and crosscutting concepts to explain novel phenomena or solve new problems. Students at this level continue to build a network of knowledge, practice, and crosscutting concepts to bridge disciplines. They can apply all of those to real-world phenomena and design problems.
Conceptual (3) (minimum goal for all students)	Performance on an item or assessment shows well-developed understanding of three-dimensional learning. Students are making connections among practices, core ideas, and crosscutting concepts in order to answer more complex questions and solve more complex problems. To get to the next level (strategic), students need to continue to build connections and be able to transfer this knowledge from the classroom to real-world phenomena.
Recognition (2)	Performance on an item or assessment shows developing understanding of three-dimensional learning. Students have built a foundational repertoire of pieces of knowledge and practices, and use academic language with greater facility. To get to the next level (conceptual), students need to continue to add knowledge about practices, core ideas, and crosscutting concepts, and then build connections among those pieces to form more complex understandings about phenomena.
Notions (1)	Performance on an item or assessment shows relatively little understanding of three-dimensional learning. Students may include some scientific vocabulary or recall of simple facts or procedures, but there is little evidence of the impact of instruction. To get to the next level (recognition), students need to develop practices, begin to incorporate academic language in their communication, and construct more pieces of knowledge that help explain phenomena.

Environments Module—FOSS Next Generation

ENVIRONMENTS – Assessment

Through our own research (ASK Project, 2003-2009) and that of others (most notably, Dylan Wiliam and Carol Dweck), FOSS developers have concluded that the most important assessment goal is to support learning in a way that helps students develop a growth mindset. That means that the focus must be on formative assessment.

It also means that we don't use traditional scoring. Instead, we code student responses to categorize the level at which students have demonstrated competence. For instance, at first glance it looks as though students are being awarded a point for an incorrect answer. In fact, their work is in the level 1 category. They have attempted to answer the question, but the answer shows little influence of instruction. You will also notice that some items can be coded no higher than a level 2, even though the student-progress model has four levels. That particular item asks students only to demonstrate that they can perform a simple practice or have a piece of knowledge (a level 2 proficiency) essential to a broader understanding. When it is important to know what pieces students have or don't have in order to determine how to help them improve, a simpler question can help identify differentiated instruction that is necessary for some students. These less-complex questions may not include all three dimensions, but are important in terms of providing diagnostic formative information.

Coding Benchmark Assessments

Answer sheets and coding guides for all benchmark assessments are available on FOSSweb. A two-page spread has been dedicated to each page of each assessment. Here we show an example for coding a multiple-answer and an open-response question.

Recording Benchmark Data

Use **Assessment Record** to record students' responses on the benchmark assessments (*Survey*, I-Checks, and *Posttest*). Follow this procedure if you plan to use the I-Checks for formative assessment.

1. Record students' responses on *Assessment Record* or on a spreadsheet. For multiple-choice and multiple-answer items, record the *letter* of the response rather than the code. (Later, if you need a total score, you can replace the letter responses with numbers.) For multiple-answer items, check the coding guide; for short-answer questions, record the word or the code, depending on the answer patterns you want to see. Do not make any marks on students' papers. (Students will do that later.)

> **NOTE**
> Coding guides for interim assessments are found on FOSSweb.

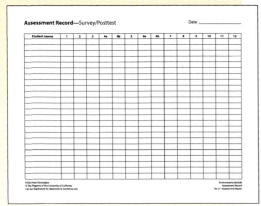

354 Full Option Science System

Benchmark Assessment

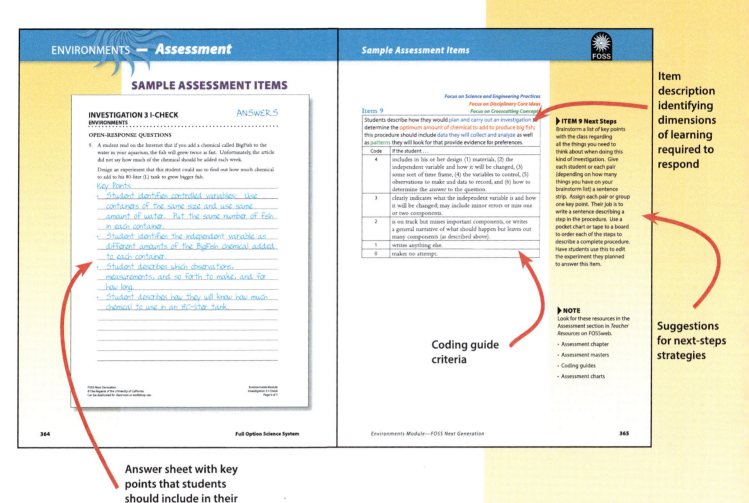

Answer sheet with key points that students should include in their responses

Coding guide criteria

Item description identifying dimensions of learning required to respond

Suggestions for next-steps strategies

2. Look for patterns of right and wrong answers, so that you can determine which questions will need follow-up reflection and discussion by students.

3. Review students' open-response answers. Record codes on *Assessment Record*. Make notes about concepts or ideas you want to bring back for students' reflection in self-assessment activities.

4. Choose three or four items from the assessment that you want to discuss with students during class time. Determine the next steps you want to introduce. (See the following section on next-step strategies.)

> **NOTE**
> If students take the benchmark assessments online, FOSSmap does most of the coding for you. You need to code open-response items and check codes for short-answer items. Then you can run a variety of class and individual diagnostic reports.

ENVIRONMENTS — Assessment

NEXT-STEP Strategies

The ASK Project (Assessing Science Knowledge) was funded by the National Science Foundation in 2003. For 6 years, the FOSS development team worked with nine centers around the United States, including more than 500 teachers and their students as research partners. Based on the evidence provided by the assessments, we learned very quickly that assessment is worth doing only if follow-up action is taken to enhance understanding. Self-assessment provides students the opportunity to be responsible for their own learning and is a very effective tool for building students' scientific knowledge and practice.

Self-assessment is more than reading correct answers to the class and having students mark whether they got the right answer. Self-assessment must provide an opportunity for students to reflect on their current thinking and judge whether that thinking needs to change. This kind of reflective process also helps students develop a better understanding of what is expected in terms of well-constructed responses.

Self-assessment requires deep, thoughtful engagement with complex ideas. It involves students in whole-class or small-group discussion, followed by critical analysis of their own work. For this reason, we suggest that you focus your probing discussions on three or four questions from an I-Check, rather than on the entire assessment. The techniques described here are meant to give you a few strategies for entering the process of self-assessment. There is no single right way to engage students in this process, but it works best when you change the process from time to time to keep it fresh. The strategies listed here are sorted into two groups: (1) strategies for whole-class feedback, and (2) strategies for individual-student and small-group feedback.

Strategies for the Whole Class

Multiple-choice discussions. Students sit in groups of three or four, depending on how many possible answers there were for a given question. You assign an answer to each student (not necessarily the answer he or she chose). Each student is responsible for explaining to the group whether the assigned answer is correct and why or why not.

Multiple-choice corners. When the class is equally split on what students have chosen as the correct response, or only a few students got the correct answer, have them meet in different corners of the room. Those who chose A go to one corner, those who chose B go to another corner, and so on. Each corner group needs to come up with

> **TEACHING NOTE**
>
> If students have taken an assessment online, print out their responses (the Student Responses Report) and use a projection system and the assessment masters or the PDF to show students the items under discussion.

Next-Step Strategies

an argument to convince the other corners that their answer is correct. As in a class debate, students are allowed to disagree with themselves if they become convinced their position is flawed or the reasoning of another group is more convincing. They then move to that corner and continue by helping their new group shape its argument. (Don't be surprised if you find all students migrating to one corner before the presentation of arguments even begins!)

Key points. Begin this strategy by discussing the item in question. After it is clear that students understand what is intended by the item prompt, call on individuals or groups to suggest key points that should be included in a complete answer. Write the key points on the board as phrases or individual words that will scaffold students' revision, rather than complete sentences they might mindlessly copy. When students return to their responses, they can number each of the key points they originally included in their answers, then add anything they missed.

Revision with color. Another way that students can revise their answers after a key-points discussion is to use colored pens or pencils and the three C's. As they read over their responses, they *confirm* correct information by underlining with a green pen; they *complete* their responses by adding information that was missing, using a blue pen; and they *correct* wrong information, using a red pen.

Review and critique anonymous student work. Use examples of student work from another class, or fabricate student work samples that emulate the problems students in your class are having. Project the work, using an overhead projector, a document camera, or an interactive whiteboard. Have students discuss the strengths and weaknesses of the responses. This is a good strategy to use when first getting students to write in their notebooks. It helps them understand expectations about what and how much to write.

Line of learning. Many teachers have students use a line of learning to show how their thinking has changed. When students return to original work (embedded or benchmark) to revise their understanding of a concept, they start by drawing and dating a line of learning under the original writing. The line of learning delineates students' original, individual thinking from their thinking after a class or group discussion has helped them reconsider and revise their thoughts.

Find the problems. Review the codes you have recorded and let students know how many items they got right and how many need to be corrected. Give them time in class to work in groups or individually to find their mistakes and correct them. They can turn in the assessment again, or write a few sentences in their notebooks explaining how their thinking has changed.

Environments Module—FOSS Next Generation

ENVIRONMENTS – Assessment

> **NOTE**
> You can make inexpensive whiteboards by using card stock and plastic sheet protectors. Students use whiteboard marking pens to write answers. Old socks make great erasers.

Group consensus/whiteboards. Have students in each group (or pairs in each group) work together to compare their answers on selected I-Check questions or key points of a notebook entry. They first create a response that the group agrees is the best answer. Groups write their responses on a whiteboard. When you give a signal, one student in each group holds up his or her whiteboard and compares answers. The class discusses any discrepant answers.

Class debate. A student volunteers an answer to an item on an assessment (usually one that many students are having trouble with or one that elicits a persistent misconception). That student is in charge of the debate. He or she puts forth an answer or explanation. Other students agree or disagree, and must provide evidence to back up their thinking. Students are allowed to disagree with themselves if they hear an argument during the discussion that leads them to change their thinking. You can ask questions to keep the discussion on track, but otherwise you should stay on the sidelines.

Critical competitor. Use the critical-competitor strategy when you want students to attend to a specific detail. You need to present students with two things that are similar in all but one or two aspects. You can use any medium: two drawings, two pieces of writing, or a combination (such as a diagram compared to a description). The point is to compare two pieces of communication or representations in some way that will help students focus on an important detail they might be missing.

Sentence frames. After completing other self-assessment activities, have students consider all the items on the assessment and write a short reflection, using sentence frames. This strategy directs students to choose one or two items that they would like to tell you more about.

I used to think ___, but now I think ___.

I should have gotten this one right, but I just ___.

I know ___, but I'm still not sure about ___.

The most important thing to remember about ___ is ___.

Can you help me with ___?

I shouldn't have gotten this one wrong, because I know ___.

I'm still confused about ___.

Next time, I will remember to ___.

Now I know ___.

Next-Step Strategies

Strategies for Individual Students and Small Groups

Feedback notes. As you read through students' notebooks, add self-stick notes with comments or questions that help guide students to further reflect on and improve their understanding.

> **NOTE**
> You can also create feedback notes for embedded assessments using FOSSmap.

Response log. Set up a response log at the back of students' science notebooks (before or after the index or glossary if those are used). Fold a notebook page in half, or draw a line down the center of the page. Have students write "Teacher Feedback" at the top of the left side of the page and "My Responses" at the top of the right side of the page. When you want a student to think about something in his or her notebook, write your note in the "Teacher Feedback" column (or students can move a self-stick note from another page to the response log). Students then respond in the right column, either addressing your comment there or telling you which page to turn to in order to see how they have responded.

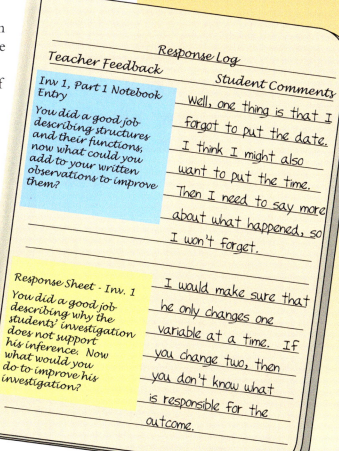

Conferences. Use silent-reading time or other times when students work independently to confer with small groups or individual students.

Centers. Set up a center at which students can continue to explore their ideas and refine their thinking. You might pair students so that a student who understands the concept well works with another student who needs some help.

Reteach or clarify a concept. Set up a modified investigation in which a small group of students works with the concept again.

Environments Module—FOSS Next Generation

ENVIRONMENTS — Assessment

FOSSMAP and Online Assessment

FOSSmap (fossmap.com) is the assessment management program designed specifically for teachers using the FOSS Program in grades 3–5. This user-friendly system allows you to open online assessments for students, to review codes for student responses, and to run reports to help you assess student learning. FOSSmap was developed at the Lawrence Hall of Science in conjunction with the Berkeley Evaluation and Assessment Research (BEAR) center at the University of California, as part of a 5-year research and development project funded by the National Science Foundation. It is based on the tools developed in the Assessing Science Knowledge (ASK) project.

Embedded-assessment data can be entered into FOSSmap to provide evidence of differentiated instruction, to run reports for formative analysis, and to print notes to provide feedback in student notebooks. It is also a tool for teacher reflection and instructional improvement from year to year.

FOSSmap allows you to give students access to the **Online Assessment** system (fossmap.com/icheck). Students log into this system to take the benchmark assessments (*Survey*, I-Checks, and *Posttest*). Responses are automatically sent to the FOSSmap teacher program where most are automatically coded. You will need to check short answers (mainly for correct answers that include inventive spelling), and to code open-response items. Students can answer open-response items on the computer or using paper and pencil, depending on how extensive their typing skills are.

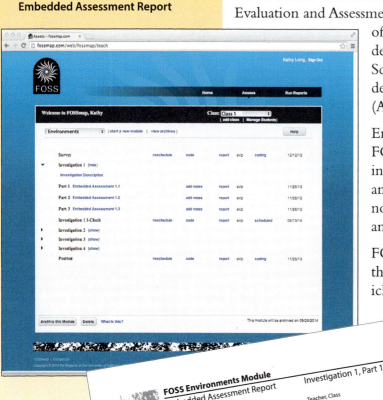

Navigation page and a sample Embedded Assessment Report

FOSSmap and Online Assessment

FOSSmap Reports

The **Embedded Assessment Report** is a record you can run for each embedded assessment your students complete. It lists the names of students who "got it" as well as those that need some individual help. It is also a good place to record things that you might want to put more emphasis on or do slightly differently the next time you teach the module.

The **Code Frequency Chart** tells you at a glance which items were problems for the class. The red-bar items are the ones you want to take back to students for self-assessment activities.

The **Class by Item Report** shows the detail of each item and students' responses. You can go directly to the problem items (identified using the Code Frequency Chart) and see what the problem is. This report displays students' names for each response, with a brief description of what each code means in terms of full or partial credit. The report helps you decide what steps need to be taken next.

The **Student Responses Report** provides a printout of individual students' responses to all items answered online (including open-response items if they were typed into the system). This report is used for student self-assessment activities.

The **Student by Item Report** (a good report to send home to parents) lists all the items on a test and shows how individual students responded to each item. It also provides the correct answer and a description of what the student knows or needs to work on, based on the evidence to be inferred from the item.

The **Class Item Codes Report** provides a spreadsheet that can be opened in any spreadsheet program. It gives you a list of the students, the maximum code for each item, and the code each student received on each item. You can use this sheet if you want to convert codes into scores in order to determine percentage correct if that is needed for giving grades. To do that, you need to subtract 1 from each code, so that you are not actually awarding a point for wrong answers. Remember though, that FOSS assessments are designed to be diagnostic and not minimum mastery, so you may need to adjust your cut points for giving ABC grades. For example, instead of 90% being an A, you may decide that 80% is a better cut point for an A.

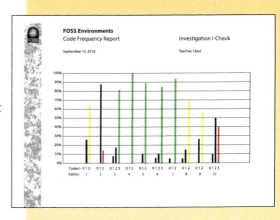

Code frequency, Class by Item, and Student by Item reports

Environments Module—FOSS Next Generation 361

SAMPLE ASSESSMENT ITEMS

INVESTIGATION 1 I-CHECK
ENVIRONMENTS

ANSWERS

4. Sanderlings are small shorebirds. They are found on sandy beaches, mudflats, lagoons, and rocky shores. A student decided to study the sanderlings and determine their preferred environment. He spent 15 minutes each day observing each environment and recorded his observations in his science notebook.

Sanderling Observations (15 minutes/day)				
	Sandy beach	Mudflat	Lagoon	Rocky shore
Monday	24	12	7	4
Tuesday	26	8	5	0
Wednesday	18	15	5	3
Thursday	21	6	4	7
Friday	25	9	11	0

Using the data in the table above, which is the Sanderling's preferred environment?

(Mark the one best answer.)

- ● **A** Sandy beach
- ○ **B** Mudflat
- ○ **C** Lagoon
- ○ **D** Rocky shore

Sanderling

5. Everything that surrounds and influences an organism is its _____.

(Mark the one best answer.)

- ○ **A** environmental factor
- ○ **B** preference
- ● **C** environment
- ○ **D** basic need

Sample Assessment Items

Item 4

Focus on Science and Engineering Practices
Focus on Disciplinary Core Ideas
Focus on Crosscutting Concepts

Students analyze the data provided to look for patterns that provide evidence to explain sanderling preferences.

Code	If the student …
2	marks A.
1	marks B, C, D, or more than one answer.
0	makes no attempt.

Item 5

Focus on Disciplinary Core Ideas

Students know an environment is everything that surrounds and influences an organism.

Code	If the student …
2	marks C.
1	marks A, B, D, or more than one answer.
0	makes no attempt.

▶ **ITEM 4 Next Steps**

Do a think aloud for the class, or have students do the talking. Read the item and look carefully at the chart. Vocalize analysis of the chart and the way it leads you to a conclusion about which location the sanderlings prefer.

▶ **ITEM 5 Next Steps**

Some students may choose A rather than C to answer this question. Discuss how the two answers are alike and different. An environmental factor is only one thing, an environment is about all the things that can affect an organism.

▶ **NOTE**

Look for these resources in the Assessment section in *Teacher Resources* on FOSSweb.

- Assessment chapter
- Assessment masters
- Coding guides
- Assessment charts

Environments Module—FOSS Next Generation

SAMPLE ASSESSMENT ITEMS

INVESTIGATION 3 I-CHECK
ENVIRONMENTS

ANSWERS

OPEN-RESPONSE QUESTIONS

9. A student read on the Internet that if you add a chemical called BigFish to the water in your aquarium, the fish will grow twice as fast. Unfortunately, the article did not say how much of the chemical should be added each week.

   Design an experiment that this student could use to find out how much chemical to add to his 80-liter (L) tank to grow bigger fish.

   Key Points
   - *Student identifies controlled variables: Use containers of the same size and use same amount of water. Put the same number of fish in each container.*
   - *Student identifies the independent variable as different amounts of the BigFish chemical added to each container.*
   - *Student describes which observations, measurements, and so forth to make, and for how long.*
   - *Student describes how they will know how much chemical to use in an 80-liter tank.*

Sample Assessment Items

Focus on Science and Engineering Practices
Focus on Disciplinary Core Ideas
Focus on Crosscutting Concepts

Item 9

Students describe how they would *plan and carry out an investigation* to determine the *optimum amount of chemical to add to produce big fish*; this procedure should include *data they will collect and analyze* as well as *patterns* they will look for that provide evidence for preferences.

Code	If the student . . .
4	includes in his or her design (1) materials, (2) the independent variable and how it will be changed, (3) some sort of time frame, (4) the variables to control, (5) observations to make and data to record, and (6) how to determine the answer to the question.
3	clearly indicates what the independent variable is and how it will be changed; may include minor errors or miss one or two components.
2	is on track but misses important components, or writes a general narrative of what should happen but leaves out many components (as described above).
1	writes anything else.
0	makes no attempt.

▶ **ITEM 9 Next Steps**
Brainstorm a list of key points with the class regarding all the things you need to think about when doing this kind of investigation. Give each student or each pair (depending on how many things you have on your brainstorm list) a sentence strip. Assign each pair or group one key point. Their job is to write a sentence describing a step in the procedure. Use a pocket chart or tape to a board to order each of the steps to describe a complete procedure. Have students use this to edit the experiment they planned to answer this item.

▶ **NOTE**
Look for these resources in the Assessment section in *Teacher Resources* on FOSSweb.

- Assessment chapter
- Assessment masters
- Coding guides
- Assessment charts

Environments Module—FOSS Next Generation 365

ENVIRONMENTS — *Assessment*